New Research in PLANT ANATOMY

Supplements to the Journals of the Linnean Society of London

Forthcoming titles

Early mammals
Symposium held in June 1970 and to be published as Supplement 1 to the Zoological Journal, Vol. 50, 1971.

Biology and chemistry of the Umbelliferae
Symposium held in September 1970 in conjunction with the Phytochemical Society and to be published as a supplement to the Botanical Journal.

Behavioural aspects of parasitic transmission
Symposium arranged with the British Section of the Society of Protozoologists and the British Society of Parasitology and to be held in July 1971.

Taxonomy and geography of higher plants in relation to evolution
Symposium organized in conjunction with the Botanical Society of the British Isles and to be held in September 1971.

New Research in PLANT ANATOMY

Edited by

N. K. B. Robson
Department of Botany, British Museum (Natural History)

D. F. Cutler and **M. Gregory**
Jodrell Laboratory, Royal Botanic Gardens, Kew

Supplement 1 to the Botanical Journal
of the Linnean Society Volume 63
1970

Published for the Linnean Society of London
by Academic Press

ACADEMIC PRESS INC. (LONDON) LIMITED
Berkeley Square House
Berkeley Square
London W1X 6BA

U.S. Edition published by
ACADEMIC PRESS INC.
111 Fifth Avenue
New York
New York 10003

Library of Congress Catalog Card No. 74-141731

ISBN: 0 12 590650 1

Made and printed in Great Britain by
William Clowes and Sons, Limited, London, Beccles and Colchester

Foreword

This is the first Symposium volume to appear as a supplement to the Botanical Journal of the Linnean Society. *In the past, symposia have been published as a part of one of the* Journals *but from now on they will usually appear only as hard-backed supplements to the current volume of the appropriate* Journal.

This volume consists of papers devoted to New Research in Plant Anatomy, some of which were read at the Symposium of that title arranged by the Plant Anatomy Group (of the Society) in honour of Dr C. R. Metcalfe, sometime Botanical Secretary of the Society. The Officers and Council are indebted to the Editors of this volume for arranging the Symposium and also for preparing this volume for publication.

The Society's Symposia are arranged by the Society, often in conjunction with other biological societies, with contributions by leading specialists. Any Fellow or representative of another biological society wishing to organise a Symposium should in the first instance contact either an Officer or the Executive Secretary of the Linnean Society at its Rooms in Burlington House, Piccadilly, London, W1V 0LQ.

The Society's Rooms provide a pleasant, centrally situated and well equipped meeting place for all-day meetings, and the Council sincerely hope that Fellows and other biological societies will avail themselves of the facilities that the Linnean Society has to offer.

DORIS M. KERMACK—Editorial Secretary

November 1970

List of contributors

AYENSU, EDWARD S. *Department of Botany, Smithsonian Institution, Washington, D.C. 20560, U.S.A.* (p. 127)

BRITTAN, N. H. *Botany Department, University of Western Australia, Nedlands, Western Australia 6009, Australia* (p. 57)

BUTH, G. M. *Department of Botany, Aligarh Muslim University, Aligarh, India* (p. 169)

BYSTROM, BARBARA G. *Space Biology Laboratory, Brain Research Institute, University of California, Los Angeles, California 90024, U.S.A.* (p. 15)

CARLQUIST, SHERWIN. *Claremont Graduate School and Rancho Santa Ana Botanic Garden, Claremont, California 91711, U.S.A.* (p. 181)

CHALK, LAURENCE. *33 Belsyre Court, Observatory Street, Oxford, England* (p. 163)

CHEADLE, VERNON I. *Department of Biological Sciences, University of California, Santa Barbara, California 93106, U.S.A.* (p. 45)

CHOWDHURY, K. A. *Department of Botany, Aligarh Muslim University, Aligarh, India* (p. 169)

CLIFFORD, H. T. *Department of Botany, University of Queensland, St. Lucia, Brisbane 4067, Queensland, Australia* (p. 25)

FAHN, A. *Department of Botany, The Hebrew University of Jerusalem, Jerusalem, Israel* (p. 51)

FINDLAY, G. W. D. *Department of Botany, Imperial College of Science and Technology, Prince Consort Road, London, S.W.7, England* (p. 71)

GREGUSS, PÁL. *Botanical Institute, The University, Szeged, Hungary* (p. 83)

HOWARD, RICHARD A. *Arnold Arboretum of Harvard University, Jamaica Plain, Cambridge, Massachusetts 02138, U.S.A.* (p. 195)

KAPLAN, DONALD R. *Department of Botany, University of California, Berkeley, California 94720, U.S.A.* (p. 101)

KUKKONEN, I. *Department of Botany, University of Helsinki, Unioninkatu 44, Helsinki 17, Suomi-Finland* (p. 137)

LEVY, J. F. *Department of Botany, Imperial College of Science and Technology, Prince Consort Road, London, S.W.7, England* (p. 71)

PHILIPSON, W. R. *Botany Department, University of Canterbury, Christchurch, New Zealand* (p. 87)

PHIPPS, ROBERT E. *Department of Botany, University of Maryland, College Park, Maryland 20742, U.S.A.* (p. 215)

RACHMILEVITZ, TALIA. *Department of Botany, The Hebrew University of Jerusalem, Jerusalem, Israel* (p. 51)

SCOTT, FLORA MURRAY. *Department of Botanical Sciences, University of California, 405 Hilgard Avenue, Los Angeles, California 90024, U.S.A.* (p. 15)

STACE, C. A. *Department of Botany, The University, Manchester M13 9PL, England* (p. 75)

STANT, MARGARET Y. *Jodrell Laboratory, Royal Botanic Gardens, Kew, Richmond, Surrey, England* (p. 147)

STERN, WILLIAM L. *Department of Botany, University of Maryland, College Park, Maryland 20742, U.S.A.* (p. 215)

SWEITZER, EDWARD M. *Department of Botany, University of Maryland, College Park, Maryland 20742, U.S.A.* (p. 215)

TOMLINSON, P B. *Fairchild Tropical Garden, Miami, Florida 33156, U.S.A. and Cabot Foundation, Harvard University, Cambridge, Massachusetts, U.S.A.* (p. 1)

VAUGHAN, J. G. *Department of Biology, Queen Elizabeth College, Campden Hill Road, London, W.8, England* (p. 35)

Introduction

This symposium was arranged by the Plant Anatomy Group of the Linnean Society of London in honour of Dr C. Russell Metcalfe, who retired from the post of Keeper of the Jodrell Laboratory, Royal Botanic Gardens, Kew in September 1969.

The immediate and willing co-operation of those called upon to contribute papers is evidence of the high regard in which Dr Metcalfe is held, not only in this country but also all over the world. Of the twenty papers published here, seven were presented at the meeting held in London at the Rooms of the Society on 18 September 1970.

During his career at Kew, Dr Metcalfe advanced the study of plant anatomy, not only by his extensive writings but also by the kindly interest he showed to visitors to the Jodrell Laboratory and the encouragement he gave them. He has always been ready to spend time with people.

The range of subjects involving plant anatomy these days is surely impressive. The papers contained in this volume indicate some of the main lines of investigation which are being followed. Some give the classical approach, still essential to our basic understanding of the plant, whilst others show the results of the application of the latest techniques.

If this volume, as it is hoped, provides a stimulus to would-be anatomists to share in the increasing interest of this type of research, it will only be doing what Dr Metcalfe has been striving to do during the past forty years.

D. F. Cutler
(*Convenor, Plant Anatomy Group*)

M. Gregory

N. K. B. Robson

The publication of this symposium volume is a notable event at a time when the morphology and anatomy of plants, though basic aspects of botanical science, have been somewhat neglected. It has, however, been my good fortune during the past twelve months to have met not only a great many plant anatomists but also workers in many diverse aspects of botany from a great many countries. I have also travelled extensively in the U.S.A., visiting botanical departments at universities and other institutions in widely separated parts of the American continent. These experiences have afforded a wonderful opportunity to take a broad look at the current botanical scene, and have enabled me to think afresh about the significance of morphology and anatomy in relation to botany as a whole. These contacts were made after I had retired from the Keepership of the Jodrell Laboratory, when my official duties were limited to teaching plant anatomy to two classes of students in the University of California at Los Angeles, thus providing plenty of opportunity to consider these matters without being involved in the more mundane problems which normally fall to the lot of a working botanist.

It seems quite plain that, perhaps without realizing it, we are gradually regaining our awareness of an obvious truth that was at one time taken as axiomatic. This is that a thorough knowledge of the form and structure of the organisms on which we are working is an essential preliminary to many different lines of enquiry. Furthermore, to be of the maximum value this morphological knowledge, besides applying in depth to the organisms with which we are immediately concerned, must also extend widely so as to enable comparisons with other plants to be made. These two approaches should serve to remind us that morphology and anatomy constitute a basic discipline which underlies or forms part of investigations concerned with physiology and ecology, about which it may consequently be necessary to develop revised and perhaps novel concepts. We should also remember that the comparative anatomy of living and fossil plants is but a part of an entity which must be comprehended as a whole before we make any real progress in understanding the broad lines of evolution. Unless integration is achieved we may sometimes be tempted to indulge in more speculation than is justified by the available facts—a process that may lead to copious writing without adding anything of real substance to our knowledge or understanding.

It is probably true to say that ontogenetic studies in a broad sense are among the commonest lines of morphological investigation at the present time. These investigations are of great importance because they cover all the problems which face us when we turn to the differentiation of tissues, a subject about which there still appears to be little understanding. Besides this, the study of ontogeny can often enlighten our comparative studies of mature plant organs by explaining the origin of differences by which taxa at and above the rank of species can be distinguished. Then again, electron microscope studies are calling for a reinterpretation of histology at much higher magnifications than those obtainable with the light microscope. Here there is a danger that we may go astray unless we start with observations under the light microscope before proceeding to explain the disclosures made available to us by electron microscopy.

The twenty articles in this volume provide significant contributions to some of the aspects of morphology and anatomy to which reference has been made, and it is indeed encouraging to be reminded that so many investigators are active in so many parts of the world at the present time. It is very rewarding to see that nearly all the authors of these contributions are known to me personally and that a high proportion of them are amongst the still greater number who have at some time worked on plant anatomy at the Jodrell Laboratory during my tenure of office there. That these articles have been published together in a single volume is particularly valuable as they collectively constitute a useful addition to botanical knowledge.

C. R. Metcalfe

Formerly Keeper of the Jodrell Laboratory,
Royal Botanic Gardens, Kew

Contents

Dichotomous branching in *Flagellaria indica* (Monocotyledones)

P. B. TOMLINSON, F.L.S.

Fairchild Tropical Garden, Miami, Florida, U.S.A.
and Cabot Foundation, Harvard University, Cambridge, Massachusetts, U.S.A.

Branching in the climbing aerial shoots of many populations of *Flagellaria indica* (Flagellariaceae) involves a bifurcation of the axis. The distichous arrangement of leaves found below the fork is continued into each new shoot without interruption. Leaves above or below the fork may have additional blades. Initiation of this bifurcation involves a division of the shoot apex, above the insertion of the youngest visible leaf primordium, into two more or less equal halves. Central cells of the corpus as well as the tunica are involved in the division. The twin shoots remain identical in size and produce equal numbers of leaves. By definition, the branching is a true dichotomy. It has a developmental, rather than an evolutionary interpretation. It is suggested that it may occur elsewhere in the monocotyledons and two other examples are already known.

CONTENTS

INTRODUCTION

In recent investigations of the morphology of a wide variety of monocotyledons, I have encountered a type of vegetative branching which differs from normal axillary branching and for which the term 'dichotomous' seems appropriate. This paper describes this kind of branching in *Flagellaria*. Such examples require careful analysis because dichotomous branching, involving an equal division of a shoot apex or apical cell, has been regarded as a unique property of lower vascular plants and, by implication, is considered a 'primitive' feature within the Tracheophyta in general. That true dichotomy does occur in angiosperms has been verified recently by Nolan (1969) in *Asclepias syriaca*. We are indebted to Nolan for a discussion of the subject and for a definition of dichotomy which is accepted in this present article. In *Asclepias* dichotomy occurs in the reproductive phase of growth; shoots bifurcate so that one branch becomes a peduncle, the other becoming a continuing main axis which repeats the bifurcation at each later node. In the monocotyledons with a similar type of branching, dichotomy is a property of the vegetative shoot. The branches may or may not remain alike.

Dichotomous branching has long been suspected in palms as it would account for the familiar equal forking of the aerial stem which is a constant feature of certain species of *Hyphaene* (Borassoideae): 20 out of 27 species in the treatment by Beccari (1924). A similar forking of the axis may also occur in species of *Hyphaene* which are normally unbranched, e.g. in *H. ventricosa* (Lewalle, 1968). The belief that this kind of branching involves an equal dichotomy of the shoot apex rests only on circumstantial evidence. Schoute (1909) examined a piece of dried mature stem which included a bifurcation, and concluded that the leaf arrangement at the fork could not have arisen by axillary branching. His material was too limited for him to be certain, however, and subsequently no one has attempted to observe early stages in the branching process. *Hyphaene* is neither very accessible nor commonly cultivated and, since branching in any one individual is very infrequent, the task of augmenting Schoute's observations is a considerable one.

More recently, using similar circumstantial evidence, it has been shown that the *Hyphaene*-type of branching is of wide occurrence in palms (Tomlinson & Moore, 1966; Tomlinson, 1967), since it is found in a number of unrelated genera in several of the major subgroups of palms. Examples are *Allagoptera* (Cocoideae), *Chamaedorea* and *Vonitra* (Arecoideae), *Nannorrhops* (Coryphoideae) and *Nypa* (Nypoideae). These examples have hitherto been overlooked because the palms are either rare in nature or unfamiliar in cultivation. In *Allagoptera* and *Nypa* the stems are subterranean and not very accessible. Until recently none of these records was convincing since a microscopical demonstration of dichotomizing shoot apices was lacking, so they have merely served to make the problem more intriguing. However, on two recent visits to New Guinea I have collected material which shows that branching of the rhizomatous stem of *Nypa* involves equal dichotomy. This will be reported in detail in a later publication.

It should be emphasized that the *Hyphaene*-type branching is a normal feature of growth in the palms cited, and in general I do not propose to discuss the numerous examples of pathologically branched specimens of palms which are common among both wild and cultivated palms and about which there is an extensive literature (e.g. Davis, 1969). *Chrysalidocarpus lutescens* is a specific example which does deserve comment. This commonly cultivated palm is multiple-stemmed (branching via basal suckers) but the aerial stem is usually unbranched. Certain individuals of this species, on the other hand, frequently have branched aerial stems, and my colleague, Dr J. B. Fisher, who is investigating such specimens, suggests that the branching, which is not axillary, may involve dichotomy of the shoot apex.

So far these examples have been confined to palms but the same kind of branching has been discovered in two monocotyledons unrelated both to each other and to the palms. One is *Flagellaria indica* (Flagellariaceae), which forms the subject of this present article, and the other is *Thalassia testudinum* (Hydrocharitaceae), which will be described elsewhere. In the later discussion in this article it is suggested that further examples should be sought in other monocotyledons.

It is important to emphasize that these discoveries are all based on careful examination of as wide a range of materials as has been possible and should not be mistaken for the work of very superficial observers who see 'dichotomous branching' in the bifurcated stems of many woody monocotyledons as a means of bolstering some dubious 'evolutionary' theories (e.g. Meeuse, 1965: 31; Greguss, 1968). In all the examples, excepting *Hyphaene*, quoted by these authors (*Aloë*, *Cordyline*, *Dracaena*, *Pandanus*, *Yucca*), branching is sympodial and bifurcation is the result of equal growth of two lateral branches, usually below a terminal inflorescence. Buds and branches in these plants are always in the normal, axillary position. This is an observation which can be verified by any student of elementary botany and by the simplest means.

The present article is restricted to an account of branching in the aerial stems of *Flagellaria indica* and may be regarded as the first unequivocable demonstration of dichotomous branching in a monocotyledon.

TAXONOMY

Flagellaria is one of three genera traditionally included in the Flagellariaceae, an assemblage which, on a wide range of evidence, seems unnatural (e.g. see Tomlinson, 1969: 81). *Flagellaria* is normally regarded as including four species, one in Tropical Africa, the other three in the Indo-Malayan region (Backer, 1951). *Flagellaria indica* L. is widely distributed from Ceylon to Polynesia, northward to Formosa and southward to New South Wales.

MORPHOLOGY

Habit

Flagellaria is a scrambling vine, usually of disturbed habitats. Its morphology is represented diagrammatically in Fig. 1. The climbing, aerial stems arise from a very regularly branched sympodial scale-bearing rhizome, which appears previously neither to have been collected nor described. Rhizome buds, which are the renewal shoots for the whole system, arise as axillary buds at the base of each erect shoot. They are almost invariably in pairs, one on each side of the parent stem, and subtended by two successive scale-leaves. Each bud grows out to make a short rhizome segment before repeating the branching pattern and then turning erect. Aerial stems have internodes up to 30 cm long and are supported by tendrillous leaves, the tendrils being the extended and thickened leaf tips coiled abaxially like a watch spring (du Sablon, 1887). The leaf blade itself is lanceolate with an indistinct petiole attached to the closed tubular sheathing base. There is a transition from scale to foliage leaves at the base of the erect shoot. The lower foliage leaves lack tendrils. Inflorescences are terminal panicles borne at the end of very long shoots. Leaf insertion is distichous throughout the vegetative parts. Foliage leaves do not subtend vegetative buds, an observation supported by microscopic investigation.

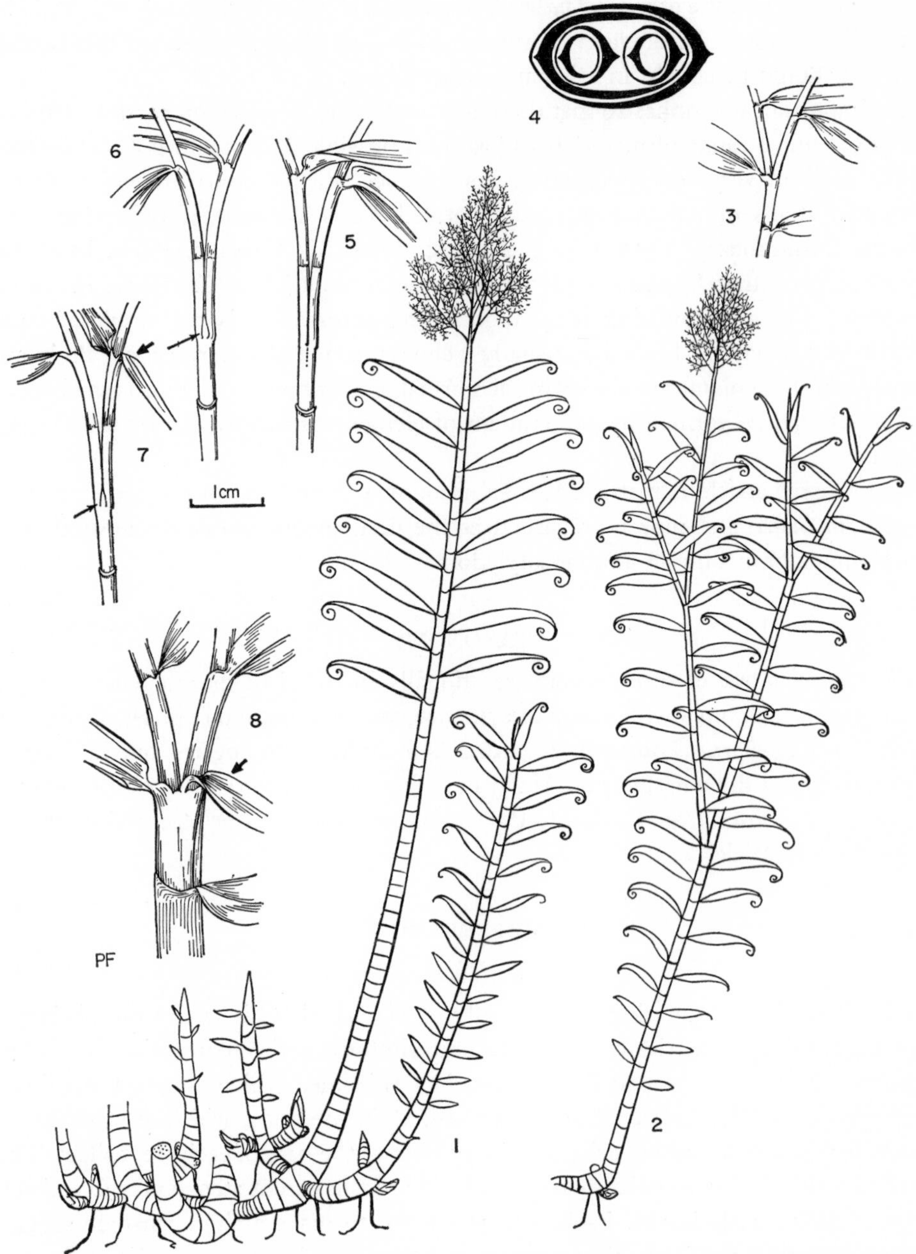

FIGURES 1 to 8. *Flagellaria indica.* Habit and branching. 1. Diagrammatic representation of habit of a plant with unbranched aerial stems, as in Fijian populations. 2. The same for a plant with regularly branched aerial stems, as in Queensland populations. In both figures, internodes and stems are much foreshortened and the interval between branches is shorter than in nature. 3. Leaf arrangement at level of forking. 4. Diagrammatic T.S. of the same. 5. Forking stem with leaf below the fork removed. 6. The same from the other side. 7. Forking stem with leaf below the fork removed. 8. Larger specimen with abnormal leaf. In Figs 6 and 7, the protuberance at the level of the fork is shown by a thin arrow. In Figs 7 and 8, the additional blades on leaves above or below the fork are shown with thick arrows. Figs 3 to 8 are from the New Guinea population sampled as 30.vi.69E.

A peculiar method of branching in these aerial stems was drawn to my attention by Mr J. T. Waterhouse of the University of New South Wales. This involves their equal forking, as illustrated diagrammatically in Fig. 2. Forking can occur at any level and may even involve the inflorescence. It appears to have escaped the notice of previous workers, which is surprising because it is quite commonly shown in herbarium specimens and in fact the illustration in the account by Backer (1951) is of a bifurcated specimen, although the leaf arrangement may not have been represented correctly. My recent (1969) travels in Fiji, New Guinea and Queensland enabled me to study plants in detail in the field. It became evident that this type of branching occurs only in certain populations. Branching was a regular feature of Australian and New Guinean plants, but was not observed in Fijian specimens. Localities for populations seen in the field are listed in the Appendix (p. 13).

Aerial branching

Branching is always in the plane of insertion of the leaves. The relation of parts at the level of branching is very constant in all populations and is represented in Figs 3 to 8. Internodes are covered by overlapping leaf bases, so in Figs 5 to 7 the leaf below the branch has been removed to reveal the fork. At the level of branching the normal distichy which exists below the fork is continued without interruption into both shoots. This is shown in Fig. 3 and in diagrammatic transverse section in Fig. 4. Sister shoots are therefore alike, but not mirror images of each other. This arrangement corresponds to the disposition of leaves along main axis and branch in normal axillary branching, but later description will show that this is incidental. Bifurcation of the axis is always in the middle of the internode and the first leaves on each of the resulting twin shoots are almost invariably situated at exactly the same level. Commonly the leaves at the level of branching are abnormal. Examples are shown in Figs 7 and 8 (thick arrows). In Fig. 7 the first leaf on the right-hand branch has two subequal blades attached on opposite sides of the same sheath and in Fig. 8 the sheath of the leaf immediately below the branch has an additional small blade inserted at right-angles to the blade in the normal position. Populations from New Guinea showed additional modifications of the axis itself, commonly with a minute woody prolongation of the axis between, but always to one side of, the fork (thin arrows in Figs 6 and 7). More extreme irregularities have been seen in which additional branches are associated with the main fork. This shows that in certain populations the bifurcation is not very regular. Subsequent description, therefore, is restricted to a Queensland population with very regular branching.

Frequency of branching

The distance between forks is very long; in the New Guinea populations, for example, there were some 30 to 50 intervening internodes, with the interval between successive forks not very constant. In the population studied near Cairns, Queensland, on the other hand, the interval was much shorter, with a range of 18–26 internodes, and repeated very regularly. By selecting apices of this order of distance beyond a previous fork the chances of obtaining early stages in branch development were made very high. For this reason, and because branching was always of the simple kind illustrated in Figs 3 to 5, subsequent description refers only to the Queensland population, which was sampled thoroughly for anatomical investigation.

MATERIAL AND METHODS

For microscopic investigation of shoot apex morphology, the growing ends of shoots selected in the above manner in the field were fixed in FAA. Subsequently in the laboratory they were trimmed and embedded in 'Paraplast' after dehydration in the tertiary butyl alcohol series in the usual way. Of 24 shoots originally gathered, 8 had dichotomies revealed during dissection and 16 were without visible dichotomy after dissection. Eight from this latter group were subsequently shown by microscopic methods to have very early stages in branching. Serial transverse and longitudinal sections were cut on a 'Jung' rotary microtome at 8 μm, stained in safranin and Delafield's haematoxylin and mounted in the usual way. In order to recognize easily early stages in branching, all shoots without visible dichotomy were sectioned transversely. For this reason I have no longitudinal sections of dividing apices. However, the outline appearance in longitudinal section of apices in which a dichotomy was subsequently revealed were reconstructed from serial transverse sections and are shown in Figs 14 to 18. For comparison, drawings were made of actual longitudinal sections of shoots in which early dissection had already revealed a dichotomy (Figs 10 and 11) together with other reconstructions of the longitudinal appearance of undivided apices, from serial transverse sections (Figs 9, 12 and 13). Photographs which form Plates 1 to 3 were taken on a Wild M-20 microscope. To assist in the analysis of the course of vascular bundles at the level of branching, serial sections were photographed on ciné film by the method described in detail by Zimmermann & Tomlinson (1965, 1966). Preliminary technical work was carried out in Auckland, New Zealand, and completed in Miami.

MICROSCOPIC OBSERVATIONS

Unbranched apices

Apices close above a previous fork and therefore in which branching was very unlikely are shown in Plate **1B–D, F** and Figs 9 to 11. The apex is more or less a hemisphere but with a marked tendency to be flattened either in the plane of distichy or perpendicular to it, probably depending on the stage within a plastochrone at which it was fixed. The apex has a 2(–3)-layered tunica of which only the outer layer remains distinct in lateral organs. The tunic layers surround a corpus of central cells with irregularly orientated walls. Apical cells of both tunica and corpus are distinguished by the intense staining of their walls (Plate 1**B–D**). Below the corpus is a rib meristem in which divisions transverse to the axis predominate so as to produce vertical files of cells. The regularity of this meristem also varies according to the stage in the plastochrone at which the apex is fixed (cf. Plate 1**C** and **D**). Early stages in the initiation of the youngest leaf primordium (P_1) are indicated by anticlinal divisions in the two outermost cell layers followed by irregular divisions in the layers immediately within. This activity is evident in both longitudinal and transverse sections. It leads

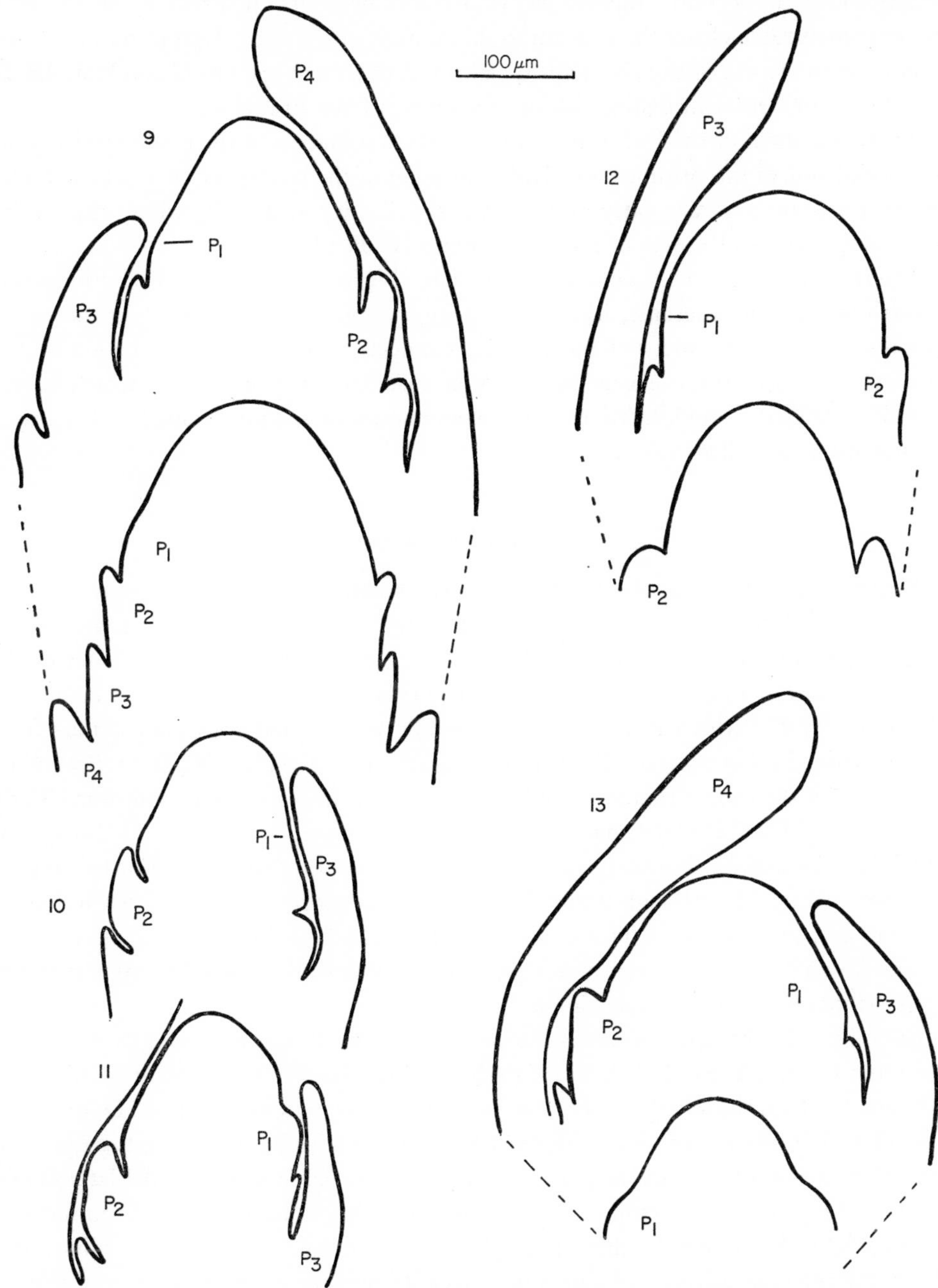

FIGURES 9 to 13. *Flagellaria indica*. Outline of undivided apices in L.S. 9 to 11 are of apices close above a recent dichotomy. 12 and 13 are remote from a previous dichotomy and possibly represent early stages in dichotomy. 10 and 11 are camera lucida drawings of twin apices from a recently bifurcated shoot cut in L.S. The remaining figures are reconstructed from serial T.S., the upper figure in the plane of leaf distichy, the lower at right angles to it. A photograph of one section from 12 is shown as Plate **1A**.

to the development of a collar-like structure which encircles the whole apex. This encircling growth is complete before the next youngest leaf is evident. Consequently in any shoot, the second youngest leaf (P_2) always encircles the axis completely. P_1 is never recognizable closer than 80 μm to the summit of the apex. Leaf initiation always occurs on the flanks of the shoot apex, as is evident from Figs 9 to 13 and Plate 1**B–D, F**. This is important in distinguishing leaf growth from branching.

Further growth of the leaf primordium involves initiation of the blade meristem on the dorsal side of the primordium. Differentiation of the tendril as the thickened apex of the blade occurs very early and is always well advanced in P_4 where the tendril commonly exceeds the rest of the blade (Plate 1**E, F**, td).

Internodal elongation occurs relatively late, beginning at about P_{12}. Elongation growth is then considerable, but does not involve an obvious intercalary meristem. Details of the development of the vascular connection between leaf and stem do not concern us at this point except to note that vascular interconnection, which occurs late, is restricted to nodes. Microscopic examination of sections confirms that leaves do not subtend axillary buds.

Branching apices

Seven of the apices sectioned showed various early stages in dichotomy, and one apex, although undivided, could be interpreted as showing incipient dichotomy (Fig. 12; Plate 1**A**). Branching merely involves a more or less equal division of the apex without any major morphological change (Plate 3**A, C–E**). It always occurs well above the level at which P_1 can be recognized in unbranched apices and differs from leaf inception in that central cells of the corpus are involved (Plate 3**C, D**, c). Branching is clearly not the transformation of a leaf primordium into a shoot primordium. Early stages (Figs 14 to 18) show that at the time of dichotomy the volume of the apex is much increased and it is greatly extended in the plane of leaf distichy. For this reason the apex in Fig. 12, although undivided, may in fact represent the very earliest stage of shoot division because of its size and shape (cf. Plate 1**A**). I had insufficient material to decide whether or not there was a gradual increase in the size of an apex between each dichotomy along any one shoot.

Subsequently the apex becomes bilobed, usually with one lobe slightly exceeding the other (Plate 3**A–F**). This is not significant since there is no constant relationship between the two; the smallest lobe may be on the same side as P_1 (Plate 3**D**) or opposite (Plate 3**C**). Whatever slight difference there may be in the size of the two apices at inception, it is evident that their developmental potential is identical. All later stages show that equality is maintained so that the twin shoots have identical numbers of leaves (Plate 2**A**). The only difference is in the size of the youngest primordium, but the developmental advance of one shoot over its partner never exceeds one plastochrone (e.g. Plate 3**B**). In later stages of development there is some mutual flattening of the sides of the shoots in contact (Plate 2**A**), but since there is no dominance of one twin, there is no chance for a keeled, prophyllar structure to develop. The median vascular bundle of the first leaves on each of the new shoots is often displaced to a pseudolateral position (Plate 2**A**, m). Such leaves are commonly distinguished from previous or subsequent leaves by an aberrant number of blades, as has already been described. There is some displacement of leaf primordia to fit the requirements for close packing so that the first leaves on daughter shoots may not lie in the plane of distichy of the parent shoot (e.g. Plate 3**B**). The only subsequent divergence in

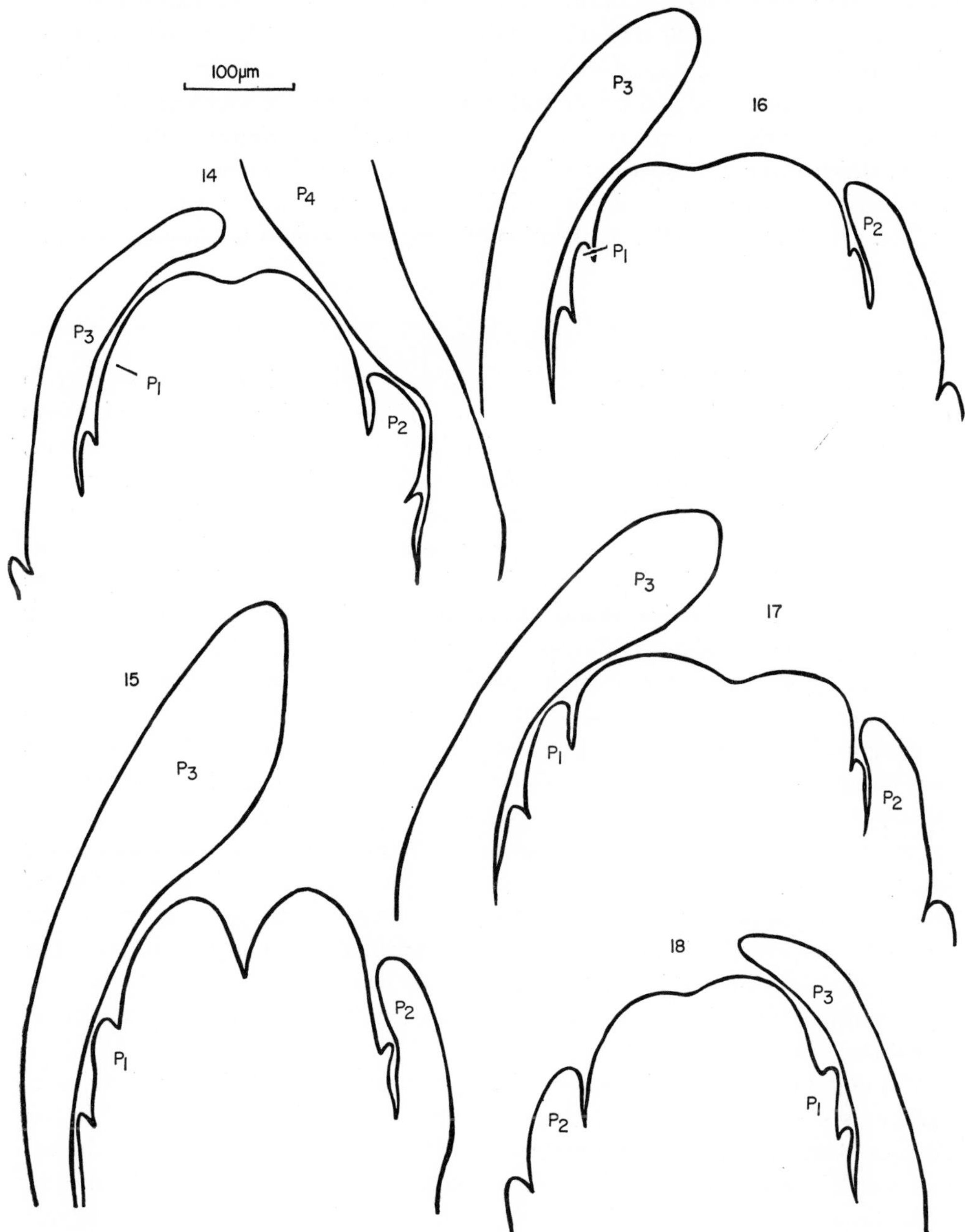

FIGURES 14 to 18. *Flagellaria indica.* Outline of dividing apices in L.S., in plane of leaf distichy. These are reconstructed from serial T.S. Single sections corresponding to each of these are shown in Plate 3 as follows: **A** = 17; **C** = 18; **D** = 16; **E** = 14; **F** = 15.

behaviour of twin shoots is that one may flower before the other, as suggested diagrammatically in Fig. 2.

Vascular system of branched shoots

An examination of sections of dichotomized shoots of all ages suggests that there is no discontinuity of growth at the time of shoot division (e.g. Plates 1**E** and 2**B**). There is no abortion or loss of meristematic tissue. This continuity at the level of branching is most clearly indicated in the distribution of vascular bundles, which all extend from the stem below the fork into one of the two shoots. This is shown in the analytical films, which made a comparison of vasculature above and below a fork very easy. Bundles in the stem below a fork are divided more or less equally between the twin shoots. The developmental interpretation of these observations is discussed below.

DISCUSSION

The morphological and microscopical evidence presented above shows that in certain populations of *Flagellaria indica* growth of the climbing aerial stems involves a kind of branching which has not previously been described in monocotyledons, although it has been suspected in palms. Clearly the branching is not axillary, although axillary branching does occur in the rhizomatous parts of *Flagellaria*. Whether one describes it as dichotomous branching or not is simply a matter of definition. Fortunately, Nolan (1969), in his account of the similar kind of branching involved in the production of reproductive shoots in *Asclepias syriaca*, provides the following definition: 'A true dichotomy of a shoot apex can be defined as a bifurcation produced when the apical center (one or more apical initials) of the apex is transformed by internal cellular divisions into two apical centers; each new apex is devoid of a subtending organ and each establishes its own morphological identity. The new apices may be equal to each other or to the parental apex in size, form and function but such equality is not a necessary condition.' The type of branching found in *Flagellaria* agrees with this definition. It is true that there is a leaf in the *position* of a subtending organ, but equally clear that this leaf does not subtend a branch, in a *developmental* sense. The qualification about subsequent behaviour of the twin shoots is not necessary for *Flagellaria* because their developmental potential is identical. The qualification is necessary in *Thalassia* because the twin shoots have a very different developmental potential although they originate simultaneously as will be described elsewhere.

It is unfortunate that Nolan extends his discussion in an evolutionary direction in order to speculate about the 'phylogenetic significance' of his observations, in the absence of any physiological understanding of the dichotomy he describes. In the monocotyledons it is more profitable to consider the developmental significance of dichotomous branching. This is appropriate in view of the recent advances in our understanding of monocotyledonous growth made as a result of analyses of the development of the vascular system in larger monocotyledons carried out in collaboration with my colleague, Dr M. H. Zimmermann (Zimmermann & Tomlinson, 1967,

1968, 1969). These studies have shown that the origin of the vascular system of monocotyledons involves a developmental process in which procambial strands associated with the growth of leaf primordia, with a 'downwardly directed' developmental orientation, link with 'upwardly directed' strands in the stem. A fundamental feature of this concept is that the 'upwardly directed' traces initially have no developmental association with the leaf with which they will ultimately link. The axial bundles (or 'vertical bundles' of the above papers) are initiated within a 'cap' of meristematic tissue close beneath the shoot apex proper and become recognizable as bundles ending 'blindly' in a distal direction long before the leaf with which they are ultimately connected via linking, downwardly-directed leaf traces, exists as a visible primordium. This is far too simple a statement (about what is obviously a very complex subject) to be understood without detailed reference to the articles cited above, but in this general expression it may provide an explanation for the existence of dichotomizing shoots in monocotyledons.

On this basis, bifurcation of the shoot of a monocotyledon is, in developmental terms, 'easy' since an equal division of the shoot apex merely redirects the ends of unconnected axial bundles towards the resulting twin growth centres. Linkage between vascular systems of leaf and stem above and below the fork is unaffected. (This, of course, offers no explanation of why the apex should bifurcate in the first instance.) Monocotyledonous growth therefore lends itself to dichotomous branching and to talk about this kind of branching in evolutionary terms completely divorced from any understanding of principles of growth can only lead to meaningless conclusions. In *Thalassia*, where dichotomy is also found, the conclusion that it represents a 'primitively persistent' feature is particularly valueless since the Hydrocharitaceae are such a specialized group of monocotyledons.

Returning to shoot growth in other monocotyledons, a type of branching which is developmentally comparable to a dichotomy is found in the inflorescence axes of many palms, in which branches are frequently adnate to the parent axis for part of, or even the whole of, the internode above their theoretical node of origin. A branch may equal or even exceed its parent axis in size. Morphologically, these branches are axillary since they are subtended by a bract (in whose axil they presumably originate). That they are developmentally comparable to the *Flagellaria*-type of dichotomy is indicated by the equal distribution of vascular bundles between each axis, as has been described for *Rhapis excelsa* (Tomlinson & Zimmermann, 1968). This reflects the precocious origin of the lateral. By a further extension of this principle, it could be reasoned that any other bifurcation of a monocotyledonous axis merely reflects precocious development of a lateral. This is arguable in *Flagellaria* where leaf arrangement does indeed conform to a situation which might arise if a lateral bud were to develop before its subtending leaf. But the developmental evidence presented in this article contradicts this. In other examples of dichotomizing monocotyledons (*Nypa*, *Thalassia*), the leaves at the level of forking are not arranged in a way which suggests precocious development of a lateral, so that there is morphological as well as developmental evidence in favour of true dichotomy. Quite obviously this topic could easily degenerate into mere philosophical speculation which would not add to our understanding of dichotomous branching as a feature of growth.

So far we have dealt developmentally with a population of *Flagellaria* with regular forking of the stem. I have made no developmental observations on the less regularly branched New Guinea populations, examples of which are illustrated in Figs 6 and 7. Probably these can be accepted as instances in which the apex divides not twice but three times, with one apex aborting early. Abnormal leaves, with additional, incompletely developed blades (Figs 7, 8), may in turn reflect aberrations induced by the double growth centre. Whether the abnormal leaf is above or below the fork may be determined by whether the bifurcation occurs early or late within the plastochrone interval. These, and other matters associated with this type of branching, lead to interesting morphogenetic speculation. Why, for example, does the first leaf on each of the twin shoots occupy a comparable position, continuing the previously established distichy? Is this explicable in terms of modern theories of phyllotaxis?

Turning aside from this line of thought, we may again reiterate the conclusion that monocotyledonous growth favours dichotomous branching, so that dichotomy is not necessarily a primitive feature of angiosperm morphology. The conclusion which should surprise us is not that dichotomous branching occurs in monocotyledons but that it has been overlooked for so long. This, however, merely reflects our relative ignorance of monocotyledonous morphology (see Tomlinson, 1970). My prediction is that when attempts are made to correct these deficiences by careful examination of the growth habits of monocotyledons (unfashionable though this kind of inquiry may be), dichotomous shoots may prove to be quite frequent in monocotyledons.

ACKNOWLEDGEMENTS

It is a particular pleasure to be allowed to include this article in a publication honouring Dr C. R. Metcalfe's services to botanical science in general and to the Linnean Society in particular, since it was his initial guidance in descriptive plant anatomy which laid the foundation for much of my own research and provided so essential a background for its extension into aspects of the developmental anatomy of monocotyledons.

For assistance in collecting *Flagellaria* in the South Pacific I am indebted to Mr J. W. Parham and Dr A. C. Smith (Fiji); Mr J. S. Womersley (New Guinea); Dr L. J. Brass and Mr Vince Moriarty (Queensland). In the Department of Botany, University of Auckland, New Zealand, I am grateful for the facilities provided by Professor V. J. Chapman and Associate Professor L. H. Millener and for the technical assistance of Miss Theresa Brownlee. Illustrations which form Figs 1 to 8 were drawn by Priscilla Fawcett, Botanical Illustrator at Fairchild Tropical Garden. Support for continuing studies on the anatomy of the monocotyledons is provided by Grant GB 5762-X from the National Science Foundation, Washington, D.C.

REFERENCES

Backer, C. A., 1951. Flagellariaceae. *Flora Malesiana Bull.* (*ser. 1*), **4**: 245–250.

Beccari, O., 1924. *Palme della tribù Borasseae* (ed. U. Martelli). Firenze.

Davis, T. A., 1969. Ramifying and twisting stems of Palmyra palm (*Borassus flabellifer*). *Principes*, **13**: 47–66.

Greguss, P., 1968. Dichotomous branching of monocotyledonous trees. *Phytomorphology*, **18**: 515–520.

Lewalle, J., 1968. A note on *Hyphaene ventricosa*. *Principes*, **12**: 104–105.

MEEUSE, A. D. J., 1965. Angiosperms—past and present. *Advg Front. Pl. Sci.*, **11**: 1–228.
NOLAN, J. R., 1969. Bifurcation of the stem apex in *Asclepias syriaca*. *Am. J. Bot.*, **56**: 603–609.
SABLON, LECLERC DU, 1887. Recherches sur l'enroulement des vrilles. *Annls Sci. nat.*, *Bot.* (*ser.* 7), **5**: 1–50.
SCHOUTE, J. C., 1909. Über die Verästelung bei monokotylen Bäumen. II. Die Verästelung von *Hyphaene*. *Recl Trav. bot. néerl.*, **6**: 211–232.
TOMLINSON, P. B., 1967. Dichotomous branching in *Allagoptera*? *Principes*, **11**: 72.
TOMLINSON, P. B., 1969. *Anatomy of Monocotyledons* (ed. C. R. Metcalfe). III. *Commelinales to Zingiberales*. Oxford: Clarendon Press.
TOMLINSON, P. B., 1970. Monocotyledons—towards an understanding of their morphology and anatomy. In R. D. Preston (ed.), *Advances in botanical research*, **3**. London: Academic Press.
TOMLINSON, P. B., & MOORE, H. E., 1966. Dichotomous branching in palms? *Principes*, **10**: 21–29.
TOMLINSON, P. B., & ZIMMERMANN, M. H., 1968. Anatomy of the palm *Rhapis excelsa*, V. Inflorescence. *J. Arnold Arbor.*, **49**: 291–306.
ZIMMERMANN, M. H. & TOMLINSON, P. B., 1965. Anatomy of the palm *Rhapis excelsa*, I. Mature vegetative axis. *J. Arnold Arbor.*, **46**: 160–178.
ZIMMERMANN, M. H. & TOMLINSON, P. B., 1966. Analysis of complex vascular systems in plants: optical shuttle method. *Science, N.Y.*, **152**: 72–73.
ZIMMERMANN, M. H. & TOMLINSON, P. B., 1967. Anatomy of the palm *Rhapis excelsa*, IV. Vascular development in apex of vegetative aerial axis and rhizome. *J. Arnold Arbor.*, **48**: 122–142.
ZIMMERMANN, M. H. & TOMLINSON, P. B., 1968. Vascular construction and development in the aerial stem of *Prionium* (Juncaceae). *Am. J. Bot.*, **55**: 1100–1109.
ZIMMERMANN, M. H. & TOMLINSON, P. B., 1969. The vascular system in the axis of *Dracaena fragrans* (Agavaceae), I. Distribution and development of primary strands. *J. Arnold Arbor.*, **50**: 370–383.

APPENDIX

Localities of populations examined, with voucher number, if available.

Fiji

Voucher numbers are those of Department of Agriculture, Suva, Fiji.
Flagellaria indica L. Mount Korambamba, Viti Levu. SUVA 16505.
Flagellaria gigantea Hook. f. Toninaiwau, Colo-i-Suva, Viti Levu. SUVA 16647.
Flagellaria sp. (cf. *F. neocaledonica*) Buca Bay, Vanua Levu. SUVA 16874.

New Guinea

Flagellaria indica L. Edie Creek Road, Wau, Morobe District. P.B.T. 30.vi.69E.; Gabensis village between Wau and Bulolo, Morobe District; no voucher. North-east of Port Moresby International Airport, Port Moresby; no voucher.

Australia

Flagellaria indica L. Six miles south of Cairns, Queensland. P.B.T. 6.vii.69A.

EXPLANATION OF PLATES

PLATE 1

A. T.S. unbranched shoot remote from a previous dichotomy, 8 μm below apex; P_{1-2} are below the level of the section. (The same apex in L.S. forms Fig. 12.)
B–D. Approximately median L.S. of 3 apices close above a recent dichotomy.
E. L.S. recently dichotomized shoot, only the left-hand apex is cut medianly.
F. Same shoot shown in **B** at higher magnification.
c, Corpus; P_{1-4}, leaf primordia in order of age, P_1 the youngest; r, rib meristem in **D**; t, tunica; td, tendril.

PLATE 2

A. T.S. recently dichotomized shoot at approximate level of insertion of P_1 on each apex.
B. L.S. recently dichotomized shoot, each apex cut more or less medianly.
m, Presumed median vascular bundle of first leaf on each daughter shoot, displaced laterally; P_{1-7}, leaf primordia in order of age, P_1 the youngest; P_L, leaf on parent axis immediately below fork.

PLATE 3

A–F. Bifurcating apices in T.S.
A. 16 μm below apex (cf. Fig. 17 which shows the same shoot in L.S.).
B. Another shoot at later stage, 40 μm below apex of tallest (lower) shoot.
C, D. Early stages in dichotomies.
C. 8 μm below apex (cf. Fig. 18). **D.** 8 μm below apex (cf. Fig. 16).
E. 16 μm below apex (cf. Fig. 14).
F. 40 μm below apex, the two new apices are now quite distinct.
c, Central corpus cells which are involved in dichotomy; P_{3-4}, leaf primordia in order of age, P_3 the youngest (P_{1-2} are below the level of the sections); P_L, leaf on parent shoot immediately below fork.

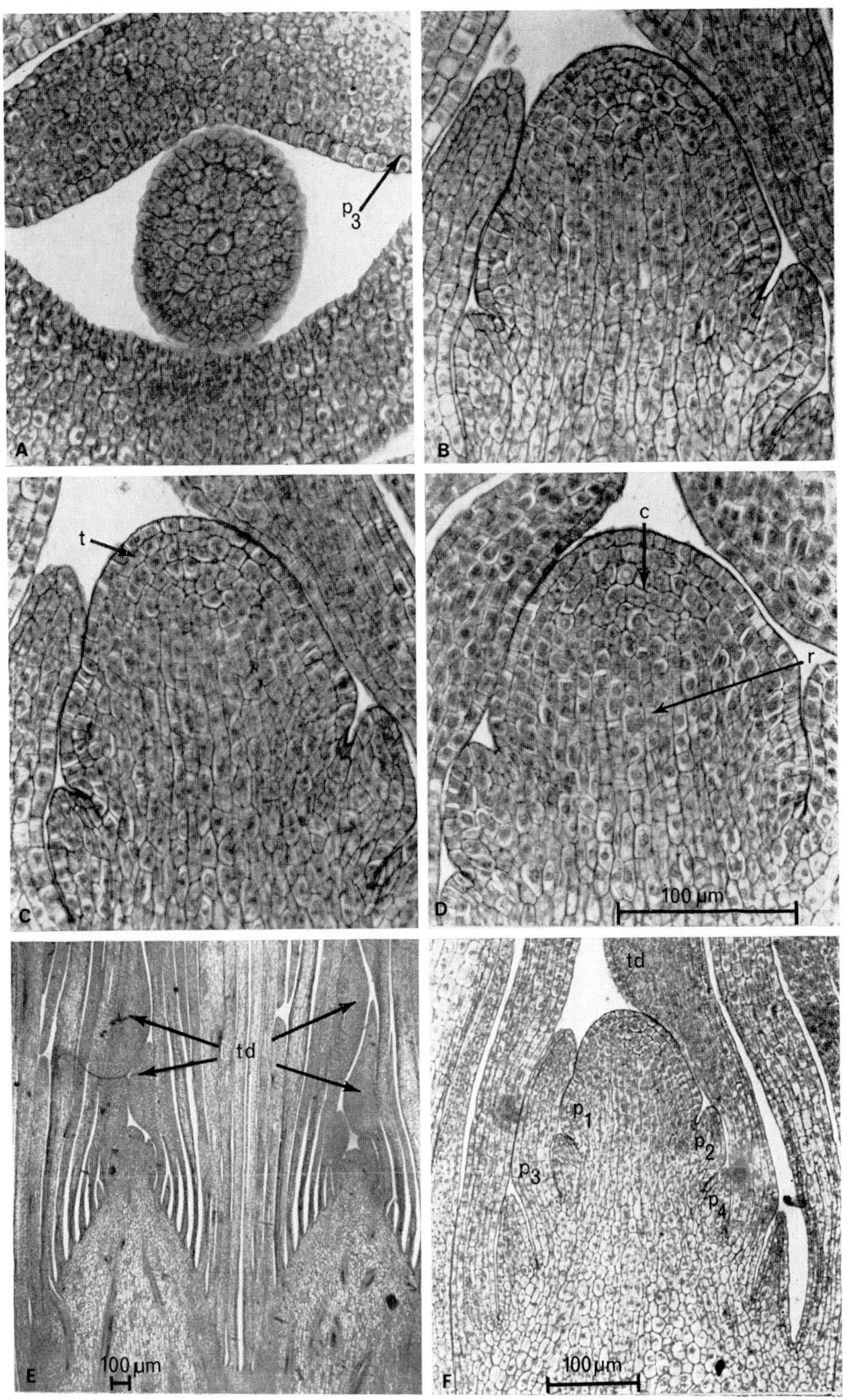
p3
A
B
t
C
c
r
100 µm
D
td
100 µm
E
td
p1
p2
p3
p4
100 µm
F

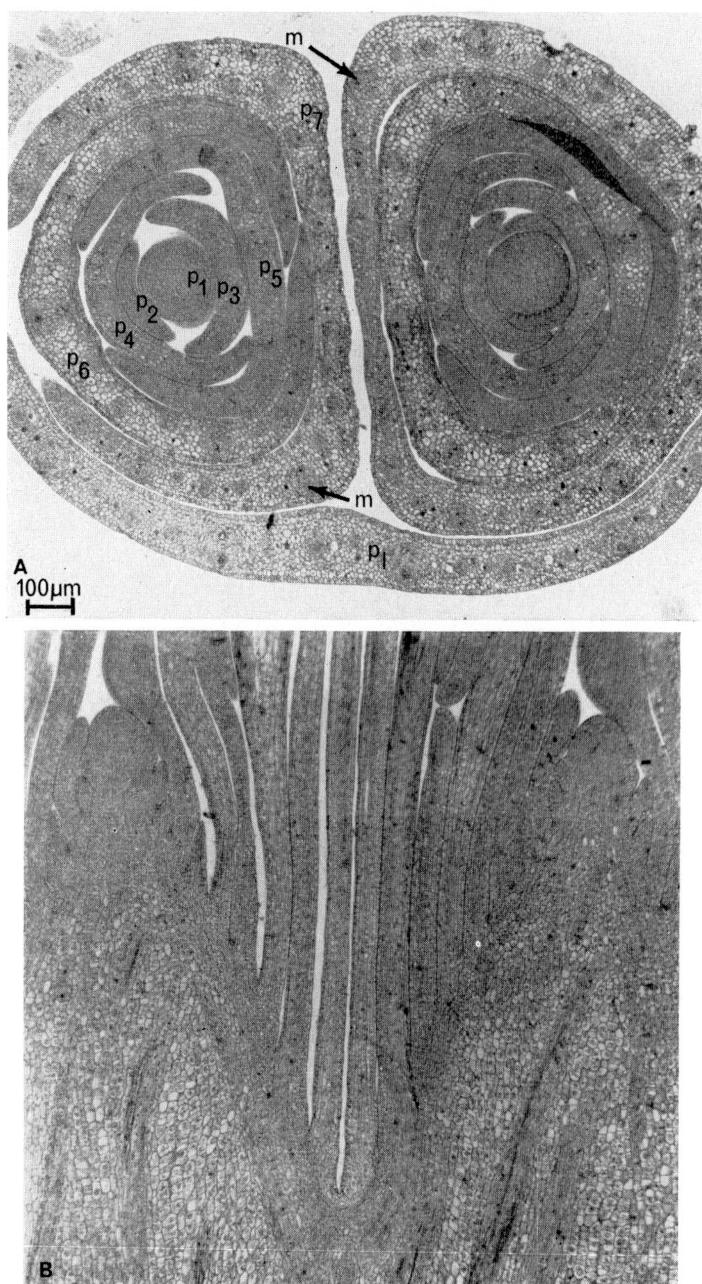
m
p7
p5
p1 p3
p2
p4
p6
m
p1
A
100μm
B

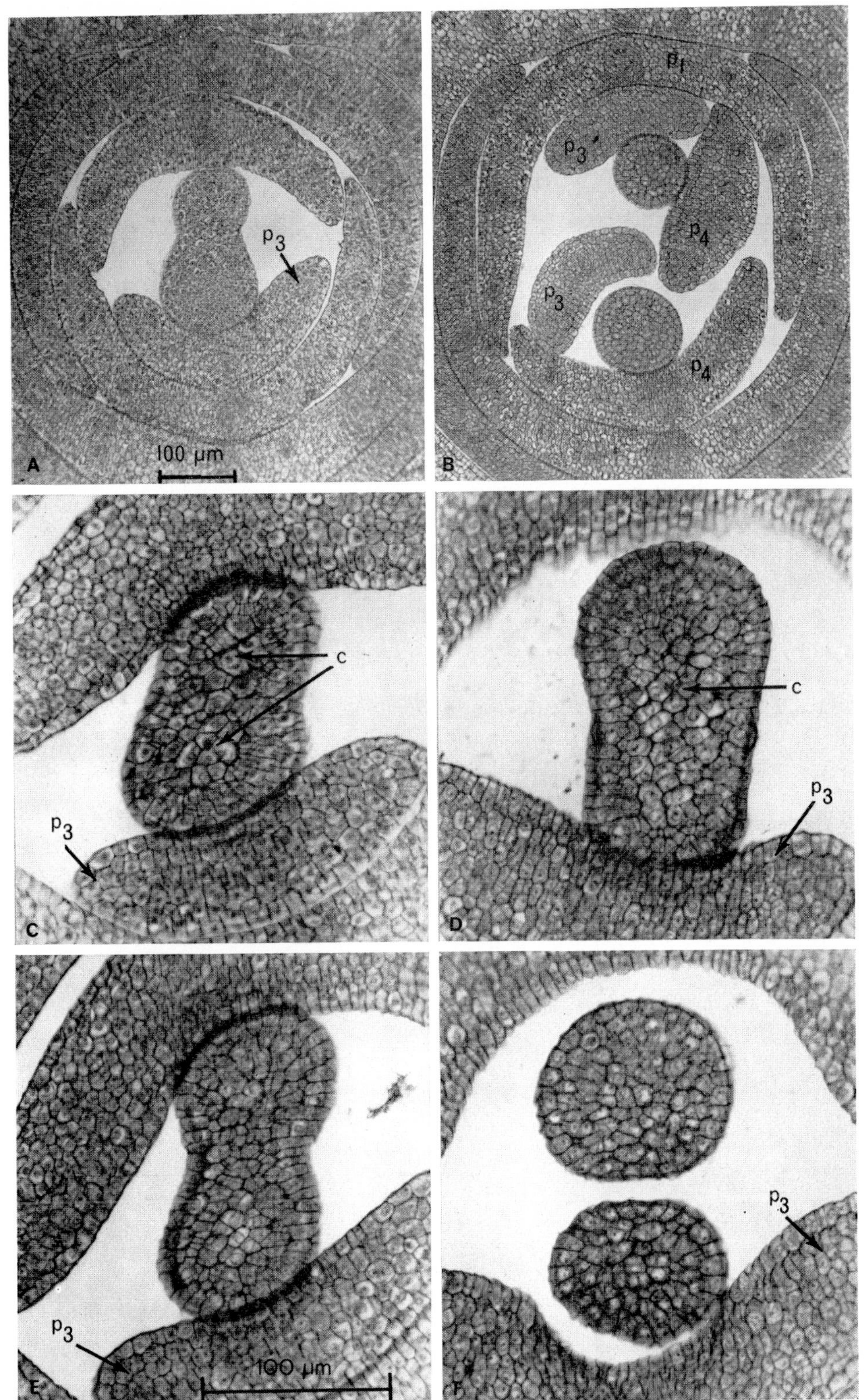

P. B. TOMLINSON

Mucilaginous idioblasts in okra, *Hibiscus esculentus* L.

FLORA MURRAY SCOTT

Department of Botanical Sciences, University of California, Los Angeles, U.S.A.

AND

BARBARA G. BYSTROM

Space Biology Laboratory, Brain Research Institute, University of California, Los Angeles, U.S.A.

The mucilage which flows freely from cut tissues of okra occurs in a diffuse network of mucilaginous idioblasts. The protoplast of the idioblast consists of a thin plasmolemma, a mass of apparently homogeneous mucilage, and a central nucleocytoplasmic core. Cytoplasmic strands radiate from the core and terminate as plasmodesmata visible in the pits of the cell wall. Aqueous sap vacuoles, typical in unspecialised cells, are not observed at any stage during growth. Chloroplasts and lesser organelles are not obvious. The cell walls of the idioblasts, like those of the contiguous parenchymatous cells, consist of cellulose, but they are reinforced in the region of the middle lamella by a suberin-like, perforate sheath. Mucilage stains with aqueous dyes such as neutral red, Janus green and others and swells immediately in contact with water. The cell contents of the idioblasts may be observed in fresh material or, more easily, in tissues which have been immersed in ethyl alcohol for 24 hours or more. The nature of the sheath may be tested with $I_2KI–H_2SO_4$.

CONTENTS

INTRODUCTION

Okra, *Hibiscus esculentus* L., family Malvaceae (Metcalfe & Chalk, 1950), is a rapidly growing annual with palmate leaves and conspicuous flowers (Plate 1). Winton & Winton (1935) comment that, in the north, the fruit is a vegetable of secondary importance, but, in the south, the plant is grown for its immature fruits, which, when used in soups, impart a peculiar mucilaginous consistency.

In the tender edible stage, the pod in cultivated varieties consists of from five to eight carpels and ranges in length from 2–16 cm. If the pod is cut while still green, mucilage flows freely from all parts. In the flowers, solitary in the leaf axils, an involucre of slender bracts encircles the conical, gamosepalous calyx, which is deciduous and splits lengthwise as the flower opens. The petals, yellow with reddish centre, and the stamen groups are in number equal to the carpels of the ovary. Hairs, abundant on the epidermis of all organs except the stamens, are mainly unicellular in younger organs, and multicellular in the older. In addition, short-stalked, capitate, multicellular oil glands occur sporadically on the various organs. Judging by their blackening on treatment with osmic acid, the contents of these glands include unsaturated fatty acids.

The occurrence and the unique structure of the mucilage-containing cells are discussed first for the carpels and thereafter for the vegetative organs, the stems and the leaves. Typical mucilage idioblasts are not observed in the younger roots. This introductory survey indicates the need for intensive investigation of mucilaginous idioblasts with light and electron microscopes and with microchemical techniques.

Mucilage, a complex of carbohydrates varying in different species, is an hygrophilous substance, which becomes slimy in contact with water. It occurs throughout the entire plant kingdom. Apart from the comprehensive account of mucilage by Tunmann & Rosenthaler (1931), the anatomy of mucilaginous cells is generally ignored in current textbooks. In higher plants, mucilage is familiar in the following situations: (1) It is a component of epidermal cell walls, as in the coating of germinating seeds. It is also evident as a film of varying thickness on the epidermis of young growing roots, including the root hairs. (2) Intercellular spaces in the specialized layers of certain seed coats, as in *Cercidium* (Scott, Bystrom & Bowler, 1962), are entirely filled with what appears to be homogeneous mucilage. No air-containing spaces are observed at any stage of growth in this mucilaginous mass. (3) In typical living cells, the vacuolar sap is aqueous, colloidal and hygrophilous and contains varying amounts of carbohydrates and proteins and, in general, traces of mucilage (Crafts, Currier & Stocking, 1949). (4) In okra, however, and possibly in other mucilaginous species, typical aqueous cell sap is not observed at any stage of growth of the mucilaginous idioblasts. The entire cell, apart from the encircling plasmolemma and a central branching core of nucleocytoplasm, is filled with a mass of mucilage, and appears to be lacking in chloroplasts and lesser organelles.

MATERIAL AND METHODS

In definitive anatomical examination, sections of fresh material are of necessity cut sufficiently thick so that at least one layer of living cells remains undamaged. The exudation of mucilage during cutting is troublesome. Segments of fresh fruits are therefore seared with a match when cut, and are immediately immersed in 95% ethyl alcohol. Within 24 hours or more, the fruits are partially hardened, and may be sectioned by hand, or by a sliding microtome, with or without the use of pith. The microtome blade and the fruit are kept moist with alcohol and the sections are mounted directly on a slide. The sections are irrigated with one or two drops of neutral red, Janus green, thionin or other aqueous stain and covered with a cover glass in the usual way. They are then observed immediately and continuously. The sections may be kept satisfactorily for several weeks in glycerin. As has been noted, mucilage is hygrophilous, and even after prolonged immersion in alcohol, it retains its capacity to absorb water. In the sections the mucilaginous idioblasts vary in appearance according to whether or not they have been damaged in cutting, permitting the outflow of the mucilage in the aqueous dye.

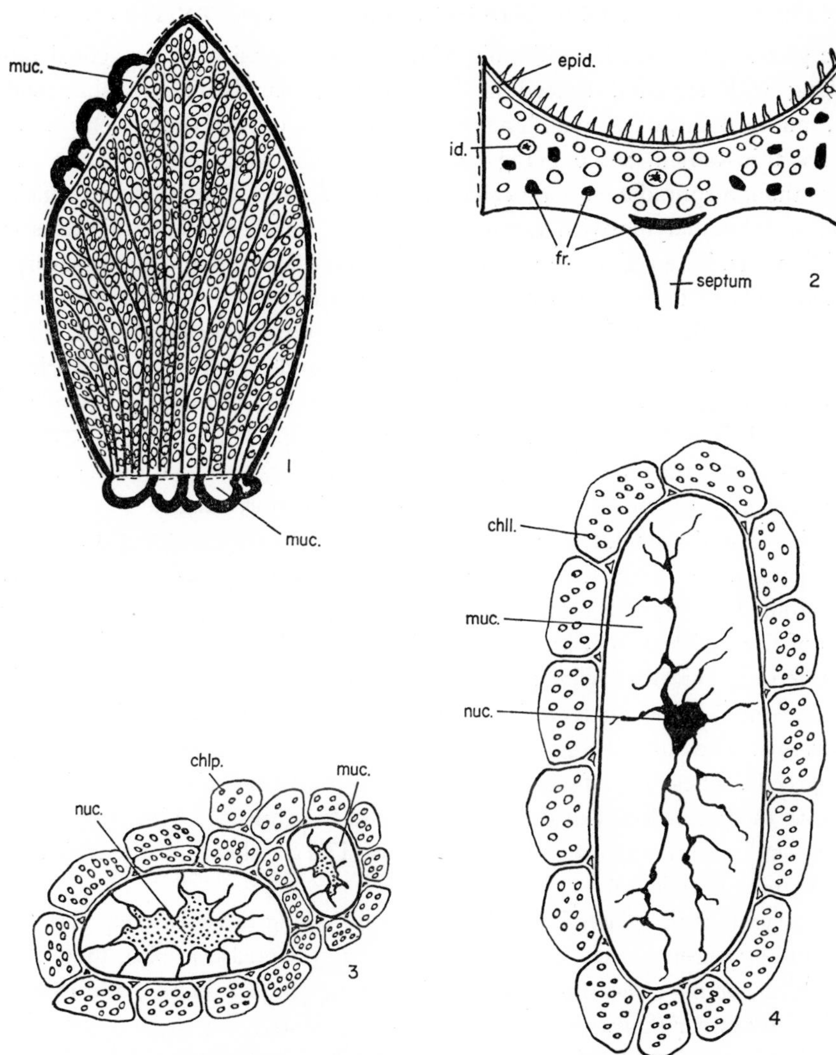

FIGURES 1 to 4. Mucilaginous idioblasts in carpels of varying size.

1. Carpel (actual size 8 × 6 mm, dissected from bud 2·5 cm long) stained with neutral red, showing venation and distribution of idioblasts and exudation of mucilage at cut surfaces (length of idioblasts about 20 μm).

2. T.S. of carpel wall (fruit 10 cm long, size of T.S. 9 mm) showing distribution of idioblasts, fibrovascular strands and interlocular septum (actual length of larger idioblasts about 250 μm); nucleocytoplasmic core is occasionally visible in some idioblasts.

3. Idioblasts (fresh fruit 3·5 cm long) stained with Janus green, showing nucleocytoplasmic core with microgranulation and cytoplasmic strands extending into cell wall (actual lengths of idioblasts 96 μm and 45 μm); surrounding cells contain chloroplasts.

4. Idioblast (fresh fruit 10 cm long) stained with neutral red, showing nucleocytoplasmic core (actual size 384 × 105 μm) with radiating strands, surrounded by cells containing chlorophyll.

The figures are diagrammatic, based on freehand and camera lucida drawings. chlp., Chloroplasts; chll., chlorophyll; epid., epidermis with hairs; fr., fibro-vascular strands; id., idioblast showing nucleocytoplasmic core; muc., mucilage; nuc., nucleocytoplasmic core.

OBSERVATIONS OF IDIOBLASTS IN FLORAL AND VEGETATIVE ORGANS

In the growing fruits, the epidermis of the carpels consists of polygonal tabular cells, along with stomata, unicellular and multicellular hairs and short-stalked, capitate, multicellular, oil glands. The parenchymatous cells of the underlying hypodermis are thin-walled, some being collenchymatous, while others, more or less numerous though sporadic, contain calcium oxalate druses. The mesocarp consists in the main of chlorenchyma with varying amounts of starch and sporadic druse-containing idioblasts. The reticulate vascular system ramifies through the mesocarp.

The cell walls of the parenchymatous elements in the body of the mesocarp are composed in the main of cellulose, and, as is usual, are pitted on all cell faces. Air-filled intercellular spaces are ubiquitous and are lined with lipid. Chloroplasts, mitochondria and lesser organelles are present throughout the tissue. The mucilage, which flows freely from freshly cut fruits (Fig. 1), is contained in specialized mucilaginous idioblasts. These cells may be solitary, but in general they are contiguous and form a diffuse network, and locally they may be grouped in short canals.

In the ridged mid-carpel, the individual xylem strands are slender and the phloem is inconspicuous. In contrast, the xylem strands of the marginal veins, from which arise the funicular strands supplying the ovules, appear in transverse section as crescents of thicker strands (Fig. 2). The mid-region of the carpel is thus a zone of comparative mechanical weakness, which marks the future plane of dehiscence along the ridges of the 3–5-locular fruit. The septa between the loculi contain median and marginal veins. The outstanding feature of the mesocarp is the network of mucilage-containing cells. These cells mainly lie parallel to the vascular strands and are contiguous with the parenchyma of the mesocarp or of the veins. The inner epidermis of the carpel loculi is similar to the outer epidermis. The basal core of the fruit consists of parenchymatous cells only.

The mucilaginous idioblasts in meristems and in young tissues are equal in size to the surrounding parenchymatous cells. Later, as they increase in size, they may remain spherical, or may become ellipsoidal in outline (Figs 3 and 4), and eventually are consistently larger than the parenchymatous elements.

The cell walls of the idioblasts, like those of the surrounding chlorenchymatous

---

FIGURES 5 to 8. Mucilaginous idioblasts in carpels of varying size.

5. **A.** Idioblasts (fruit 3·5 cm long) stained with neutral red, showing free-floating mucilage (actual diameter about 220 μm); note variation in outline of nucleocytoplasmic cores and their microgranulation. **B, C.** Nucleocytoplasmic cores in exuded mucilage (actual sizes about 64 and 180 μm); note fantastic outlines and microgranulation.

6. Idioblast (fruit 2 cm long) during treatment with $I_2KI–H_2SO_4$, surrounded by parenchymatous cells containing chlorophyll (actual diameter 64 μm); note suberin-like sheath and cellulose in cell wall, nucleus evident in nucleocytoplasmic core and ubiquitous microgranulation.

7. Idioblast (fruit 8 cm long) during treatment with $I_2KI–H_2SO_4$ (actual length 190 μm); note suberin-like sheath in cell wall, nucleocytoplasmic core with cytoplasmic strands ending in visible plasmodesmata in pits in cell wall, and ubiquitous microgranulation.

8. Sheaths (fruit 8 cm long) isolated during treatment with $I_2KI–H_2SO_4$ (actual lengths 192 and 240 μm); note occasional pits and microgranulation.

cell., Cellulose in cell wall; chll., chlorophyll; muc., mucilage; nuc., nucleocytoplasmic core; nucleus, nucleus and nucleolus of nucleocytoplasmic core; sub., suberin-like sheath.

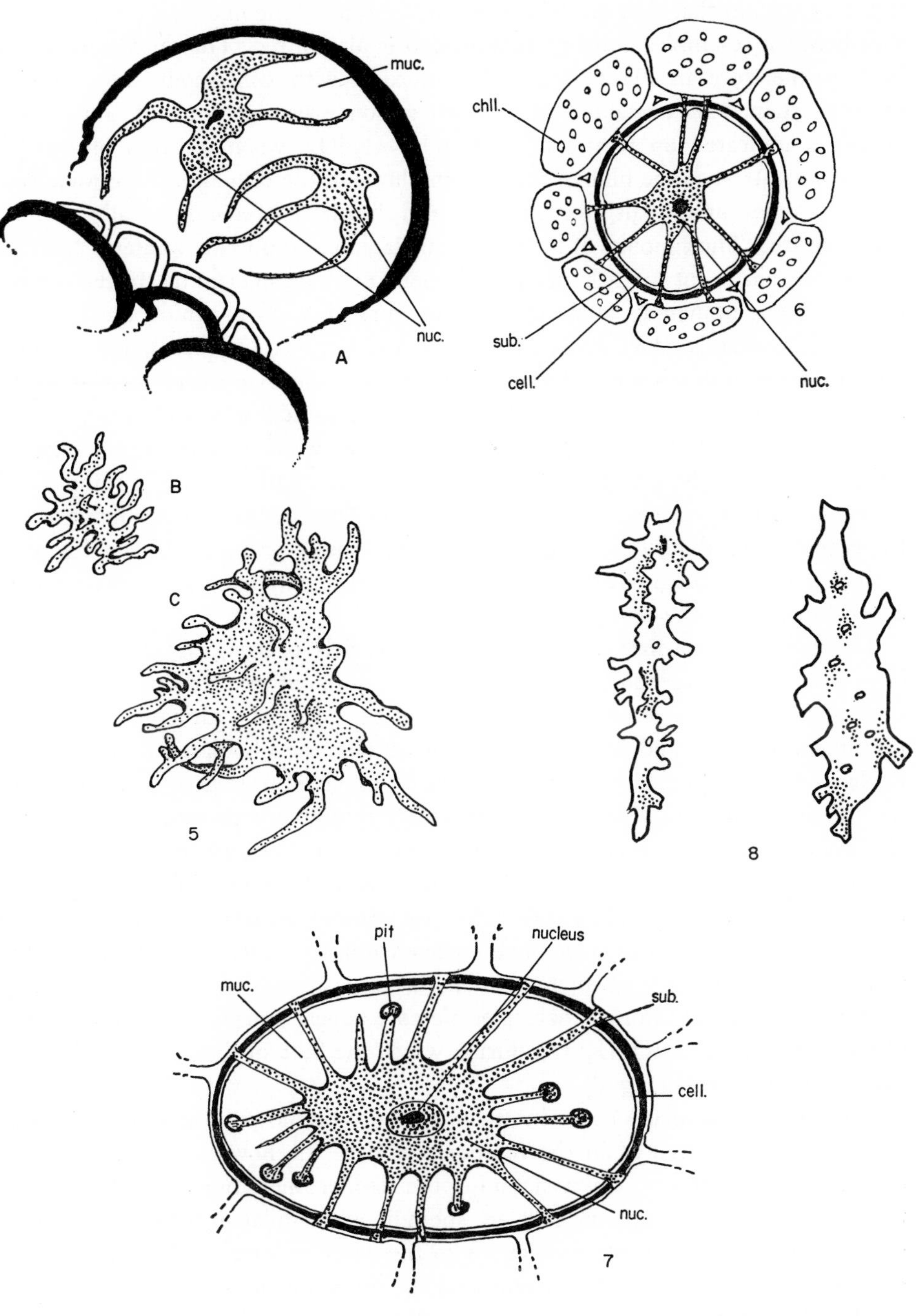

FIGURES 5 to 8

cells, consist in the main of cellulose, and are pitted on all cell faces. They are distinctive, however, in that they are reinforced in the zone of the middle lamella by a lipid sheath which resembles cutin or suberin and is also pitted. The mucilaginous idioblast is therefore in plasmodesmatal connexion with contiguous cells. In fresh sections, stained or unstained, the sheath is somewhat birefringent. Its chemical nature is indicated on treatment with I_2KI–H_2SO_4, when it turns brownish in colour in contrast to the blue of the swelling layer of the cellulose of the wall. When the cellulose eventually disintegrates, the lipid sheath may remain partially in situ, or may float free. The contents of the intact mucilaginous idioblasts, whatever their size, consist of translucent turgid mucilage, enveloped in a plasmolemma. In the centre of the cell lies a branching nucleocytoplasmic core (Fig. 5) in which the nucleus is visible. The strands which radiate from this core terminate as plasmodesmata in the pits of the cell wall (Figs 6 and 7). In general the nucleocytoplasmic core stains first and most deeply in aqueous stains. It is roughly ellipsoidal in form but is variable in outline because of the radiating cytoplasmic strands which emerge from flanges and taper off towards the pits in the cell wall. The entire surface of the protoplast, the interface between the plasmolemma and the mucilage, appears to be microgranular; the granules are about 0·5 μm in diameter and closely packed. This microgranular pattern persists throughout the length of the radiating strands into the plasmolemma and is mirrored in the lipid sheath.

The mesocarp is a firm tissue composed of closely packed, turgid cells, chlorenchymatous or mucilaginous, spherical or ellipsoidal in outline. Striking changes occur immediately after sectioning and treatment of the sections with aqueous stains. As the mucilage swells and flows out in all directions, the idioblastic sheaths and the nucleocytoplasmic branching cores are more or less freed from their sites and stain in the thinning mucilage. The sheaths apparently resemble cutin and suberin in their elastic properties. When cut, they shrink and may at once assume fantastic, blunt spiny outlines. The spines indicate the former sites of the intercellular spaces adjacent to the contiguous surrounding cells, while the plasmodesmata and pits were located in the intervening curving areas. The nucleocytoplasmic cores are varied in outline, and when floating free their microgranular surface is at once obvious. These changes in form in sheath and in nucleocytoplasmic core appear equally clearly on treatment of the sections with I_2KI–H_2SO_4, during which the bulk of the cellulose of the wall eventually disappears (Fig. 8).

Mucilaginous idioblasts occur in other organs of the flower: the bracts, the sepals and the petals, as well as in the receptacle and in the flower stalk. They are not observed in the stamens. The venation of the slender linear bracts is reticulate as in the carpels. The mucilaginous cells are sporadic and similar in form to those in the carpels and are more or less contiguous to the vascular strands. The gamosepalous calyx is likewise reticulate in venation, and the numerous mucilaginous idioblasts lie nearly adjacent to the veins. Calcium oxalate druses are sporadic and numerous between the veins. The petals are delicate in texture. The epidermis consists in the main of polygonal tabular cells along with stomata and hairs. The striking feature of the mesophyll of the petal, as of the mesocarp of the carpel, is the concentration of idioblasts flanking the veins in the proximal areas of the petals (Fig. 9). Distally in the younger petals where the idioblasts have not yet differentiated there is an abundance of calcium oxalate crystals. The idioblasts of the petals are in general smaller but are

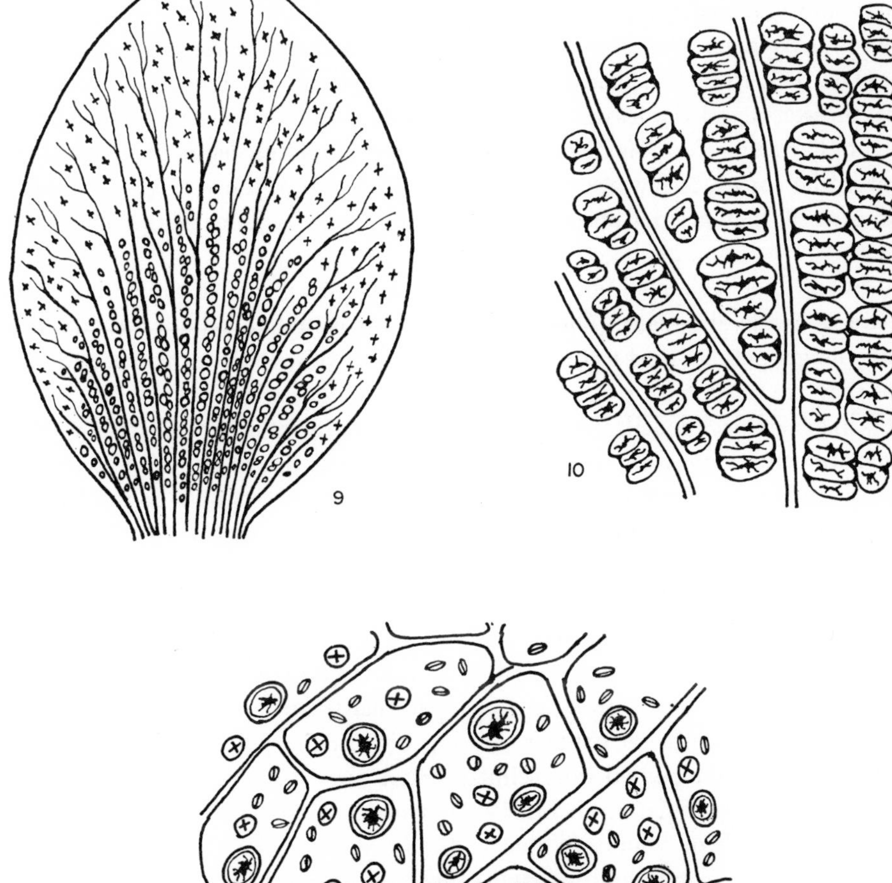

FIGURES 9 to 11. Mucilaginous idioblasts in petal and leaf.

9. Petal (bud 1·5 cm long; actual size of petal 5 × 3 mm) stained with neutral red, showing venation and distribution of mucilaginous idioblasts (circles) and druses (crosses).

10. Part of petal (Fig. 9) stained with neutral red, showing venation and idioblasts in which nucleocytoplasmic cores are visible (actual size of larger idioblasts about 90 × 64 μm).

11. Leaf (small, about 2 × 2 cm) seen in surface view, showing venation and distribution of mucilaginous idioblasts, druses and stomata in areoles (actual diameter of mucilaginous idioblasts about 43 μm).

similar to those of the carpels in form (Fig. 10), in their capacity for staining with aqueous dyes, and in their reaction to treatment with $I_2KI–H_2SO_4$. Similar mucilaginous idioblasts are present in the receptacle and in the flower stalk.

Dehiscence of the fruit occurs when the mature pods are buff in colour, and the tissues are dry and air-filled. The last layer of the pericarp to split along the ridges is the paper-white, fibrous, porous endocarp, the pores of which appear to indicate the previous sites of stomata.

Since mucilaginous idioblasts occur in the parenchyma of the receptacle and of the flower stalk, it is to be expected that they are distributed throughout the vegetative organs, the stems and the leaves (Fig. 11). Mucilage flows freely from these organs when cut. The stem of okra is dicotyledonous in type, and mucilaginous idioblasts are present in the cortex. A few are present in the solid pith of the younger internodes, but, in general, they are not observed in the hollow pith cylinder of the older internodes.

Druses and mucilaginous idioblasts occur in the parenchyma of the petioles. The latter are as a rule abundant. They form an almost complete cylinder in the distal zone of younger petioles next to the base of the blade, while in the proximal and older regions they are separated by intervening parenchymatous cells.

Roots in general lack mucilaginous idioblasts. They are not observed in the younger meristematic and elongating regions. In older zones, however, after secondary thickening and cork formation has occurred, sporadic idioblasts may be present near the transition zone of root and shoot.

DISCUSSION AND CONCLUSIONS

In general, cells in the apical primordia of shoot and root and also in the primordia of leaf, flower and fruit are polyhedral. The cell walls, consisting in the main of cellulose, are thin and pitted on all cell faces, so that the protoplasts throughout the tissue are in plasmodesmatal contact. During growth the polyhedral cells become rounded, and minute air-filled intercellular spaces appear, increase in size, and are lipid-lined from their initiation (Scott, 1964). The deposition of this lipid seems to be the first visible result of wounding, that is, the breaking of the plasmodesmata. The subsequent increase in cell volume entails increase in protoplasm, and the development of sap vacuoles, with their aqueous content of sugars and other ergastic substances. The specific factors in the biophysical and biochemical environment which determine whether a cell will differentiate as a xylem or a phloem element, or will remain as a more or less unspecialized parenchyma cell, are in general unknown.

The differentiation of mucilaginous idioblasts in okra seems to be unique. The youngest cells observed in the developing primordia do not appear to contain aqueous sap vacuoles at any time during growth. From the time of cell initiation, the protoplast appears to be filled with mucilage only, except for a cytoplasmic envelope, the plasmolemma, and a central, branching nucleocytoplasmic core. The protoplast is apparently devoid of chloroplasts, mitochondria and even the lesser organelles characteristic of other cells. It is in plasmodesmatal connexion with contiguous cells through the pitted cellulose walls. At an early stage of development, lipid material becomes evident in the zone of the middle lamella and initiates the cutin-suberin-like pitted sheath of the mature idioblast. It is hoped that electron microscope studies now in progress will show unique ultrastructural differences of these idioblasts from other cells.

When mucilage is formed in mass in intercellular spaces (Scott *et al.*, 1962) it appears as a rule to be homogeneous. In *Cercidium*, however, a cutin-suberin-like lipid layer is differentiated in this mass, parallel to, but slightly distant from the surrounding cell walls. It is suggested that in this mass of mucilage a gradient in oxygen content may be a factor in the differentiation of this suberin-like lipid layer. This falls in line with the hypothesis of Priestley & Woffenden (1922) on the conditions essential for suberin formation. In okra, investigation of the biochemical conditions in the zone of the middle lamella of the idioblast cell wall still awaits detailed study.

Tissue culture today is an important tool in the study of development in the higher plant. Comparison of the culture of roots, stems and leaves of various ages in okra might throw light on the environmental conditions essential for continued growth. When and where do idioblasts appear? If idioblasts are included in the original culture explants, do they proliferate into more mucilage-containing cells, or do they, on division, give rise to parenchymatous cells with their normal complement of chloroplasts?

The physiological role of mucilage in the plant as a whole appears to be undetermined. Is mucilage merely a surplus byproduct in the tissues, or does it function as reserve material significant in seed production? The abundance of mucilage requires consideration in relation to the prevalence of calcium oxalate druses throughout the same tissues. It is observed that, as a whole, the phloem appears to be somewhat inconspicuous. It seems possible that experimentation with tracers applied to the living plant at various stages of growth might yield significant results.

From the standpoint of evolution, can mucilaginous species be considered as relics of evolution, or, on the other hand, do they indicate an advance to physiologically newer strains?

In conclusion, the study of mucilage remains an outstanding anatomical problem.

REFERENCES

CRAFTS, A. S., CURRIER, H. G. & STOCKING, C. R., 1949. *Water in the physiology of plants.* Waltham, Mass.: Chronica Botanica Co.

METCALFE, C. R. & CHALK, L., 1950. *Anatomy of the Dicotyledons.* Oxford: Clarendon Press.

PRIESTLEY, J. H. & WOFFENDEN, L. M., 1922. Physiological studies in plant anatomy. V. Causal factors in cork formation. *New Phytol.*, **21**: 252–268.

SCOTT, F. M., 1964. Lipid distribution in intercellular space. *Science*, N.Y., **203**: 194–195.

SCOTT, F. M., BYSTROM, B. G. & BOWLER, E., 1962. *Cercidium floridum* seed coat, light and electron microscope study. *Am. J. Bot.*, **49**: 821–833.

TUNMANN, O. & ROSENTHALER, L., 1931. *Pflanzenmikrochemie*, 2nd ed. Berlin: Gebr. Borntraeger.

WINTON, A. L. & WINTON, K. B., 1935. *Structure and composition of foods.* New York: Wiley.

EXPLANATION OF PLATE

PLATE 1

Flower of okra. Photograph courtesy Neil Flynn.

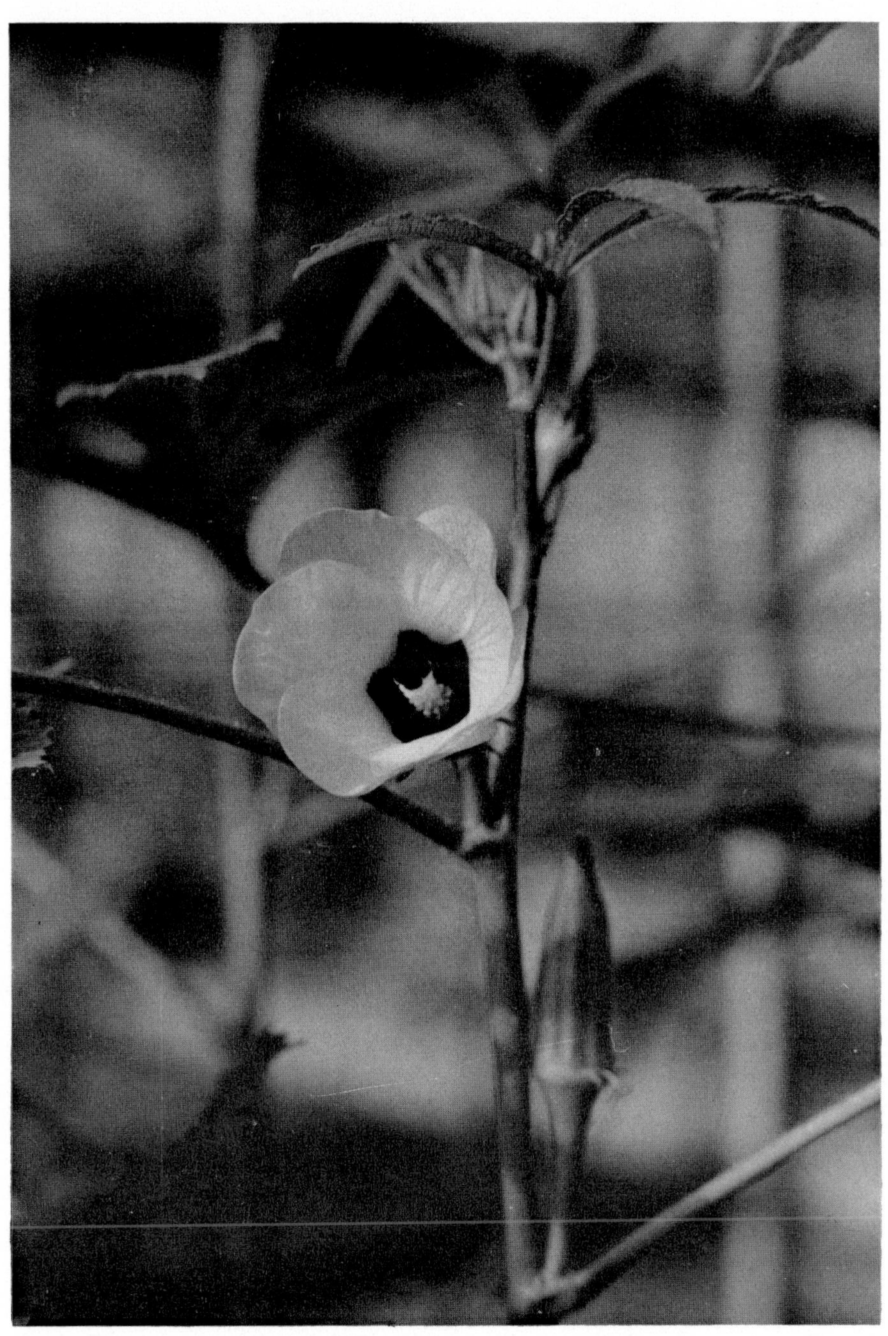

F. M. SCOTT AND B. G. BYSTROM

(*Facing p.* 24)

Monocotyledon classification with special reference to the origin of the grasses (Poaceae)

H. T. CLIFFORD, F.L.S.

Botany Department, University of Queensland, St Lucia, Australia

The interrelationships of the monocotyledons were investigated using one or two genera from each family. For each genus a wide range of attributes was employed and these were subject to two distinct classificatory programmes. Both analyses revealed an association between the Poaceae, Cyperaceae, Arecaceae and Flagellariaceae.

A more detailed analysis of the first three of these families employing primarily reproductive and anatomical characters indicated the relationships between the grasses and palms to be stronger than between either family and the sedges. As most of the attributes used in the second analysis are considered to be conservative from an evolutionary viewpoint, the overall similarity of the grasses and palms over the range of attributes scored was considered to reflect a common ancestry.

Confirmatory evidence from present day members of the two families is considered, and speculation is made as to the possible evolution of the grasses from the palms.

CONTENTS

INTRODUCTION

The interrelationships of the monocotyledons are obscure and are likely to remain so until the fossil record of the group is better documented. Meanwhile, their interrelationships must be inferred from the study of comparative morphology and anatomy, for these are the principal data available. Several taxonomists using such data have classified this subclass, and whilst there is a considerable measure of agreement amongst them as to the family limits, the divergences of opinion concerning the relationships of these are considerable. Since classifications which purport to reflect phylogenetic trends depend not only upon the attributes considered but also upon the supposed evolutionary importance of these attributes, it was decided to attempt a preliminary classification of the monocotyledons using numerical-taxonomic methods. The clusters of taxa produced computationally cannot reflect the author's views on phylogeny except in that he chose the attributes to be investigated and the model(s) to be used for generating the groups of taxa.

Such a numerical classification is otherwise fully objective and provides a base against which to compare some of the classifications proposed in recent years. Because the writer is interested in grasses and sedges particular attention will be accorded genera supposedly related to these plants in the classifications considered.

MATERIAL AND METHODS

Initially, data were assembled for 30 attributes for one or two genera for all the families of monocotyledons recognized by Takhtajan (1959) and Hutchinson (1959), together with the genera *Anarthria* and *Ecdeiocolea*. These two additional genera have been included because each has recently been raised to familial rank (Cutler & Shaw, 1965) and both were formerly included in the Restionaceae, a family placed near to the sedges and/or grasses in most classifications. The genera studied are arranged alphabetically in order of families in Appendix I and the attributes scored are tabulated in Appendix II.

The data matrix was then subject to two classificatory programs on the control data 3600 computer of the C.S.I.R.O. Division of Computing Research, Canberra. One of the programs, MULTBET, is based upon an information statistic (Lance & Williams, 1967*a*) and results in a reasonably intense clustering with well separated groups, which may be achieved at the expense of some misclassification; the other program is based on the Mean Square Distance between clusters (Burr, 1968; Lance & Williams, 1967*b*) and produces a more diffuse clustering but with less misclassification.

RESULTS OF ANALYSES

Dendrograms representing the manner in which the genera were clustered using each of the programs employed are shown in Figs 1 and 2.

In these figures the units of the vertical scale are of no particular importance but merely record increasing dissimilarity. Likewise, the absolute sequence of numbers along the vertical axis is of little importance, each figure may be regarded as a mobile in which the members may rotate at any junction. The composition of each primary group of genera, corresponding to the numbering along the horizontal axis, is given below each figure.

DISCUSSION OF RESULTS

Reference to Fig. 1 reveals that the sample of monocotyledons as clustered by the MULTBET program comprises four major groups of genera. One of these (Nos 1–6) stands out strongly from the remainder and includes the majority of genera with petaloid perianths, except for those included in No. 7. The other three groups are about equally dissimilar to one another and are ecologically well differentiated. One of them (Nos 12–15) is a group of largely wind-pollinated genera; another (No. 7) is the often bird-pollinated, tropical, rhizomatous 'ginger-plants'; and the third (Nos 8–11) of these groups embraces the majority of the aquatic monocotyledons.

Within several of the groups there are genera that are clearly misplaced, for example *Smilax* in No. 12. Nevertheless, the groups appear to be reasonable overall and, in all those families represented by two genera rather than one, both representatives are found to be closely associated. The generic pairs involved are *Epidendrum-Pterostylis* (Orchidaceae), *Poa-Panicum* (Poaceae), *Cyperus-Gahnia* (Cyperaceae), *Restio-Hypolaena* (Restionaceae), *Juncus-Luzula* (Juncaceae), *Zantedeschia-Pothos* (Araceae), *Haemodorum-Anigozanthos* (Haemodoraceae), *Xanthorrhoea-Lomandra* (Xanthorrhoeaceae), *Calamus-Arecastrum* (Arecaceae), *Moraea-Patersonia* (Iridaceae), *Cymodocea-Zannichellia* (Zannichelliaceae).

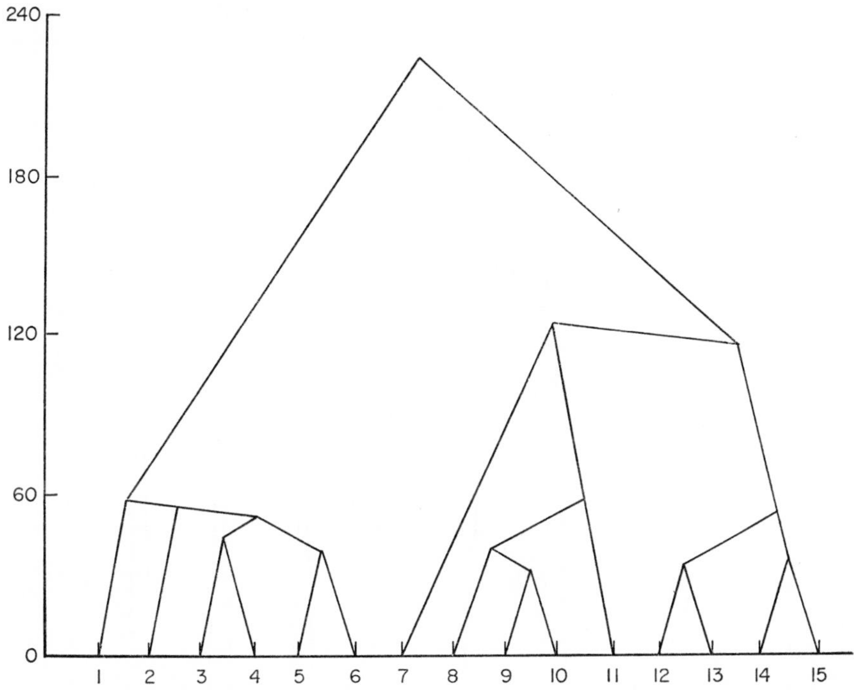

FIGURE 1. The distribution of a representative sample of monocotyledon genera as clustered by the classificatory program, MULTBET. The genera in each of the 15 groups are listed below.

1. *Apostasia, Burmannia, Corsia, Epidendrum, Isophysis, Moraea, Patersonia, Pterostylis, Thismia, Xyris.* 2. *Juncus, Luzula, Thurnia, Trichopus, Vellozia.* 3. *Agave, Aloe, Aphyllanthes, Croomia, Dianella, Dioscorea, Lomandra, Petrosavia, Philesia, Rapatea, Tacca, Tecophilaea, Trillium.* 4. *Asparagus, Petermannia, Pothos, Ruscus, Xanthorrhoea, Zantedeschia.* 5. *Anigozanthos, Cartonema, Haemodorum, Hypoxis, Mayaca, Philydrum, Pontederia.* 6. *Alstroemeria, Billbergia, Carludovica, Commelina, Crinum, Geosiris, Gilliesia, Stenomeris.* 7. *Alpinia, Canna, Lowia, Maranta, Musa, Strelitzia.* 8. *Halophila, Thalassia, Vallisneria.* 9. *Lilaea, Naias, Potamogeton, Scheuchzeria, Triglochin.* 10. *Cymodocea, Posidonia, Ruppia, Zannichellia, Zostera.* 11. *Alisma, Aponogeton, Butomus, Limnocharis.* 12. *Abolboda, Anarthria, Eriocaulon, Smilax, Sparganium, Triuris, Typha.* 13. *Centrolepis, Ecdeiocolea, Hypolaena, Restio.* 14. *Lemna, Pandanus.* 15. *Arecastrum, Calamus, Cyperus, Flagellaria, Gahnia, Panicum, Poa.*

Because of its efficiency in dealing with pairs of genera known to be closely related, the classification produced by MULTBET has been accepted as producing a reasonable scheme. Its failure to associate genera such as *Smilax* with its liliaceous relations may be due to one or more of the following—the incompleteness of the data, the non-inclusion of data used to erect other classifications, the inclusion of data not used in other classifications or the nature of the numerical model itself. It could be that some of the structures regarded as homologous in the various families are in reality not so. The difficulty in determining homologies between organs regarded as perianth members led to the exclusion of these structures from the collection of data. It is also possible that the apparent misplacement of genera in this classification is merely a reflection of their former misplacement.

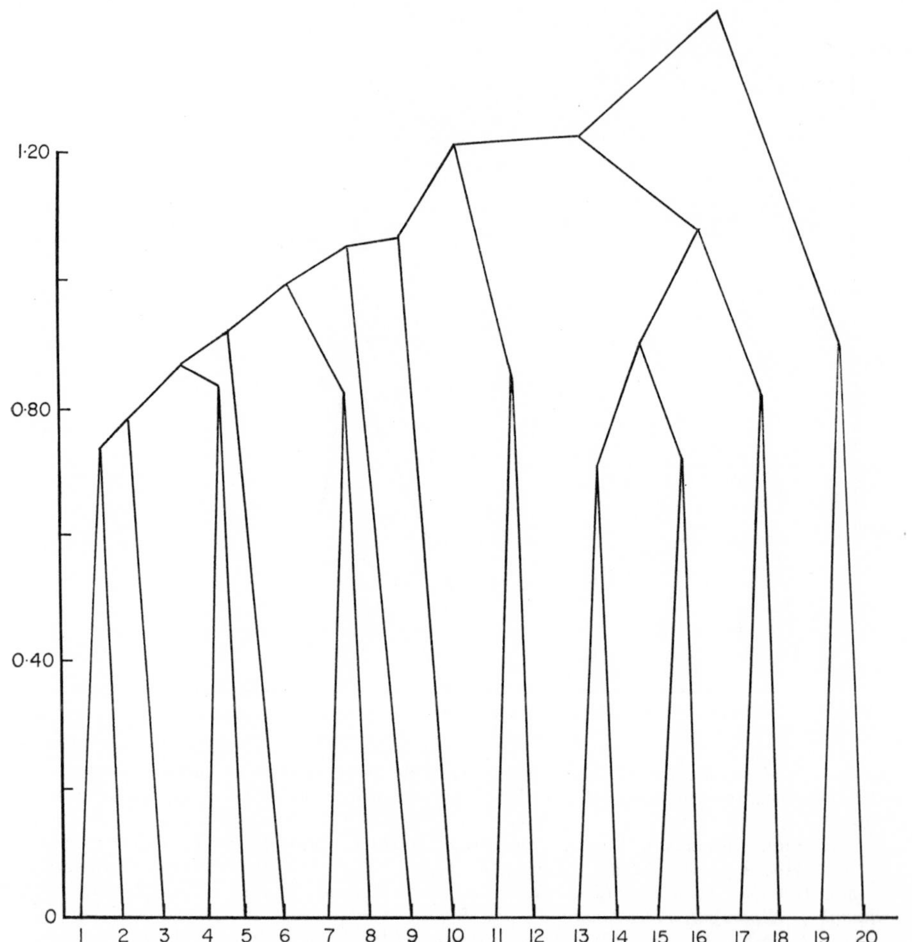

FIGURE 2. The distribution of a representative sample of monocotyledon genera as clustered by the classificatory program, M.S.D. The genera in each of the 20 groups are listed below.
1. *Agave, Aloe, Alstroemeria, Anigozanthos, Apostasia, Asparagus, Billbergia, Burmannia, Carludovica, Crinum, Dianella, Dioscorea, Gilliesia, Haemodorum, Hypoxis, Isophysis, Lomandra, Moraea, Patersonia, Petermannia, Petrosavia, Pontederia, Pterostylis, Ruscus, Stenomeris, Tacca, Tecophilaea, Trillium, Xanthorrhoea, Zantedeschia.* 2. *Abolboda, Cartonema, Commelina, Eriocaulon, Mayaca, Smilax, Xyris.* 3. *Aphyllanthes, Croomia, Philesia.* 4. *Anarthria, Calamus, Cyperus, Ecdeiocolea, Flagellaria, Gahnia, Juncus, Luzula, Panicum, Poa, Rapatea, Sparganium, Thurnia, Typha.* 5. *Arecastrum.* 6. *Lemna, Pandanus.* 7. *Trichopus, Vellozia.* 8. *Epidendrum.* 9. *Pothos.* 10. *Alpinia, Canna, Lowia, Maranta, Musa, Strelitzia.* 11. *Aponogeton, Butomus, Triuris.* 12. *Alisma, Limnocharis.* 13. *Zostera.* 14. *Ruppia, Zannichellia.* 15. *Cymodocea, Lilaea, Naias, Posidonia, Triglochin.* 16. *Potamogeton, Scheuchzeria.* 17. *Corsia, Geosiris, Thismia, Vallisneria.* 18. *Halophila, Thalassia.* 19. *Centrolepis, Hypolaena, Restio.* 20. *Philydrum.*

The less strongly clustering Mean Square Distance program (M.S.D.) united the genera in much the same way as MULTBET though the groupings are not so sharply delineated. With the exception of the ginger-plants (No. 10) those genera with conspicuous petaloid perianths are mostly located in Nos 1–3, whilst the predominantly wind-pollinated families are grouped together as Nos 4–6 and the water-families as Nos 11–18. As might have been anticipated with this program, *Smilax* is better placed than before and is much more closely associated with the lilies, being in an adjacent group well separated from the wind-pollinated genera.

As the purpose of the analyses was primarily to determine the positions of the grasses and sedges both relative to each other and to the remaining monocotyledons, particular attention will be given to the wind-pollinated groups within which the grasses and sedges are included in both classifications. The genera included with the grasses *Poa* and *Panicum* are indicated below for each classification:

MULTBET *Cyperus*, *Gahnia*, *Arecastrum*, *Calamus*, *Flagellaria*
M.S.D. *Anarthria*, *Cyperus*, *Gahnia*, *Ecdeiocolea*, *Flagellaria*, *Juncus*, *Rapatea*, *Sparganium*, *Thurnia*, *Typha*, *Calamus*, *Luzula*

A comparison of these two groupings reveals that with the exception of *Arecastrum* all the genera associated with the grasses in MULTBET are also associated with them in the M.S.D. classification. From this it may be concluded that the Cyperaceae (*Gahnia* and *Cyperus*), Flagellariaceae (*Flagellaria*) and Arecaceae (*Arecastrum* and *Calamus*) are the families most closely associated with the Poaceae in terms of these numerical **analyses.**

Additionally, the M.S.D. program associated with the grasses the genera *Anarthria*, *Ecdeiocolea*, *Juncus*, *Luzula*, *Thurnia*, *Rapatea*, *Sparganium* and *Typha*. In most classifications these would be regarded as a heterogeneous assemblage of plants. For example Hutchinson (1959) places the first five of these genera in his Juncales, *Rapatea* in the Xyridales, and *Sparganium* and *Typha* in the Typhales.

Nevertheless, whilst the group is apparently so diverse in terms of previous classifications, it has an ecological unity, in that its members are largely wind-pollinated and most of the plants grow in swampy habitats. Neither of these attributes was included in the original data and so the results reinforce the suggestion that the plants are related.

In neither of the above classifications were the Restionaceae *sen. str.* closely associated with either the grasses or the sedges.

GENERAL DISCUSSION

Neither of the classificatory programs employed above produces a reliable hierarchical structure of the taxa and it is certain that the significance of ranks within existing classifications depends upon their authors, some permitting more and some less variation within families and orders than others. It is therefore impossible to compare directly the numerical and conventional classifications. Yet, the two systems are not unrelated in that each is based upon comparisons involving large numbers of attributes.

Moreover, whilst the numerical classifications are entirely phenetic, the majority of other classifications are a mixture of phenetic and phylogenetic schemes. Assuming that overall similarity is an indication of a common ancestry for the members of a group, when the similarity is based upon a wide range of attributes it would appear from the above analyses that the grasses, sedges and *Flagellaria* are closely related, near to these come the palms, and related to all four but not so closely are the Juncaceae (inc. *Thurnia*), *Anarthria*, *Ecdeiocolea*, *Rapatea* and the Typhales.

Such a grouping of genera is nearly in accord with the proposals of Deyl (1955), who included all the above genera associated with the grasses and sedges in two orders, the Arecales and Juncales. To these orders he gave separate origins but he commented (*loc. cit.* p. 88) that *Typha* (Arecales) shows remote affinities with the grasses (Juncales).

Unlike many other phylogenetically-based classifications of the monocotyledons—Hutchinson (1959), Takhtajan (1959), Cronquist (1968)—which treat the aquatic monocotyledons as a separate line of development within the subclass, Deyl regards them as the forerunners of several separate evolutionary lines.

The Arecales he envisages as having been derived from the Aponogetonaceae-Potamogetonaceae complex; the Juncales from amongst the Zannichelliaceae-Lemnaceae-Naiadaceae-Lilaeaceae complex of families. These two groups of water-families are regarded as related and so the orders derived from them would owe a good deal to their common origin aside from any parallel or convergent evolution that may have occurred.

It would be expected that, if the grasses and sedges are as closely related both to each other and to the palms and *Flagellaria* as is suggested by the results of the MULTBET analysis, then this would be confirmed by a consideration of further data. To allow for diversity within families they will be compared, overall and not only with respect to the genera employed in the analyses, for a further set of attributes. The family Flagellariaceae is here regarded as comprising the genera *Flagellaria* and *Joinvillea*.

The further attributes that were considered are primarily reproductive or anatomical ones and are most likely to be conservative in an evolutionary sense. Where not known to the writer the data were taken from the following reference works—Davis (1966), Martin (1946), Corner (1966) or Tomlinson (1969).

The lack of information concerning the Flagellariaceae is unfortunate, but it is hoped to remedy this situation next flowering season. Meanwhile the family will be ignored, which is unfortunate, for both *Flagellaria* and *Joinvillea* have been previously likened to the grasses (Smithson, 1956).

From an inspection of Table 1 it appears that, whilst the sedges are about equally similar to the palms and the grasses, the palms and grasses are more similar to one another than is either to the sedges. In particular, in both the palms and grasses but not in the sedges, a nucellar cap structure is present; the primary root (radicle = coleorrhiza) is of limited growth; anthocyanins occur as pigments; and the peripheral endosperm may be meristematic. In most grasses and some palms the leaf epidermis consists of alternate long and short cells whereas those of the sedges are usually equal in length, whilst furthermore the seeds of several grass and palm genera are strongly grooved on the chalazal side whereas no sedge seeds are known of this shape. From the foregoing it is inferred that the sedges have a different origin from the grasses and will therefore not be considered further here.

Comparing the grasses and the palms the present day genera differ markedly in the extreme reduction of the grass flower. Amongst the grasses *Streptochaeta* has the most fully developed perianth composed of six spirally arranged members. In the palms six-numbered perianths are common and a spiral arrangement of parts is especially evident in the petaline whorl. The genus *Streptochaeta* is generally regarded as a primitive grass and so may point to a linking of the grasses and palms.

Table 1. A comparison of the Poaceae and three other families with respect to several anatomical, embryological and phenological attributes

Attribute	Poaceae	Cyperaceae	Arecaceae	Flagellariaceae
Embryo sac development	Polygonum	Polygonum	Polygonum Allium	?
Peripheral endosperm	Meristematic or non-meristematic	Non-meristematic	Meristematic	?
Embryogeny	Asterad	Onagrad	Onagrad Asterad	?
Antipodals	Few to many or absent	Few	Few	?
Nucellar cap	Well developed	Absent	Well developed	?
Endosperm	Solid to liquid	Solid	Solid to liquid	Solid
Embryonic growth	Definite	Definite	Indefinite	?
Seeds	Grooved or not grooved	Not grooved	Grooved or not grooved	Not grooved
Microspore tetrads	Isobilateral T, linear	Isobilateral tetrahedral decussate	Isobilateral tetrahedral T, or linear	?
Embryo	Lateral	Capitate, linear	Linear	Lateral
Formation of micropyle	Inner integument (if present)	Inner integument	Both integuments	?
Dioecism	+	+	+	−
Floral development	Mostly protandrous	Mostly protogynous	Mostly protandrous	?
Primary root growth	Very limited	Unlimited	Limited	?
Microhairs	±	±	±	−
Epidermal cells long and short	+	−	±	±
Anthocyanins	Present	Absent	Present	?

Theories on the origin of ovarian structures are several but, if that of Meeuse (1966) is taken as a basis for discussion, in both the palms and the grasses the ovary has an ecarpellate origin, with greater reduction in the grasses than the palms. Such a situation would be in accord with the grasses having evolved from the palms.

The lack of a coleorrhiza (or primary root of very limited growth) in the palms would be a conspicuous embryological difference between the two families if it could be maintained that this structure is never present in the palms. However, its occurrence in the family is suggested by the figure of an *Archontophoenix* embryo shown in cross section in Corner (1966: Fig. 98). This embryo appears to possess a coleorrhiza with a vertically disposed endogenous adventitious root as in grasses.

Some of the strongest lines of evidence for or against relationships between groups are to be found in studies of comparative embryology. Unfortunately the available data are insufficient to make detailed comparisons between the two families. Nevertheless seedlings are readily obtained and these reveal a remarkable similarity in morphology between the young plants of bamboos and those of many palms. In such seedlings the typical palm leaf is replaced by a simple laminate leaf or one in which the lamina is missing.

Accepting the palms and grasses to be related, even in the absence of satisfactory fossil evidence it is interesting to speculate on which is the older and hence which may be a derivative of the other. As fossil palms are reported as early as the Triassic (Corner, 1966) whereas the earliest reliably identifiable grasses come from the Upper Cretaceous (Roshevitz, 1969), it is possible for the grasses to have evolved from the palms. This view was long ago proposed without further amplification by Mattei & Tropea (1909), who unfortunately made the statement only in a footnote to another paper dealing with extra-floral nectaries in grasses.

If divergence of the grasses and palms occurred as early as the Mesozoic then it would not be surprising that connecting links between them no longer survive. However, amongst living genera there may well still be plants which show intermediary characters when looked at from this viewpoint and efforts should be directed towards seeking them out. Above all, closer attention should be given to the fossil record because this must remain the final arbiter in all phylogenetic discussions.

ACKNOWLEDGEMENTS

It is a pleasure to record my indebtedness to Charles Metcalfe for discussing this and other subjects during my stay at the Jodrell Laboratory in 1965. In the course of these discussions he revealed a knowledge of botanical sciences even deeper than that revealed in his extensive writings. His unfailing courtesy, unassuming manner and profound scholarship attracted a wide spectrum of visitors to his laboratories including non-anatomists such as myself.

REFERENCES

BURR, E. J., 1968. Cluster sorting with mixed character types. *Aust. Comp. J.*, **1**: 97–99.

CORNER, E. J. H., 1966. *The natural history of palms*. London: Weidenfeld & Nicholson.

CRONQUIST, S., 1968. *The evolution and classification of flowering plants*. London & Edinburgh: Thos. Nelson.

CUTLER, D. F. & SHAW, H. K. A., 1965. Anarthriaceae and Ecdeiocoleaceae: two new monocotyledonous families, separated from the Restionaceae. *Kew Bull.*, **19** (3): 489–499.

DAVIS, G. L., 1966. *Systematic embryology of the angiosperms*. New York: John Wiley.

DEYL, M., 1955. The evolution of the plants and taxonomy of the monocotyledons. *Sb. nár. Mus. Praze*, **11B** (6): 3–143.

HUTCHINSON, J., 1959. *The families of flowering plants*, **2**. Oxford: Clarendon Press.

LANCE, G. N. & WILLIAMS, W. T., 1967*a*. A general theory of classificatory sorting techniques. *Comp. J.*, **9** (4): 373–380.

LANCE, G. N. & WILLIAMS, W. T., 1967*b*. Mixed-data classificatory programs I. Agglomerative systems. *Aust. Comp. J.*, **1**: 15–20.

MARTIN, A. C., 1946. The comparative internal morphology of seeds. *Am. Midl. Nat.*, **36**: 513–660.

MATTEI, G. E. & TROPEA, C., 1909. Ricerche e studî sul genere *Eragrostis*, in rapporto ai nettarii estranuziali. *Contrzioni Biol. veg.*, **4**: 205–286.

MEEUSE, A. D. J., 1966. *Fundamentals of phytomorphology*. New York: Ronald Press.
ROSHEVITZ, R. J., 1969. Evoluçãs e sistemática das Gramineas. *Bolim. Inst. Bot.*, **5**: 1–20 (trans. by T. Sendulsky from original Russian published in 1946).
SMITHSON, E., 1956. The comparative anatomy of the Flagellariaceae. *Kew Bull.*, **11**: 491–501.
TAKHTAJAN, A., 1959. *Die Evolution der Angiospermen*. Jena: Gustav Fischer.
TOMLINSON, P. B., 1969. *Anatomy of the monocotyledons* (ed. C. R. Metcalfe), **3**. *Commelinales-Zingiberales*. Oxford: Clarendon Press.

APPENDIX I

The genera studied arranged alphabetically according to family. Families as in Hutchinson (1959) or Takhtajan (1959).

Abolbodaceae	*Abolboda*
Agavaceae	*Agave*
Alismataceae	*Alisma*
Aloeaceae	*Aloe*
Alstroemeriaceae	*Alstroemeria*
Amaryllidaceae	*Crinum*
Anarthriaceae	*Anarthria*
Aphyllanthaceae	*Aphyllanthes*
Aponogetonaceae	*Aponogeton*
Apostasiaceae	*Apostasia*
Araceae	*Pothos*, *Zantedeschia*
Arecaceae	*Arecastrum*, *Calamus*
Asparagaceae	*Asparagus*
Bromeliaceae	*Billbergia*
Burmanniaceae	*Burmannia*
Butomaceae	*Butomus*
Cannaceae	*Canna*
Cartonemataceae	*Cartonema*
Centrolepidaceae	*Centrolepis*
Commelinaceae	*Commelina*
Corsiaceae	*Corsia*
Cyclanthaceae	*Carludovica*
Cyperaceae	*Cyperus*, *Gahnia*
Dioscoreaceae	*Dioscorea*
Ecdeiocoleaceae	*Ecdeiocolea*
Eriocaulaceae	*Eriocaulon*
Flagellariaceae	*Flagellaria*
Geosiridaceae	*Geosiris*
Gilliesiaceae	*Gilliesia*
Haemodoraceae	*Haemodorum*, *Anigozanthos*
Halophilaceae	*Halophila*
Hydrocharitaceae	*Vallisneria*
Hypoxidaceae	*Hypoxis*
Iridaceae	*Moraea*, *Patersonia*
Isophysidaceae	*Isophysis*
Juncaceae	*Juncus*, *Luzula*
Juncaginaceae	*Triglochin*
Lemnaceae	*Lemna*
Lilaeaceae	*Lilaea*
Liliaceae	*Dianella*
Limnocharitaceae	*Limnocharis*
Lowiaceae	*Lowia*
Marantaceae	*Maranta*
Mayacaceae	*Mayaca*
Musaceae	*Musa*
Naiadaceae	*Naias*
Orchidaceae	*Epidendrum*, *Pterostylis*
Pandanaceae	*Pandanus*
Petermanniaceae	*Petermannia*
Petrosaviaceae	*Petrosavia*
Philesiaceae	*Philesia*
Philydraceae	*Philydrum*
Poaceae	*Poa*, *Panicum*
Pontederiaceae	*Pontederia*
Posidoniaceae	*Posidonia*
Potamogetonaceae	*Potamogeton*
Rapateaceae	*Rapatea*
Restionaceae	*Hypolaena*, *Restio*
Roxburghiaceae	*Croomia*
Ruppiaceae	*Ruppia*
Ruscaceae	*Ruscus*
Scheuchzeriaceae	*Scheuchzeria*
Smilacaceae	*Smilax*
Sparganiaceae	*Sparganium*
Stenomeridaceae	*Stenomeris*
Strelitziaceae	*Strelitzia*
Taccaceae	*Tacca*
Tecophilaeaceae	*Tecophilaea*
Thalassiaceae	*Thalassia*
Thismiaceae	*Thismia*
Thurniaceae	*Thurnia*
Trichopodaceae	*Trichopus*
Trilliaceae	*Trillium*
Triuridaceae	*Triuris*
Typhaceae	*Typha*
Velloziaceae	*Vellozia*
Xanthorrhoeaceae	*Xanthorrhoea*, *Lomandra*
Xyridaceae	*Xyris*
Zannichelliaceae	*Cymodocea*, *Zannichellia*
Zingiberaceae	*Alpinia*
Zosteraceae	*Zostera*

APPENDIX II

The attributes scored and the states recognized for each.

Gynoecium	pistils	free, fused
	placentation	basal, axile, parietal, pendulous, marginal, scattered
	number of placentas	merism
	position of ovary	inferior, subinferior, superior
	ovules per placenta	1, more than 1
	ovule shape	orthotropous, otherwise
	fruit	dehiscent, indehiscent
Seed	aril	present, missing
	endosperm	scarce, abundant
		ruminate, otherwise
	perisperm	developed, missing
	embryo	axis straight, strongly bent
	radicle	enlarged, not enlarged
Fruit	at maturity	dehiscent, indehiscent
Anthers	dehiscence	extrorse, lateral, introrse
	attachment	sessile, basifixed, dorsifixed
Pollen	development	simultaneous, successive
	grains	3-nucleate, 2-nucleate single, compound
	aperture	missing, present
	aperture shape	linear, circular
	grain symmetry	radial, bilateral
	tapetum	amoeboid, glandular
Nectaries	—	present, absent
Leaves	phyllotaxy	2-ranked, spiral
	ligule	present, absent
	bases	sheathing, not sheathing
	venation	reticulate, convergent, pinnate
Stomata	guard cells	typical, graminoid
		present, absent
	subsidiary cells	absent, 2, more than 2

Seed anatomy and feed microscopy

J. G. VAUGHAN, F.L.S.

Queen Elizabeth College, University of London, England

An account is given of the application of light microscopy to the identification of seed residues in animal feeds. It is shown that with the introduction of new fruits and seeds as potential feed constituents there is scope for further original investigations in feed microscopy.

CONTENTS

INTRODUCTION

Some years ago I asked Dr C. R. Metcalfe to suggest a field of research in plant anatomy, the result of which might be of applied interest to industry and commerce. In his capacity as Keeper of the Jodrell Laboratory, Dr Metcalfe had frequently received samples of rape and mustard seed residues from commercial sources for botanical determination of the seed incorporated in them. Such residues might well be included in animal rations and, for a variety of reasons, botanical authenticity is important. During the industrial processing of rape and mustard seeds for oil, the material is reduced to fine fragments, but variation in the structure of the testa allows limited identification with the light microscope. Rapes and mustards belong to the genera *Brassica* and *Sinapis*, the taxonomy of which is notoriously difficult to resolve. This no doubt was the basis of Dr Metcalfe's suggestion for further work. I should like to thank him for his suggestion that I should investigate the seed anatomy of *Brassica* and *Sinapis* in that, not only have I found the problems in these genera most interesting, but his suggestion has also provided a stimulus for investigations of many types of seeds of commercial importance.

Studies of seed anatomy and taxonomy are relatively meagre in comparison with investigations of vegetative anatomy (Carlquist, 1961; Barton, 1967). The last comprehensive account of seed anatomy was produced practically 50 years ago (Netolitzky, 1926). As in the case of anatomical studies of woods, many seed types have been investigated because of their economic importance.

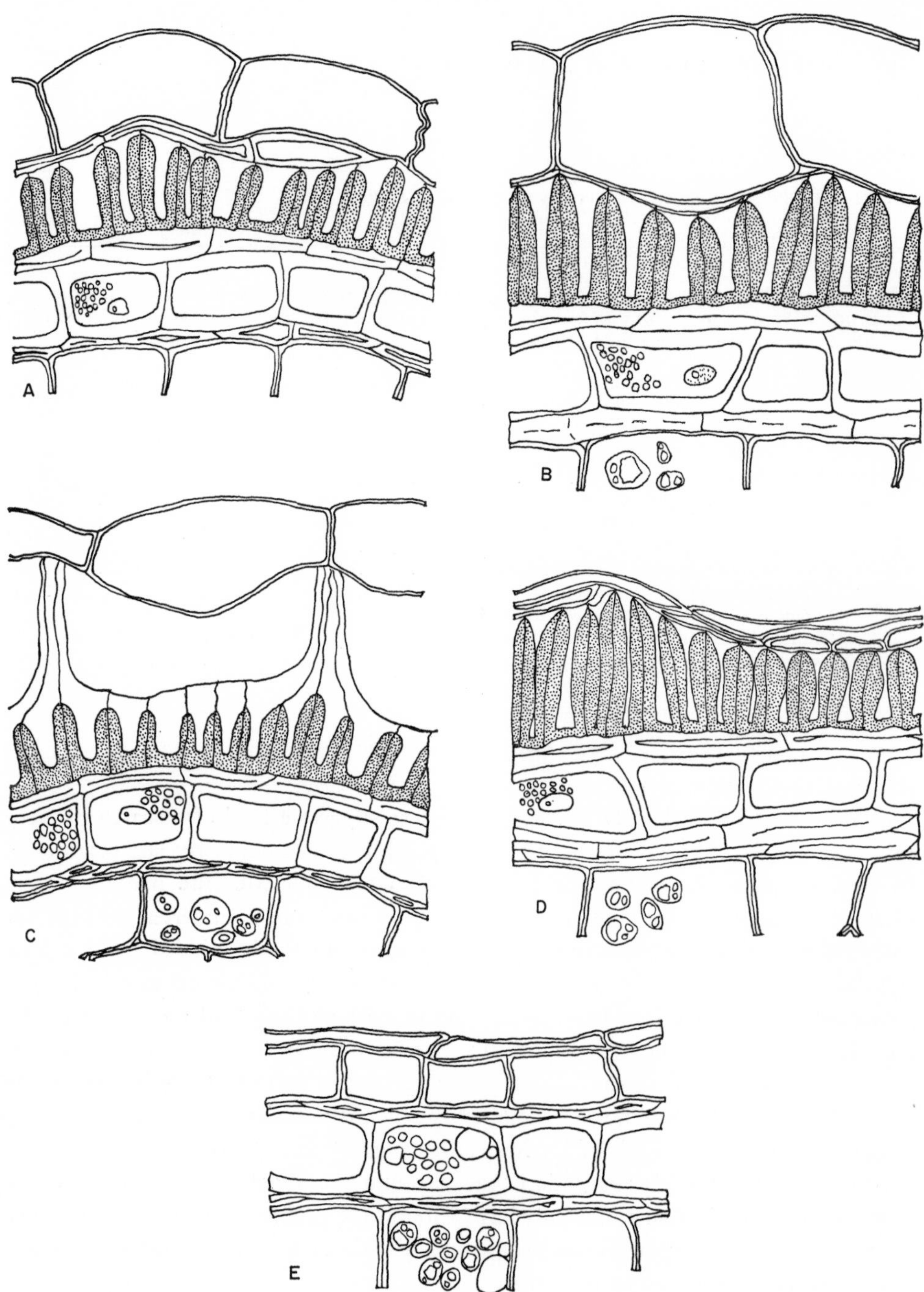

FIGURE 1. T.S. of testa, endosperm and outer portion of cotyledon of various cruciferous seeds, ×600: **A**, *Brassica carinata*; **B**, *B. juncea* (European); **C**, *B. nigra*; **D**, *B. juncea* (Indian); **E**, *Crambe hispanica*.

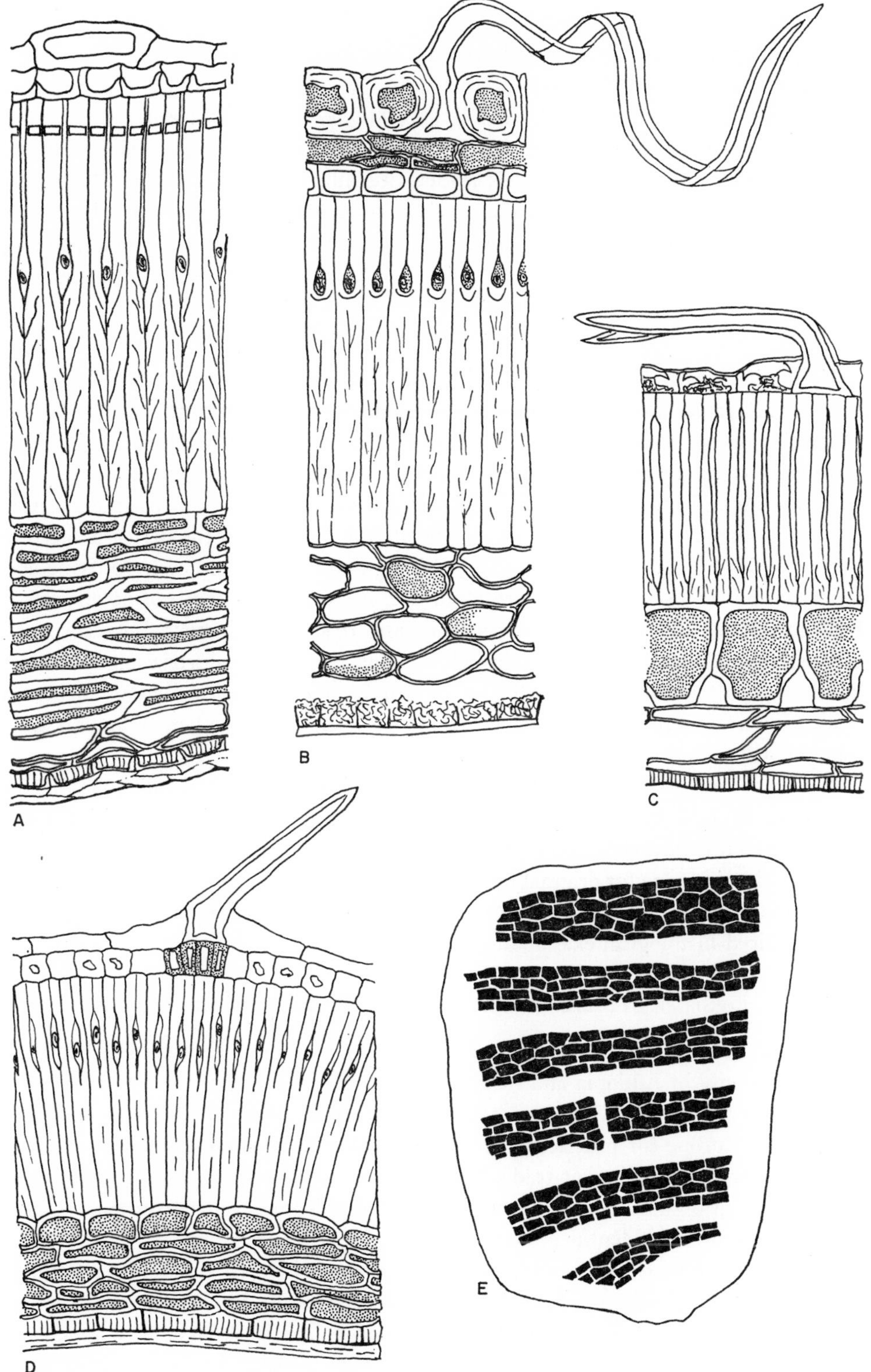

FIGURE 2. **A–D**, T.S. of testa of various malvaceous seeds, ×300: **A**, *Hibiscus esculentus*; **B**, *Gossypium* sp.; **C**, *Hibiscus cannabinus*; **D**, *Abutilon avicennae*; **E**, Surface view of *A. avicennae* chalaza, ×150.

For some time it has been realized that the light microscope could be used to identify seed fragments in products that originated from the industrial processes of oil extraction and milling (Collin & Perrot, 1904; Winton, 1906; Bussard & Brioux, 1925; Morris, 1928; Moeller & Griebel, 1929; Parkinson & Fielding, 1930; Winton & Winton, 1932, 1935; Gassner, 1955; Juillet, Susplugas & Courp, 1955). This type of identification has normally been the preserve of the specialist, but with the present day emphasis on quality control in the animal feed industry, there is a strong likelihood that the technique will become more widely adopted in industrial laboratories.

ROUTINE FEED MICROSCOPY

Feed microscopy is usually practised in the industrial laboratory by chemists whose training has rarely included much study of botany. In addition to this problem, large quantities of materials have to be analysed in a relatively short time. It is highly improbable that any industrial analyst could spare time for the cutting of sections, and therefore identification is normally made on the surface features of fragments. After the processing of fruits and seeds, it is normally the structure of the pericarp and testa that provides features for identification, although endosperm and embryo tissue contents, particularly starch grains, are often useful. As regards the routine analysis of feed materials, the different constituents are readily distinguishable except for cereals, *Brassica* species and certain palms. This helps the industrial analyst and it has been the task of the specialist to interpret and emphasize obvious structural features which enable rapid identification. Plate 1 shows some plant fragments commonly encountered in feed microscopy.

New plant materials are constantly being submitted to the feed industry for consideration as feed constituents. Their usage depends on various nutritional and economic factors but, although feed microscopy is a relatively old discipline, it means that there is still scope for original investigation in the field of industrial seed anatomy.

The following are some example of seeds and fruits which, for commercial reasons, have required histological examination in recent years.

Abyssinian Mustard or Ethiopian Rape (Brassica carinata *Braun*)

Of the cultivated *Brassica* species, *B. carinata* has the most restricted distribution It is grown only in Ethiopia and neighbouring territories as an oil-seed plant and a vegetable (Vaughan, 1957). Since the Second World War, *B. carinata* seed has appeared in Europe for processing in oil mills and also as an adulterant of *B. nigra* seed meant for the condiment trade. Seed residues of *B. carinata* are highly suspect as a constituent of animal feeds because of their allyl isothiocyanate content, so some means of microscopic identification is required.

The testa of *Brassica* seeds may consist of (i) an epidermal layer, one cell thick, which may be mucilaginous; (ii) a subepidermal layer; (iii) a palisade layer, one cell thick, of tumbler-shaped and possibly pigmented cells; (iv) a layer of parenchyma cells, some of which may contain pigment. Closely associated with the testa is an aleurone layer (endosperm). Diagnostic features of the *B. carinata* testa (Vaughan, 1956) are the mucilaginous epidermis, a reduced subepidermis and palisade cells of varying heights which present a reticulation in surface view. These features allow identification of *B. carinata* from all other commercial *Brassica* seeds except for the European form of *B. juncea* (Fig. 1**A**–**D**). Distinction between these two species seems impossible on testa structure alone, but this is unimportant from the commercial point of view because both residues produce allyl isothiocyanate. It is, however, possible to distinguish *B. carinata* from the Indian form of *B. juncea*, since the latter has no obvious epidermis or subepidermis (Fig. 1**D**).

When *B. nigra* seed is adulterated with *B. carinata*, difference in seed size is not an absolute criterion for the degree of adulteration and recourse must be made to the anatomical structure of the testa. Although several histological differences have been recorded, it seems that the most distinctive feature of the *B. nigra* testa, as compared with European *B. juncea*, is the elongated thin outer radial walls of the palisade cells (Fig. 1**C**).

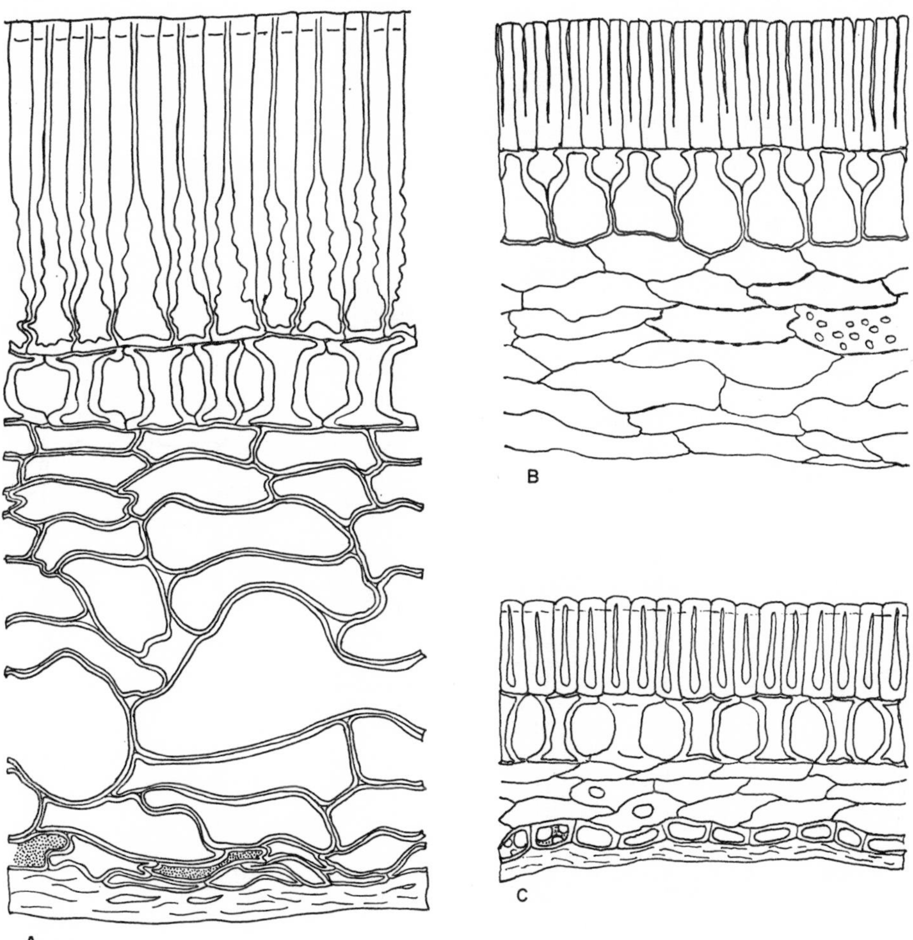

FIGURE 3. T.S. of testa of various leguminous seeds, ×300: **A**, *Vicia faba*; **B**, *Cyamopsis tetragonolobus*; **C**, *Glycine max*.

Crambe hispanica *L.*

Crambe was first developed as an oil-seed crop in the U.S.S.R. in the 1930's and its fixed oil has considerable industrial potential (Anon., 1966). Figure **1E** shows that the *Crambe* testa possesses the usual cruciferous features but its palisade cells are thin-walled and, in transverse section, are elongated transversely, a feature which easily distinguishes the seed from *Brassica* species (Vaughan, 1963).

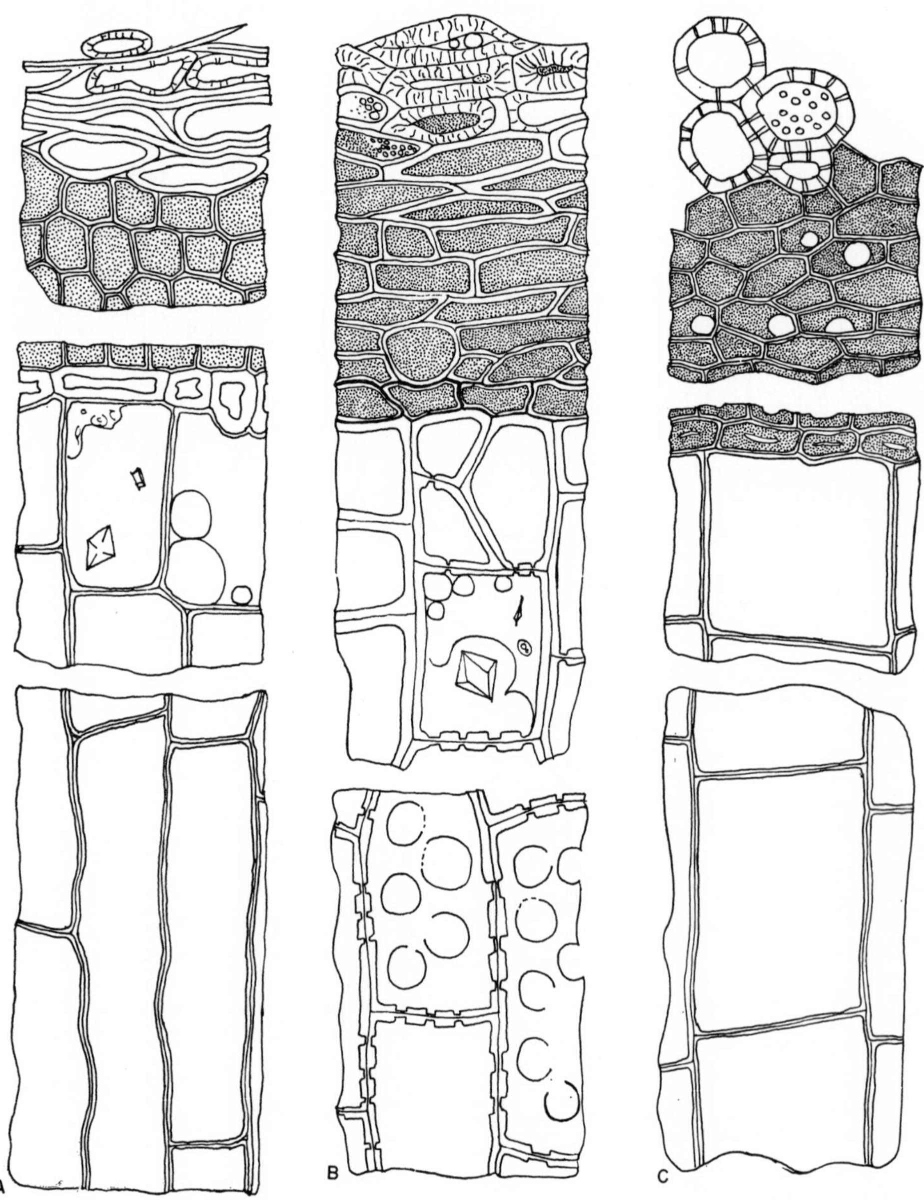

FIGURE 4. T.S. of testa and endosperm of various palm kernels, ×300: **A**, *Cocos nucifera*; **B**, *Elaeis guineensis*; **C**, *Acrocomia* sp.

Chinese Jute (Abutilon avicennae *Gaertn.*)

In recent years, *Abutilon avicennae* seed residue has appeared in European commerce. The use of *Abutilon* as an oil-seed plant seems to be fairly recent and most records of the species refer to its value as a fibre crop (Eckey, 1954).

The few species of malvaceous seeds important in commerce include Cotton (*Gossypium* spp.), Kenaf (*Hibiscus cannabinus* L.) and Okra (*Hibiscus esculentus* L.). Probably the most striking histological feature of the malvaceous testa is the layer of palisade cells, in transverse section with a radial diameter up to 150 μm (Fig. 2**A**–**D**). Distinction between the economic species is best made on the outer layers of the testa. As a practical point of identification, pieces of chalaza (Fig. 2**E**) are conspicuous in *Abutilon* seed residues and these show radiating zones of pigmented cells.

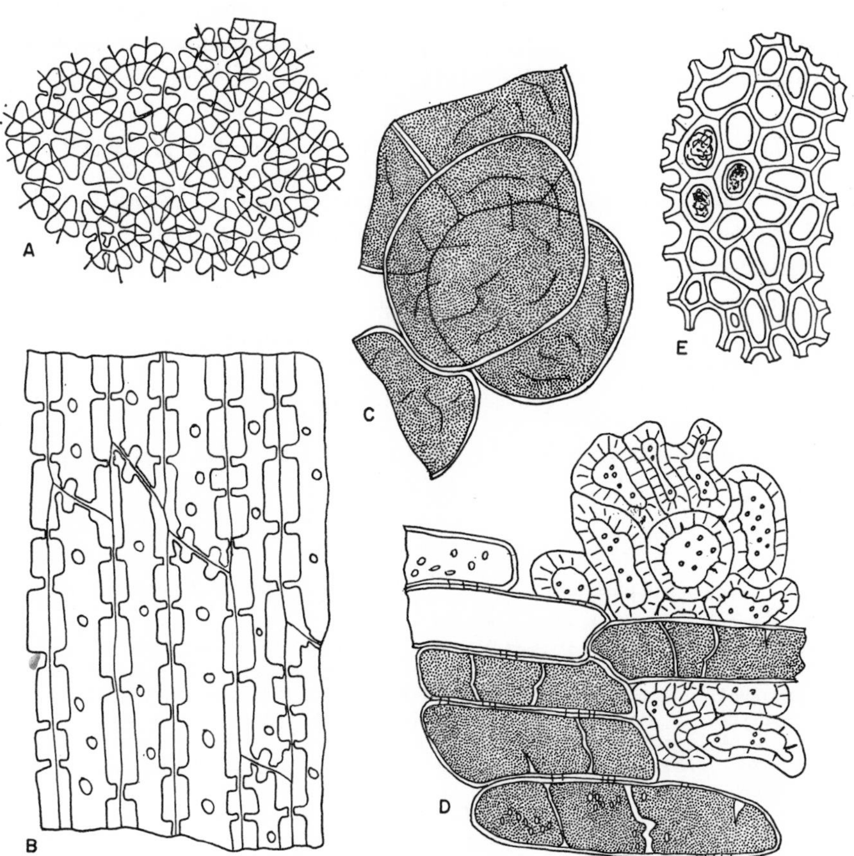

FIGURE 5. *Trachycarpus excelsus* meal fragments, ×300: **A**, T.S. endosperm; **B**, L.S. endosperm; **C**, mesocarp; **D**, testa and endocarp; **E**, epicarp.

Guar (Cyamopsis tetragonolobus (*L.*) *Taub.*)

Guar seeds are an important source of vegetable gum and the seed residue is often utilized as a constituent of animal feeds. The testa of leguminous seeds shows comparatively little differentiation that can be used for taxonomic purposes except for the epidermal and particularly the subepidermal layers. The cells of the subepidermal layer of Guar are 'bottle'-shaped, which contrasts with the 'hour-glass' cells of Soya (*Glycine max* Merr.) and the 'dumb-bell'-shaped cells of Broad Bean (*Vicia faba* L.) (Fig. 3**A**–**C**).

Paraguay Cocopalm (Acrocomia totai *Mart.*)

Most palm products in the animal feed industry are the kernal residues of Oil Palm (*Elaeis guineensis* Jacq.) and Coconut (*Cocos nucifera* L.). South America is relatively rich in oil-yielding palm species, other than Oil Palm and Coconut. Kernel residues of South American palms are not common in European commerce but do occasionally appear. Such a residue is that of Paraguay Cocopalm (Vaughan, 1960). It is usual to distinguish between palm products on the character of the endosperm cells. The endosperm cells of *Acrocomia* are relatively short and have smooth walls, to be contrasted with those of *Elaeis* and *Cocos* (Fig. 4**A**–**C**).

Chinese Hemp or Chinese Palm (Trachycarpus excelsus *H. Wendl.*)

Of recent years a Chinese Palm (*Trachycarpus excelsus* H. Wendl.) product has appeared in commerce under the name of Chinese hemp meal or Chinese palm kernel. It has a very low protein content. The plant has normally been grown for bark fibre, the processed fruit being a new product. Again, the endosperm cells assist in identification and, in this case, mesocarp tissue is present (Fig. 5**A**–**E**).

ACKNOWLEDGEMENT

I am indebted to Miss J. A. Rest for technical assistance during the preparation of this paper.

REFERENCES

Anon., 1966. *Prod. Res. Rep. U.S. Dep. Agric.*, **95**.
Barton, L. V., 1967. *Bibliography of seeds.* New York: Columbia University Press.
Bussard, L. & Brioux, C., 1925. *Tourteaux.* Paris & Liège: Libraire Polytechnique Ch. Beranger.
Carlquist, S., 1961. *Comparative plant anatomy.* New York: Holt, Rinehart & Winston.
Collin, E. & Perrot, E., 1904. *Les residus industriels.* Paris: A. Joanin.
Eckey, E. W., 1954. *Vegetable fats and oils.* New York: Reinhold.
Gassner, G., 1955. *Mikroskopische Untersuchung pflanzlicher Nahrungs und Genussmittel.* Stuttgart: Gustav Fischer.
Juillet, A., Susplugas, J. & Courp, J., 1955. *Les oleagineux et leurs tourteaux.* Paris: Paul Lechevalier.
Moeller, G. & Griebel, C., 1929. *Mikroskopie der Nahrungs und Genussmittel.* Berlin: Springer.
Morris, T. N., 1928. *Microscopic analysis of cattle foods.* Cambridge: University Press.
Netolitzky, F., 1926. *Anatomie der Angiospermen-samen.* In K. Linsbauer, *Handbuch der Pflanzenanatomie,* **10**. Berlin: Borntraeger.
Parkinson, S. T. & Fielding, W. L., 1930. *The microscopic examination of cattle foods.* London: Headley Brothers.
Vaughan, J. G., 1956. The seed coat structure of *Brassica integrifolia* (West) O. E. Schulz var. *carinata* (A. Br.). *Phytomorphology*, **6**: 363–367.
Vaughan, J. G., 1957. Ethiopian rape seed. *Fertil. Feed. Stuffs J.*, **46**: 191.
Vaughan, J. G., 1960. Structure of *Acrocomia* fruit. *Nature, Lond.*, **188**: 81.
Vaughan, J. G., 1963. The testa structure of *Crambe hispanica* L. (Cruciferae). *Kew Bull.*, **16**: 393–394.
Winton, A. L., 1906. *The microscopy of vegetable foods.* New York: John Wiley.

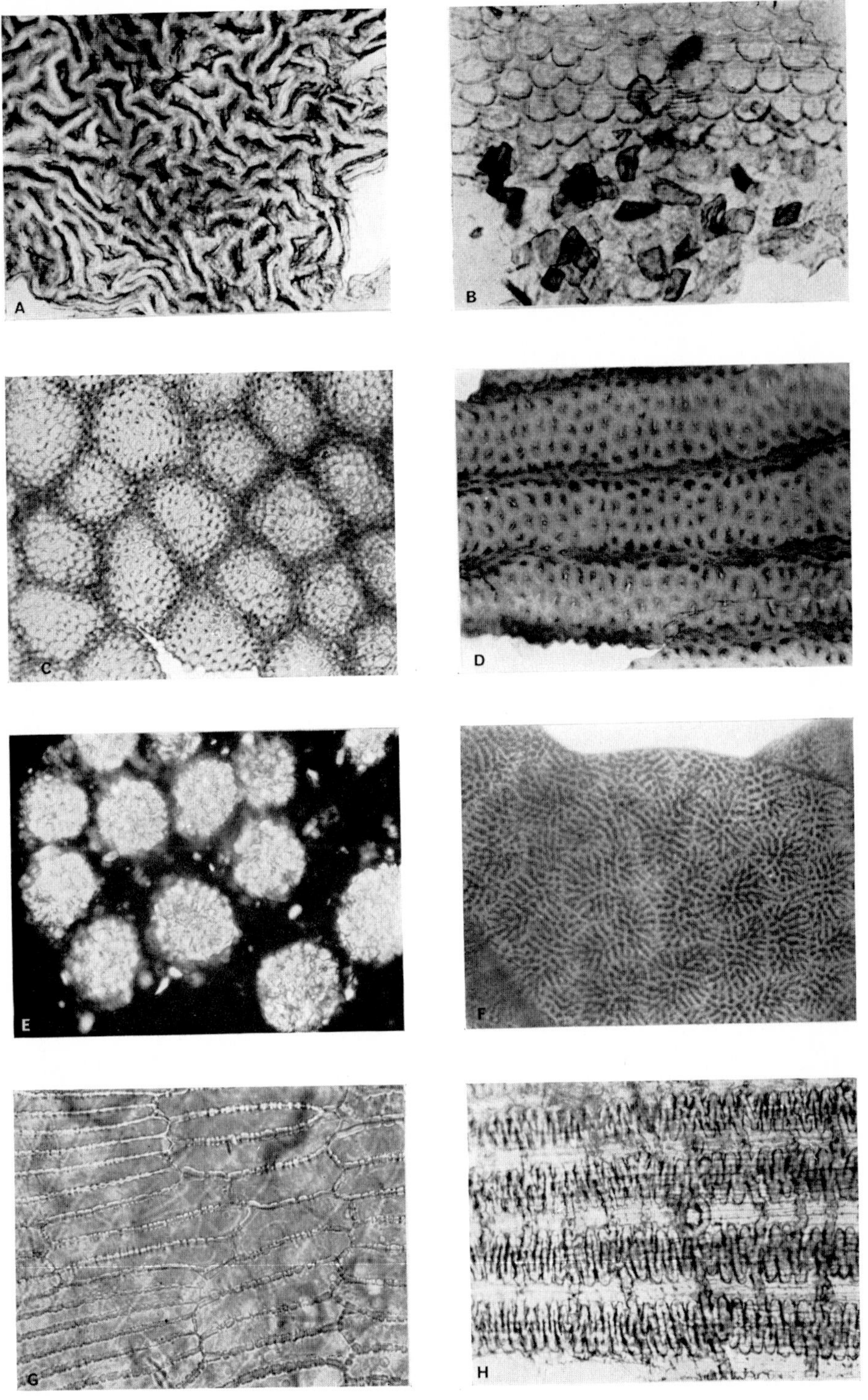

J. G. VAUGHAN

(*Facing p.* 43)

WINTON, A. L. & WINTON, K. B., 1932. *The structure and composition of foods.* **I.** *Cereals, starch, oil seeds, nuts, oils, forage plants.* New York: John Wiley.
WINTON, A. L. & WINTON, K. B., 1935. *The structure and composition of foods.* **II.** *Vegetables, legumes, fruits.* New York: John Wiley.

EXPLANATION OF PLATE

PLATE 1

Surface views of common plant fragments found in animal feed constituents.
A. Outer epidermis of Cotton (*Gossypium*) testa, ×175.
B. Round cells, fibres and pigment cells of Linseed (*Linum*) testa, ×140.
C. Palisade layer, showing reticulation, of Indian mustard (*Brassica juncea*) testa, ×115.
D. Palisade layer, showing ribs, of Field Pennycress (*Thlaspi arvense*) testa, ×115.
E. Outer epidermis of Sesame (*Sesamum*) testa showing calcium oxalate crystals in polarized light, ×390.
F. Outer epidermis of *Chenopodium* testa, ×150.
G. Cross cells of wheat (*Triticum*) bran, ×240.
H. Chaff of rice (*Oryza*), ×140.

Vessels in Pontederiaceae, Ruscaceae, Smilacaceae and Trilliaceae

VERNON I. CHEADLE

Department of Biological Sciences, University of California, Santa Barbara, California, U.S.A.

The distribution of vessels in the plant and the level of specialization of vessels as measured by nature of perforation plates are described. These features tend to confirm the retention of family status for Smilacaceae, to indicate variation in the genera of Ruscaceae, to provide evidence that Pontederiaceae is probably not primitively aquatic, and to show Trilliaceae to be primitive in its tracheary elements and thus to be similar to Uvularieae of Liliaceae.

CONTENTS

INTRODUCTION

The families whose vessels are here reported are each relatively small in number of genera and species. Together with Liliaceae and Tecophilaeaceae the four comprise Liliales of Hutchinson (1959). Vessels in Tecophilaeaceae have been described (Cheadle, 1969) and those of Liliaceae are currently under study. The origin, distribution, evolution, and use of vessels in taxonomy of monocotyledons have been reported by me in a number of early general papers (e.g. Cheadle, 1944, 1953). Additional information has appeared infrequently thereafter and, moreover, is in each paper restricted to one (e.g. Cheadle, 1960, 1963) or a few (Cheadle, 1968) families. The basic conclusions have been confirmed and reconfirmed as more material is examined. Those conclusions are as follows: Vessels originated from tracheids, arose in roots first and thence upward in the shoots, and specialized in the same sequence. Likewise, in each organ vessels arose in the late metaxylem first and then in succession in the earlier formed xylem and followed the same sequence in specialization. Vessel members having long scalariform perforation plates with many perforations are most primitive, those with transversely placed plates with a single perforation are most specialized.

MATERIALS AND METHODS

The plant parts available were macerated and usually mounted unstained in diluted glycerine. A few were stained in safranin and mounted in diaphane. Others were sectioned, stained usually in haematoxylon and safranin, and mounted in clarite. The camera lucida drawings are at identical magnifications and thus provide clear indication of size differences.

The following species were available. Names are chiefly from the *Index Kewensis* (Hooker & Jackson, 1893) and I have vouchers for those with CA numbers. Thirty-nine species in 15 of the 18 genera in Hutchinson's (1959) four families were examined.

Pontederiaceae

Eichhornia azurea Kunth M-442
E. crassipes Solms M-443
Heteranthera dubia (Jacq.) MacM. M-811
H. limosa (Sw.) Willd. M-812
H. reniformis Ruiz & Pav. M-813
Monochoria hastata (L.) Solms (UC Berkeley 49224) M-865
M. korsakowii Reg.-Mack. (UC Berkeley 53487) M-689
M. korsakowii Reg.-Mack. (UC Berkeley 584529) M-870
M. plantaginea Presl (UC Berkeley M162889) M-866
M. vaginalia (Burm. f.) Presl (UC Berkeley MOO9214) M-867
M. vaginalia (L.) Presl var. *pauciflora* (Bl.) Merr. (UC Berkeley 1346463) M-868
Pontederia cordata Larrañaga M-444, M-602, M-808
P. hastata L. (UC Berkeley 244621) M-864
Reussia subovata (Seub.) Solms M-817

Ruscaceae

Danae racemosa (L.) Moench
Ruscus aculeatus L. M-459
R. hypoglossum L. CC-570
Semele androgyna Kunth CC-573

Smilacaceae

Heterosmilax sp.
H. erythrantha Baill. ex Gagnep. (UC Berkeley M954193) M-846
H. gaudichaudiana Maxim. (UC Berkeley 300824) M-847
H. gaudichaudiana Maxim. (UC Berkeley MO11179) M-848
Rhipogonum album R. Br. CA-171
R. elseyanum F. Muell. CA-172
R. fawcettianum F. Muell. ex Benth. CA-280
R. scandens Forst. M-461
Smilax australis R. Br. CA-214
S. glyciphylla Sm. CA-284
S. herbacea L. M-463, M-604, M-747, M-748, M-749, M-750
S. panamensis Morong M-464
S. rotundifolia L. M-466, M-521

S. walteri Pursh M-467
S. sp. M-746
Trilliaceae
Medeola virginiana Merrill M-476
Paris hexaphylla Cham. (UC Berkeley 132289) M-849
P. incompleta M. Bieb. (UC Berkeley 396410) M-850
P. polyphylla Sm. (UC Berkeley 388878) M-851
P. quadrifolia L. (UC Berkeley 507419) M-852
P. tetraphylla A. Gray (UC Berkeley M309137) M-853
Scoliopus bigelovii Torr. CC-574
Trillium cernuum L. M-477
T. grandiflorum Salisb. M-478

RESULTS

The full spectrum of variations in vessel perforation plates appears in the illustrations. In only one species examined (*Semele androgyna*) are vessels with simple perforation plates present (Fig. 1**A**), and they are restricted to the late metaxylem of roots, where they occur in association with more infrequent scalariform plates having few (Fig. 1**B**) to many perforations. Otherwise, where vessels occur at all, they are likely to have scalariform plates of rather typical form (Fig. 1**C, E**) with relatively few (Fig. 1**I**) to many perforations (Fig. 1**E**). More unusual forms of perforation plates that I term basically scalariform appear in Fig. 1**D** and **G**. These may be said to have numerous branching bars separating the perforations. The perforation plates of some vessel members depart still further from the basic scalariform type and may be more net-like (Fig. 1**H**) in appearance or in a somewhat mixed condition, as depicted in Fig. 1**J**.

Vessels (primitive) occur regularly in the stems of Smilacaceae examined, but apparently only in the late metaxylem. They were identified in the late metaxylem of the inflorescence axes of three of the four species for which these organs were available and in seven of the eleven species seen. (The leaves of the three species of *Rhipogonum* examined seem to lack vessels.) I could not identify vessels in either the inflorescence axes or leaves of *Heterosmilax erythrantha*. Neither do they occur in rhizomes in any genus.

I had only the stem of *Danae racemosa* for study, but it has vessels (primitive) in both early and late metaxylem. Among other genera of Ruscaceae, primitive vessels occur in the roots of *Ruscus*, but are lacking elsewhere. In *Semele androgyna*, highly specialized vessels (single perforations) occur in the late metaxylem, and primitive ones elsewhere in the root and in the stem (Fig. 1**D**). Some perforation plates in the late metaxylem (large cells) are more advanced (Fig. 1**H**) in having few perforations in a nearly transverse plate.

Among species examined in Pontederiaceae, primitive vessels were identified in early and late metaxylem of roots, with some in the late metaxylem having few (five or fewer) perforations in nearly transversely placed perforation plates. It appears that some vessels also occur in the stems of *Eichhornia crassipes* (Fig. 1**K**) and *Heteranthera*

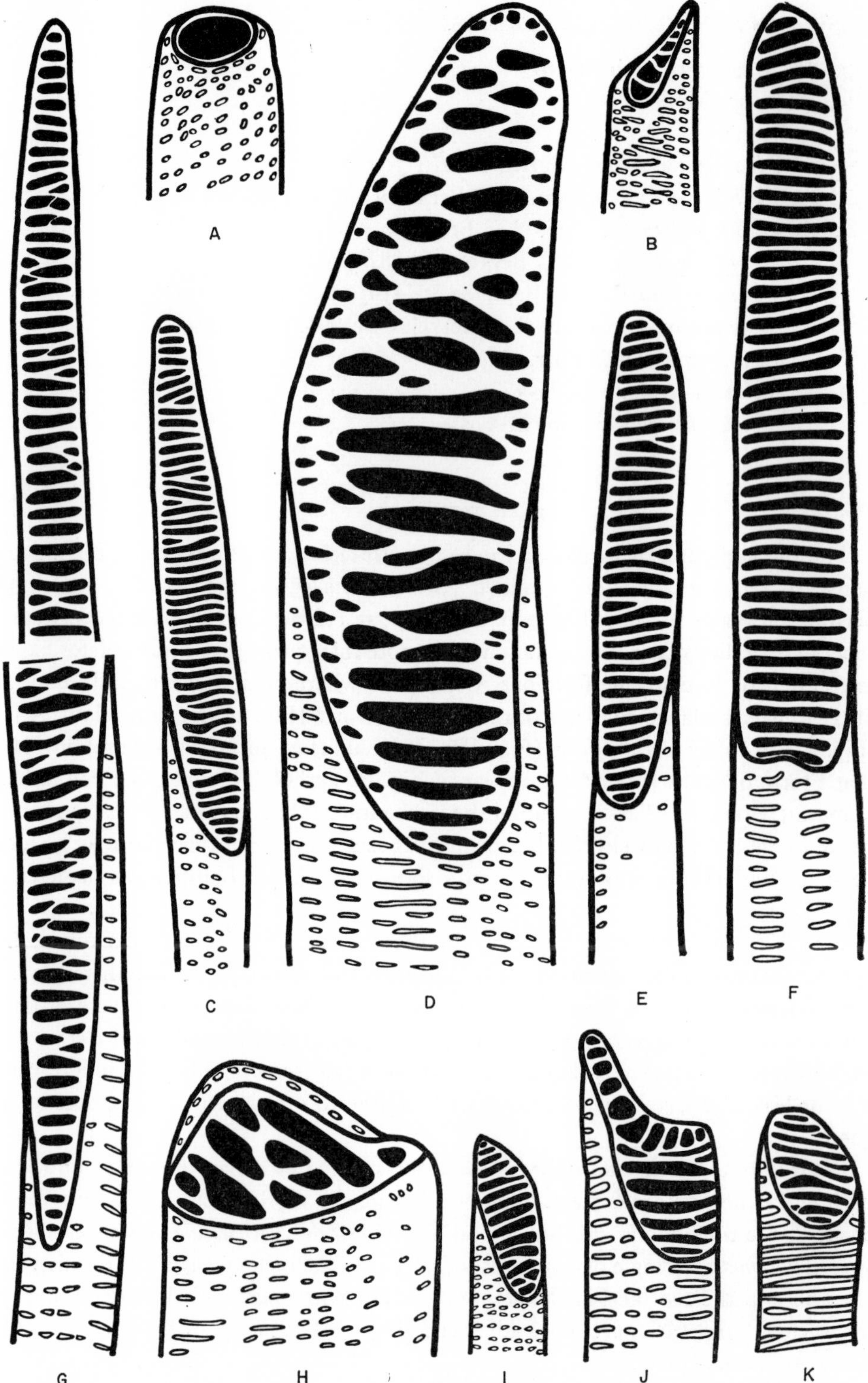

FIGURE 1. Parts of vessel members from late metaxylem of roots (**A–C, F, G, I, J**) and stems (**D, E, H, K**). Perforations shown in black. Imperforate pits closely associated with perforations (**H**) or on side walls are shown clear. (All ×420.) **A, B, D, H,** *Semele androgyna*; **C,** *Ruscus hypoglossum*; **E,** *Smilax glyciphylla*; **F,** *Rhipogonum elseyanum*; **G,** *Scoliopus bigelovii*; **I,** *Reussia subovata*; **J,** *Pontederia cordata*; **K,** *Eichhornia crassipes.*

limosa, and maybe as well in *Pontederia cordata*. They are seemingly absent elsewhere in the species available for study.

Primitive vessels were recorded for early and late metaxylem of roots in Trilliaceae, but only tracheids occur elsewhere.

Table 1 shows the comparative specialization of tracheary elements in the late metaxylem of the four families. The xylem generally is so relatively primitive that differences between early- and late-formed metaxylem are usually very minor and these are not shown. Where differences do occur (e.g. in roots of *Semele androgyna* or stems of Smilacaceae), the late metaxylem is always more specialized, a situation unfailingly uniform throughout the monocotyledons.

Table 1. Specialization* in late metaxylem of roots, stems and leaves of Pontederiaceae, Smilacaceae, Ruscaceae and Trilliaceae

	No. of genera	No. of species	Roots	Stems	Leaves
Pontederiaceae	5	12	1·00	0·25	0·00
Ruscaceae	3	4	2·00	0·50	0·00†
Smilacaceae	3	14	1·00	1·00	0·64
Trilliaceae	4	9	1·00	0·00	0·00

* Specialization based on kinds or lack of perforation plates: 1 represents scalariform (multiperforate) perforation plates only; 2, mostly scalariform; 3, simple (uniperforate) and scalariform; 4, mostly simple; and 5, simple only. Zero represents tracheids only. See text.

† Leaves in Ruscaceae reduced to scales.

It is clear from Table 1 that if any differences occur among the organs of a given family, the roots are always most highly specialized and the leaves least. This is true for any species and hence is true for any family. The value five represents highest specialization, so it is obvious that in none of the families are vessels in any organ very specialized. All four families are relatively primitive with respect to tracheary elements. Trilliaceae is most primitive in its vessels; Smilacaceae seems most advanced because it has vessels throughout (excepting rhizomes) in most species even though the perforation plates are primitive; Ruscaceae is perhaps next to Smilacaceae, for it has (in *Semele androgyna*) specialized vessels in roots, but lacks vessels in stems of two of four species and in leaves of all three species seen; and Pontederiaceae seems more nearly like Trilliaceae except for rather primitive vessels in stems of perhaps two species.

DISCUSSION AND CONCLUSIONS

These four families do not exhibit much specialization of vessels. It may be understandable that vessels in Pontederiaceae are relatively unspecialized, even in roots, and are generally lacking in the shoot system, for these are aquatic plants. It is noteworthy that roots in *Eichhornia* have very well developed vessels, including rather short scalariform perforation plates with few bars occurring frequently in the late metaxylem of roots. Some of the vessel elements in *Pontederia* are also relatively short with few bars. These observations would lead one to support Cronquist's (1968: 318) conjecture about the aquatic habitat being a 'secondary' development in Pontederiaceae. (I must confess I do not know what monocotyledons reflect the 'primary' condition of aquatic habitat; maybe none does, although this conjecture does violence to Cronquist's views on the origin of monocotyledons. I think I know a great many aquatics in the monocotyledons, on the other hand, that do not reflect a primary aquatic condition.) Hamann (1966) suggests a juxtaposition for Philydraceae and Pontederiaceae. As far as vessels are concerned, it seems to me the two families could be happily so associated, although vessels in Pontederiaceae I have seen are somewhat more specialized than those of Philydraceae.

Among Ruscaceae, vessels occur in the stems of *Danae* and *Semele*, but not in *Ruscus*, whose tracheary elements are small. I do not understand this distribution. Hutchinson (1959: 621) regards Ruscaceae as truly a family or perhaps as a very advanced tribe of Liliaceae; at least he considers it a possibly close relative of *Asparagus*. The vessels of *Asparagus* (in four species examined) have highly specialized vessels in the roots and primitive (very long vessel members) ones in the stem. *Asparagus* then resembles *Danae* and *Semele*, but not *Ruscus*. Cronquist (1968: 374) subsumes Ruscaceae under his Liliaceae.

As noted earlier, vessels (always primitive) occur throughout roots and stems of Smilacaceae examined. Only the largest tracheary elements in inflorescence axes and in leaves are vessels. Vessels were identified in leaves of all available species of *Smilax* and in one species of *Heterosmilax*; they seem to be absent in *Rhipogonum*. In view of the fact that I have not been able to identify vessels in any of the leaves of some 128 species of Hutchinson's (1959) Liliaceae examined, perhaps it is best, as Cronquist (1968: 358) suggests, to let *Smilax* and its cohorts have family status.

Hutchinson (1959: 606) states 'further development' of Uvularieae (in his Liliaceae) is to Trilliaceae. Vessel distribution and degree of specialization are similar in the two families: restricted to roots and primitive.

ACKNOWLEDGEMENTS

I am in debt to many botanists in Australia and South Africa with whom I collected. The herbarium at the University of California, Berkeley gave me much needed material from herbarium sheets. Individuals at Davis and Santa Barbara gave technical aid, and Hatsume Kosakai was cheerfully helpful in making the drawings and in otherwise preparing for the writing. I acknowledge again a Fulbright Scholarship in Australia and support from Natural Science Foundation Grant GB-5506X.

REFERENCES

CHEADLE, V. I., 1944. Specialization of vessels within the xylem of each organ in the Monocotyledoneae. *Am. J. Bot.*, **31**: 81–92.

CHEADLE, V. I., 1953. Independent origin of vessels in the monocotyledons and dicotyledons. *Phytomorphology*, **3**: 23–44.

CHEADLE, V. I., 1960. Vessels in grasses: kinds, occurrence, taxonomic implications. *Jl. S. Afr. biol. Soc.*, **1**: 27–37.

CHEADLE, V. I., 1963. Vessels in Iridaceae. *Phytomorphology*, **13**: 245–248.

CHEADLE, V. I., 1968. Vessels in Haemodorales. *Phytomorphology*, **18**: 412–420.

CHEADLE, V. I., 1969. Vessels in Amaryllidaceae and Tecophilaeaceae. *Phytomorphology*, **19**: 8–16.

CRONQUIST, A., 1968. *The evolution and classification of flowering plants*. Boston: Houghton Mifflin Co.

HAMANN, U., 1966. Embryologische, morphologisch-anatomische und systematische Untersuchungen an Philydraceen. *Willdenowia, Beih.*, **4**: 1–178.

HOOKER, J. D. & JACKSON, B. P. (*et al.*), 1893. *Index Kewensis* (including supplements 1–13). Oxford: Clarendon Press.

HUTCHINSON, J., 1959. *The families of flowering plants*. **II**. *Monocotyledons*, 2nd ed. Oxford: Clarendon Press.

Ultrastructure and nectar secretion in *Lonicera japonica*

A. FAHN, F.L.S.

AND

TALIA RACHMILEVITZ

Department of Botany, The Hebrew University of Jerusalem, Israel

The nectar-secreting cells of *Lonicera japonica* Thunb. are short hairs, which are present in a delimited area of the inner epidermis of the corolla tube. The ultrastructure of these cells was studied at various ontogenetic stages. In the development of the secretory cells there is first an increase in the vacuome and later, with the approach of the stage of secretion, the volume of cytoplasm increases while the volume of the vacuome decreases. The ribbed cuticle becomes detached from the wall at the top of the secretory hair leaving a large enclosed space, and at the same time a thick inner layer of wall protuberances develops. In the young secretory cells vesicles appear throughout the cytoplasm. Later a lamellar endoplasmic reticulum (ER) starts to appear and reaches its maximal development during the time of nectar secretion. At this stage the edges of the ER cisternae, which have a parallel arrangement, are swollen and vesicles appear to bud off from them. Both at early and late stages ER vesicles appear to approach the cell wall and their membranes appear to fuse with the plasmalemma.

It is suggested that, up to the stage of secretion, vesicles, most of which apparently develop from the Golgi bodies, supply substances to the inner layer of wall protuberances, while at the stage of secretion the vesicles that bud off from the ER release the sugar that forms the nectar.

CONTENTS

INTRODUCTION

Nectaries, which have attracted the interest of scientists since Linnaeus, have been the subject of many histological investigations (e.g. Caspary, 1848; Bonnier, 1879; Behrens, 1879; Schniewind-Thies, 1897; Daumann, 1928, 1930*a*, *b*, 1935, 1965; Frey-Wyssling, 1933; Brown, 1938; Fahn, 1953; Frei, 1955).

Various views based on histological and physiological observations have been suggested concerning the process of nectar secretion by plants. At least, it is clear that nectaries secrete sugar supplied by the phloem (Frey-Wyssling & Agthe, 1950; Agthe, 1951; Zimmermann, 1953). Many investigators have put forward the suggestion that active mechanisms take place in the nectary tissue (e.g. Helder, 1958;

Frey-Wyssling & Häusermann, 1960; Frey-Wyssling, Zimmermann & Maurizio, 1954; Lüttge, 1961, 1962).

A few cytological studies have recently been carried out with the aid of the electron microscope (Schnepf, 1964*a*, *b*; Mercer & Rathgeber, 1962; Eymé, 1966, 1967). We undertook a further electron microscopic investigation of the nectar-secreting cells. Our purpose was to learn which cell organelles are involved in nectar secretion.

MATERIAL AND METHODS

Flower nectaries of *Lonicera japonica* Thunb. growing in the Hebrew University garden were used for this investigation. Flower buds and open flowers at various stages of development were examined. The various stages could be distinguished morphologically (Plate 1**A**) according to the size of the bud, the extent of opening and the colour of the flower.

At the stage at which the flower bud has reached its maximal size, but is still closed, small droplets of nectar can already be found. Immediately after the flower opens intensive nectar secretion starts. This secretion decreases about a half a day later. Flower colour turns light yellow and nectar secretion stops one day after the flower opens.

Portions of the flowers containing the secreting cells were fixed in 5% glutaraldehyde in phosphate buffer (pH 7·5), post-fixed with 2% OsO_4, dehydrated in ethyl alcohol and embedded in Epon 812. The embedded material was sectioned with an L.K.B. ultramicrotome and the sections were used for both light and electron microscope observations. The 3–4 μm thick sections examined under the light microscope were stained with various stains, viz. safranin, ruthenium red, Sudan IV and Nile blue. With Nile blue the cuticle stained red and the cell walls blue.

The sections prepared for the electron microscope were post-stained with uranyl acetate and lead citrate.

LIGHT MICROSCOPY

The *Lonicera japonica* nectary occupies a strip about 1 mm wide and 15 mm long on the inner lower side of the basal part of the corolla tube. This secretion area is covered with one-celled epidermal hairs which secrete nectar (Plate 1**B**) (cf. Bonnier, 1879; Frei, 1955). Each of these epidermal hairs consists of a 'base' and a 'head' (Plate 1**C, D, E**). The 'base' lies between the ordinary cells of the epidermis and is narrow. The 'head' is large and ovate or spherical in shape. A cuticle covers both the ordinary and the secreting epidermal cells. This cuticle appears first on the secretory hairs and then on the ordinary epidermal cells. The cuticle upon the 'head' of the hair has an elaborate structure; it consists of a network of thick ribs which enclose very thin areas. The thin areas of the cuticle, therefore, represent deep depressions (Plate 1**E**).

At the stage in which the young secretory cell has reached its mature size the cuticle upon the 'head' still adheres to the cell wall and a large vacuole occurs in the protoplast (Plate 1**D**). Later, in the stages that precede secretion, the vacuoles become very small, the cuticle upon the 'head' becomes detached from the cell wall and a large crescent-shaped space is formed between them (Plate 1**E**). The cell wall close to the developing crescent-shaped space becomes relatively thick, as may be seen well when it is stained with ruthenium red. After the cessation of secretion the cuticle becomes folded (Plate 1**F**), and it finally adheres to the cell wall which was its original position before secretion.

ELECTRON MICROSCOPY

Plate 2 represents the earliest developmental stage in which the secretory cells can be distinguished. At this stage the cuticle is not yet developed and the cell wall is relatively thin. The cytoplasm in these cells is dense and rich in ribosomes, and it contains Golgi bodies, few mitochondria and plastids. In scattered places in the cytoplasm it is possible to distinguish ER cisternae. The vacuoles at this stage are relatively small.

Plate 3 represents the stage in which the secretory cell has already reached its characteristic shape and size. The ribbed cuticle is completely developed and has started to withdraw from the wall of the upper part of the cell. By this stage an inner layer of wall protuberances (Pr) starts to develop, and it is also characterized by the appearance of many vesicles, part of which are coated by ribosomes (Plate 3**B**, CV). Some of these vesicles are attached to the plasmalemma (Plate 3**B**, arrows), and we suspect that their membranes fuse with the plasmalemma by a process of reverse pinocytosis (Plate 3**A**, arrow). Lamellar ER, which is seen in later developmental stages of the nectar-secreting cells, does not occur here, there is no change in the number of Golgi bodies and the vacuome has increased in volume (Plates 1**D** and 3**A**).

Plates 4 and 5 are of secretory cells of flower buds, two to three days before they would have opened. The cuticle can be seen to have completely withdrawn from the upper part of the cell and, as a result, a large space is formed between the cuticle and the wall. The inner layer of wall protuberances has increased in thickness. At this stage the cytoplasm contains, in addition to a large number of vesicles, rough lamellar ER and many more mitochondria than in earlier stages (Plate 4**B**). There appears to be no change in the number of Golgi bodies, the volume of the vacuome has decreased and starch grains occur in the plastids.

Plate 6 is of secretory cells of a flower bud just before it was about to open. Here the inner layer of wall protuberances has reached its maximal thickness. The main change observed at this stage is in the amount and arrangement of the ER: there is a distinct increase in the amount of the ER, its cisternae are arranged parallel to one another and the edges of the cisternae are swollen. In Plate 6**B** connections between the swollen ends of the ER cisternae and wall protuberances can be seen (arrow).

Plates 7 and 8 are of secretory cells of open flowers at the stage of maximum secretion. The characteristic feature of this stage is the extreme development of the ER. The parallel-arranged ER occupies the greater part of the cytoplasm and the edges of the ER cisternae are swollen as in the previous stage. Near these swollen edges there are many vesicles, whose topography and arrangement resemble those of the budding-off of the Golgi vesicle (Plate 7**A**). These vesicles, which are often coated with ribosomes, are seen close to and in contact with the wall protuberances (Plate 8**B**, arrow), and sometimes the swollen edges are continuous with these wall protuberances (Plate 7**B**, arrow). Starch grains no longer appear in the plastids.

At the end of the stage of secretion the space between the cuticle and the cell wall starts to diminish until the cuticle adheres more or less to the wall. Concurrently many changes occur in the cell wall and the cytoplasm starts to disintegrate (Plate 9**A, B**). The wall protuberances merge so that the whole wall appears more or less uniform and includes only disconnected remnants of cytoplasm; the plasmalemma, which previously lined all the wall protuberances, now lines the smoothed inner surface of the wall so that the surface of the cytoplasm decreases greatly. The process of disintegration of the cytoplasm continues for several days, during which the central vacuole increases in size until it occupies most of the cell. As these processes occur after secretion ceases, we shall not describe them in detail in this article.

DISCUSSION

The examination of the various developmental stages of the secretory cells of *Lonicera japonica* shows that the main change as the time of secretion approaches occurs in the ER. The occurrence of a well-developed ER in nectar-secreting cells was recorded by Mercer & Rathgeber (1962) and Schnepf (1964*a*, *b*). In *Lonicera* at the stage of secretion the ER occupies most of the volume of the cytoplasm, and the parallel arrangement of the ER cisternae, which is seen at the time of secretion, makes possible the packing of maximal amount of ER in a minimal volume. During various stages of development, vesicles are seen to be in contact with the plasmalemma (Plates 3 and 8), and we suggest that they release their contents by reverse pinocytosis. At the stage of secretion this process is at its maximum, and the vesicles appear to develop from the swollen edges of the ER cisternae by budding. In some cases the swollen edges themselves were seen to be continuous with the wall protuberances (Plates 6**B** and 7**B**). Although most of the vesicles at the stage of secretion are of ER origin, we cannot exclude the possibility that part of the vesicles develop from the Golgi bodies, especially in the early developmental stages of the secretory cells (Plates 2**B**, 3**B** and 4**A**).

A thick inner layer of wall protuberances is formed during the development of the secretory cell, and it may be assumed, therefore, that vesicles contribute material to the growth of the wall. If the function of the vesicles is only to add material to the wall, they would be expected to disappear after the layer of wall protuberances has reached its maximal thickness. Our observations indicate, however, that the greatest number of vesicles detaches from the ER and leaves the cytoplasm after the cell wall has reached its maximal thickness, i.e. at the stage of maximal sugar secretion. This fact leads to the suggestion that vesicles take part both in wall formation and in sugar secretion. A similar view was mentioned by Eymé (1966), who attributes the function of secretion to vesicles of the Golgi bodies as well as to vesicles which originate from the ER. At the stage of secretion in *Lonicera japonica* most of the vesicles are of ER origin. We assume, therefore, that the ER plays a main role in the secretion of sugar.

In the early stages, when the inner layer of wall protuberances develops, the vesicles may be primarily of Golgi origin, in which case the main source of substances for the growth of the inner wall layer would come from the Golgi bodies.

Mercer & Rathgeber (1962), who studied the nectar-secreting multicellular hairs of a species of *Abutilon*, discussed the possibility that the transport of the 'prenectar' from the phloem to the apical cell of the hair may be along the internal space of the ER. These authors, however, point out the difficulty in accepting this view because the volume of the ER appears to occupy not more than 15% of the cytoplasm. Such an ER volume is, according to them, unable to transport the nectar at the rate secreted by the *Abutilon* hairs. In *Lonicera*, on the other hand, the ER at the secretion stage occupies a great part of the cytoplasm of the secretory cells. Although accurate measurements were not made, it seems possible to us that the secreted sugar is transported through the ER. In the cells which are situated between the phloem and the secretory cells the ER is sparse but the number of these cells per secretory cell is large. It should also be noted that the secreting cells are connected with their neighbouring cells by many plasmodesmata.

REFERENCES

AGTHE, C., 1951. Über die physiologische Herkunft des Pflanzennektars. *Ber. schweiz. bot. Ges.*, **61**: 240–274.

BEHRENS, W. J., 1879. Die Nectarien der Blüten. *Flora, Jena*, **62**: 2–11, 17–27, 49–54, 81–90, 113–123, 145–153, 233–240, 241–247, 305–314, 369–375, 433–457.

BONNIER, G., 1879. *Les nectaires*. Thèses Fac. Sci. Univ. Paris.

BROWN, W. H., 1938. The bearing of nectaries on the phylogeny of flowering plants. *Proc. Am. phil. Soc.*, **79**: 549–595.

CASPARY, R., 1848. *De Nectariis*. Elberfeldas.

DAUMANN, E., 1928. Zur Biologie der Blüte von *Nicotiana glauca* Grah. *Biologia gen.*, **4**: 6–8.

DAUMANN, E., 1930*a*. Das Blütennektarium von *Nepenthes*. Beiträge zur Kenntnis der Nektarien I. *Beih. bot. Zbl.*, **47**: 1–14.

DAUMANN, E., 1930*b*. Das Blütennektarium von *Magnolia* und die Futterkörper in der Blüte von *Calycanthus*. *Planta*, **11**: 108–116.

DAUMANN, E., 1935. Die systematische Bedeutung des Blütennektariums der Gattung *Iris*. *Beih. bot. Zbl.*, **53**: 525–625.

DAUMANN, E., 1965. Das Blütennektarium bei den Pontederiaceen und die systematische Stellung dieser Familie. *Preslia*, **37**: 407–412.

EYMÉ, J., 1966. Infrastructure des cellules nectarigènes de *Diplotaxis erucoides* D.C., *Helleborus niger* L. et *H. foetidus* L. *C.r. hebd. Séanc. Acad. Sci., Paris*, **262**: 1629–1632.

EYMÉ, J., 1967. Nouvelles observations sur l'infrastructure de tissus nectarigènes floraux. *Botaniste*, **50**: 169–183.

FAHN, A., 1953. The topography of the nectary in the flower and its phylogenetic trend. *Phytomorphology*, **3**: 424–426.

FREI, E., 1955. Die Innervierung der floralen Nektarien dikotyler Pflanzenfamilien. *Ber. schweiz. bot. Ges.*, **65**: 60–115.

FREY-WYSSLING, A., 1933. Über die physiologische Bedeutung der extrafloralen Nektarien von *Hevea brasiliensis*. *Ber. schweiz. bot. Ges.*, **42**: 1.

FREY-WYSSLING, A. & AGTHE, C., 1950. Nektar ist ausgeschiedener Phloemsaft. *Verh. schweiz. naturf. Ges.*, **130**: 175–176.

FREY-WYSSLING, A. & HÄUSERMANN, E., 1960. Deutung der gestaltlosen Nektarien. *Ber. schweiz. bot. Ges.*, **70**: 150–162.

FREY-WYSSLING, A., ZIMMERMANN, M. & MAURIZIO, A., 1954. Über den enzymatischen Zuckerumbau in Nektarien. *Experientia*, **10**: 490–492.

HELDER, R. J., 1958. The excretion of carbohydrates (nectaries). *Encycl. Plant Physiol.*, **6**: 978–990.

LÜTTGE, U., 1961. Über die Zusammensetzung des Nektars und den Mechanismus seiner Sekretion. I. *Planta*, **56**: 189–212.

LÜTTGE, U., 1962. Über die Zusammensetzung des Nektars und den Mechanismus seiner Sekretion. III. Mitteilung. Die Rolle der Rückresorption und der specifischen Zuckersekretion. *Planta*, **59**: 175–194.

MERCER, F. V. & RATHGEBER, N., 1962. Nectar secretion and cell membranes. *5th Int. Conf. Electron Microsc.*, WW 11. New York: Academic Press.

SCHNEPF, E., 1964*a*. Zur Cytologie und Physiologie pflanzlicher Drüsen. 4 Teil. Licht- und elektronenmikroskopische Untersuchungen an Septalnektarien. *Protoplasma*, **58**: 137–171.

SCHNEPF, E., 1964*b*. Zur Cytologie und Physiologie pflanzlicher Drüsen. 5 Teil. Elektronenmikroskopische Untersuchungen an Cyathialnektarien von *Euphorbia pulcherrima* in verschiedenen Funktionszuständen. *Protoplasma*, **58**: 193–219.

SCHNIEWIND-THIES, J., 1897. Beiträge zur Kenntnis der Septalnektarien. *Bot. Zbl.*, **69**: 216–218.

ZIMMERMANN, M., 1953. Papierchromatographische Untersuchungen über die pflanzliche Zuckersecretion. *Ber. schweiz. bot. Ges.*, **63**: 403–429.

ABBREVIATIONS USED IN PLATES

Bc	bacteria
C	cuticle
CV	vesicles coated with ribosomes
ER	endoplasmic reticulum
EV	vesicles budding off from the swollen edges of the ER cisternae
GB	Golgi bodies
M	mitochondria
N	nucleus
NM	nuclear membrane
P	plastids
Pr	wall protuberances
S	space between cell wall and cuticle
SC	secretory cell
SE	swollen edges of ER cisternae
SM	flower at the stage of maximal secretion
St	starch grains
V	vesicles
Vc	vacuoles
W	cell wall

EXPLANATION OF PLATES

PLATE 1

A. Inflorescence.
B. Cross-section of a flower tube when still in bud (×70).
C. Portion of a cross-section of a flower tube at the stage of secretion (×190).
D. A secretory cell with the cuticle still attached to the cell wall (×700).
E. A secretory cell at the stage of secretion, the cuticle is detached from the cell wall (×900).
F. Shows the folding of the cuticle of a secretory cell which has just ceased to secrete (×600).

PLATE 2

Electron micrographs of secretory cells of very young flower buds: **A,** ×10,500; **B,** ×22,000.

PLATE 3

Electron micrographs of a young secretory cell which has reached mature size: **A,** ×30,000; **B,** ×38,000.

PLATE 4

Electron micrographs of secretory cells of flower buds, two to three days before they would have opened: **A,** ×21,000; **B,** ×3300.

PLATE 5

As in Plate 4: **A,** ×15,000; **B,** ×25,000.

PLATE 6

Electron micrographs of secretory cells of a flower bud just before it was about to open: **A,** ×7200; **B,** ×14,000.

PLATE 7

Electron micrographs of secretory cells of flowers at the summit of secretion: **A,** ×22,000; **B,** ×21,000.

PLATE 8

As in Plate 7: **A,** ×16,000; **B,** ×14,000.

PLATE 9

Electron micrographs of secretory cells after they have ceased to secrete: **A,** ×7700; **B,** ×12,500.

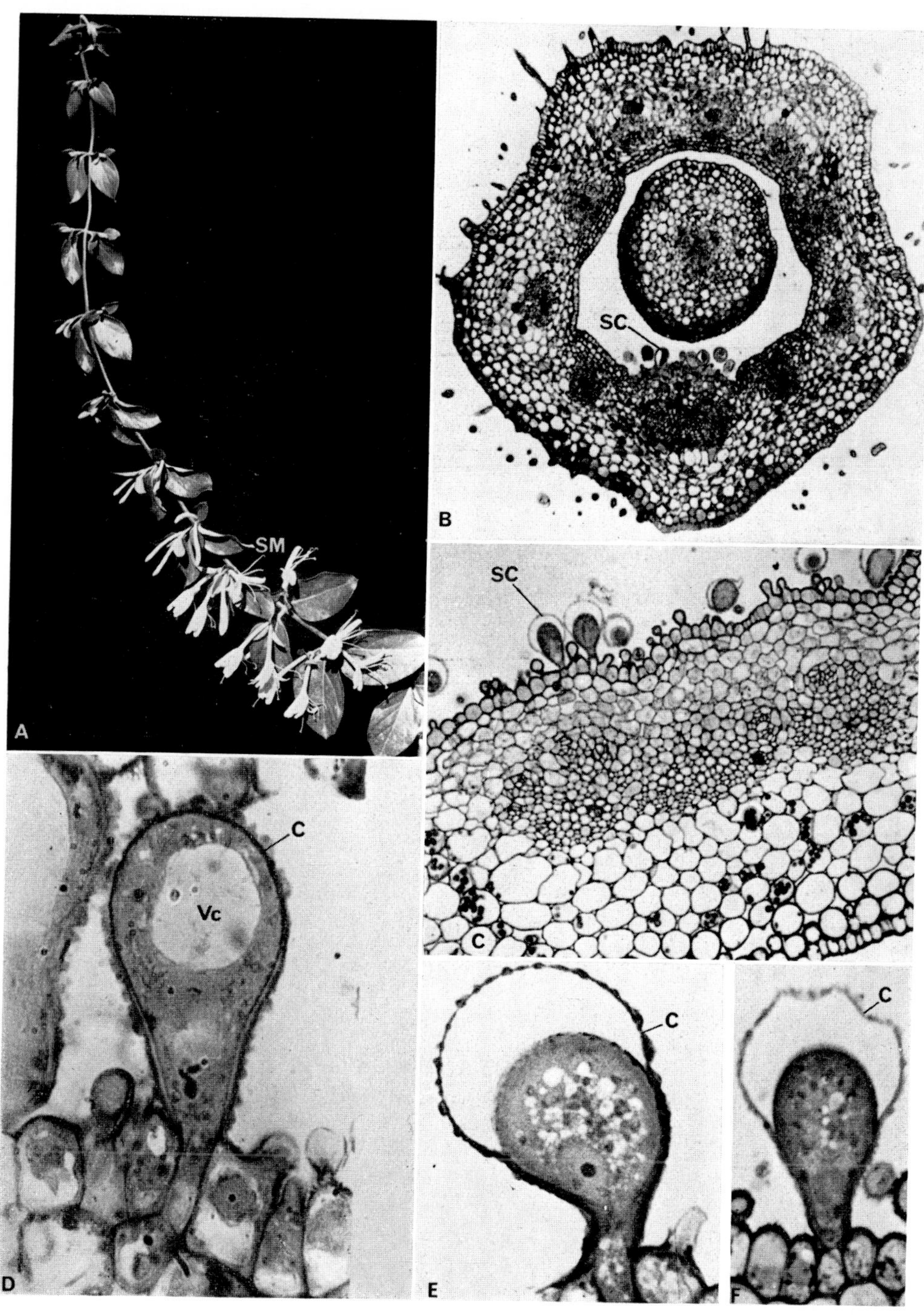

A. FAHN AND T. RACHMILEVITZ

(*Facing p.* 56)

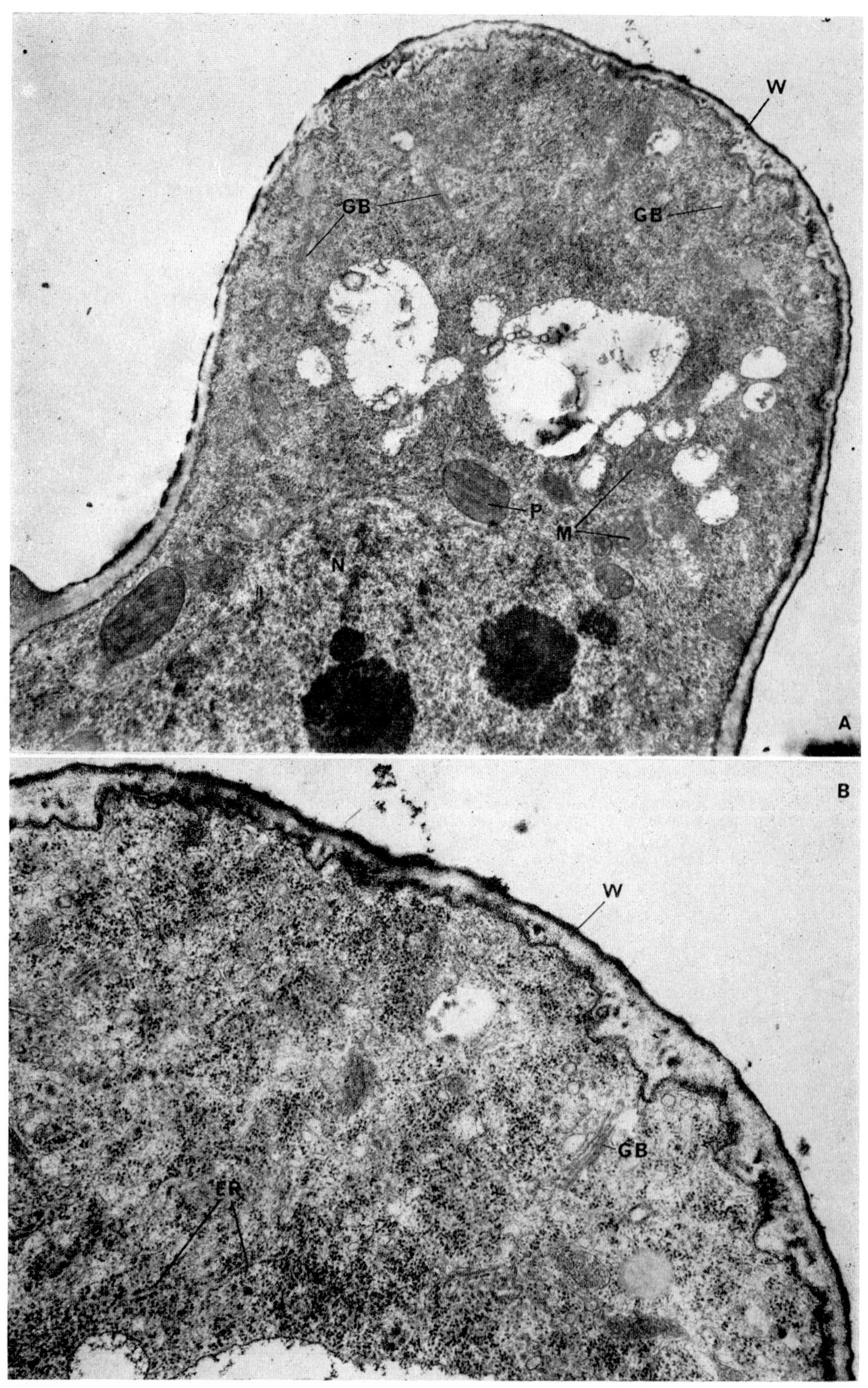

A. FAHN AND T. RACHMILEVITZ

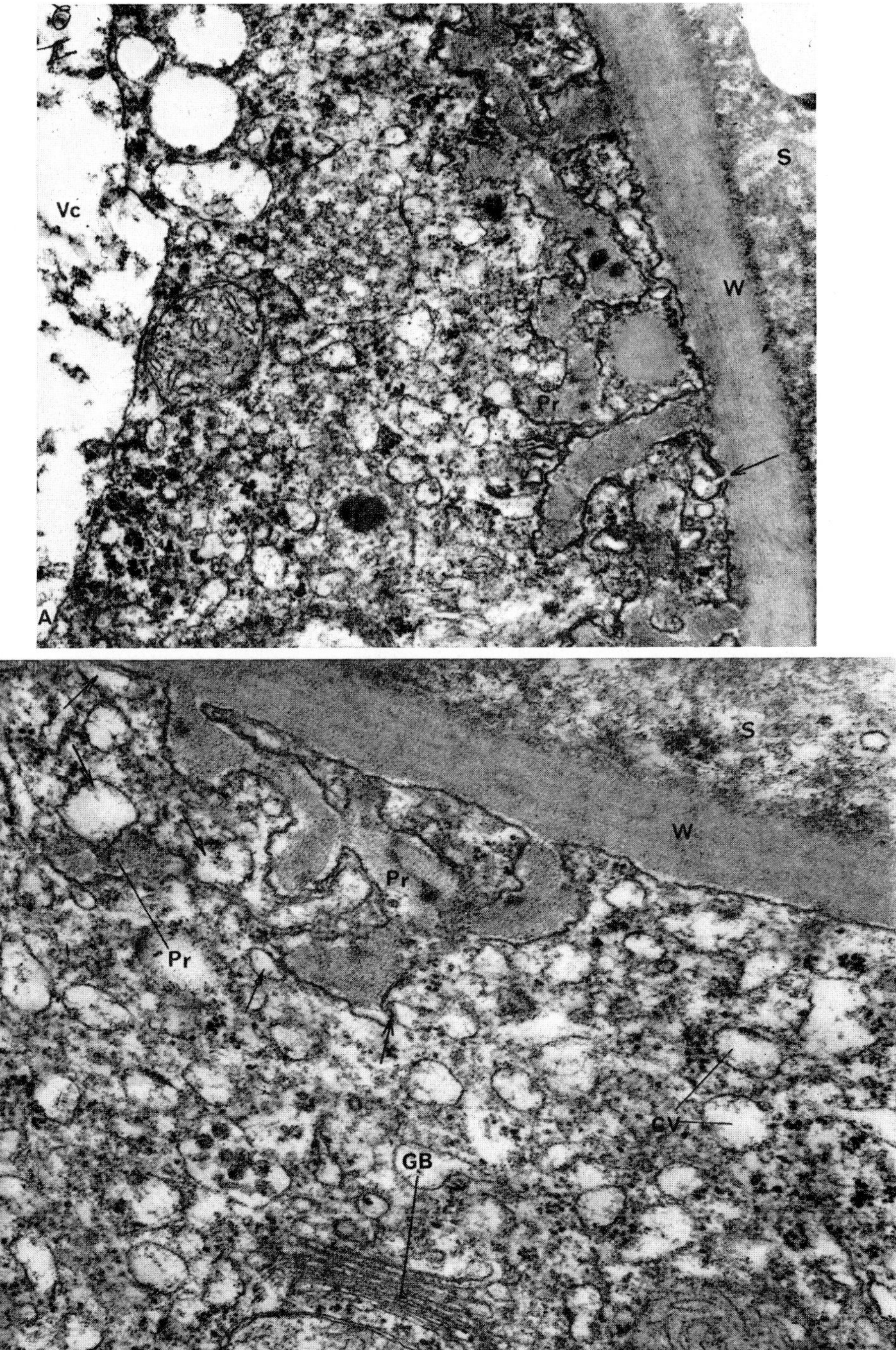

A. FAHN AND T. RACHMILEVITZ

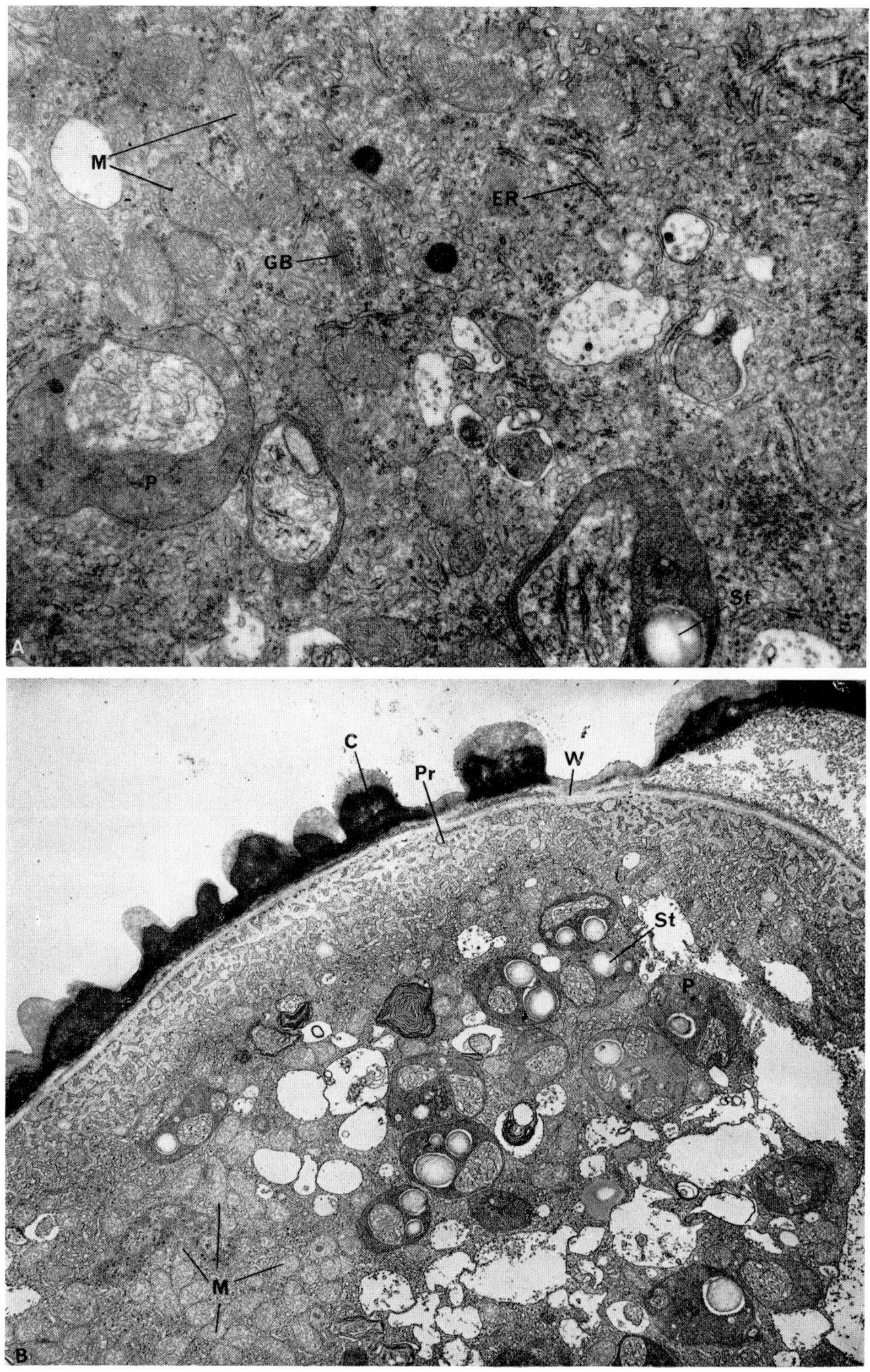

A. FAHN AND T. RACHMILEVITZ

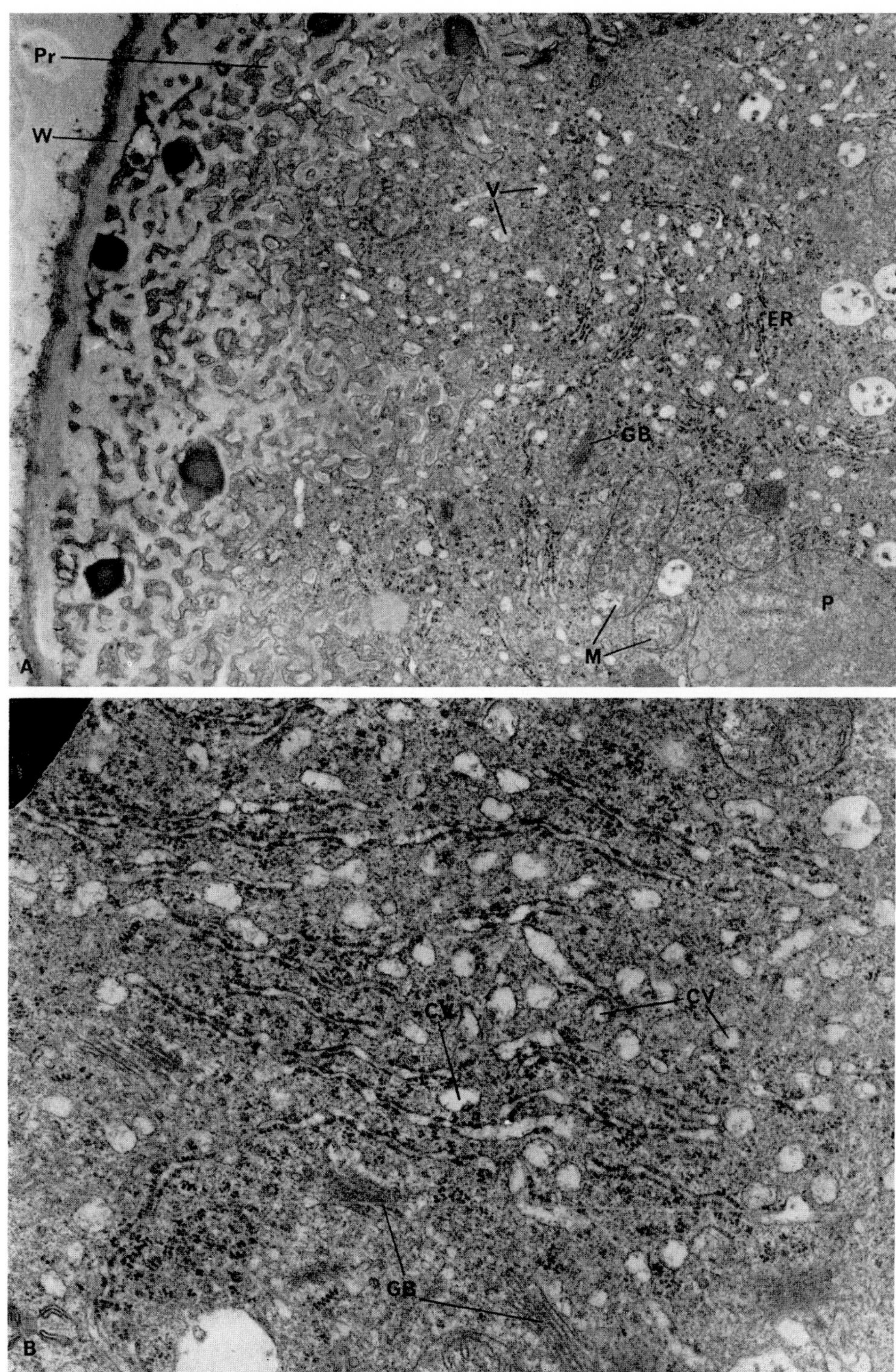

A. FAHN and T. RACHMILEVITZ

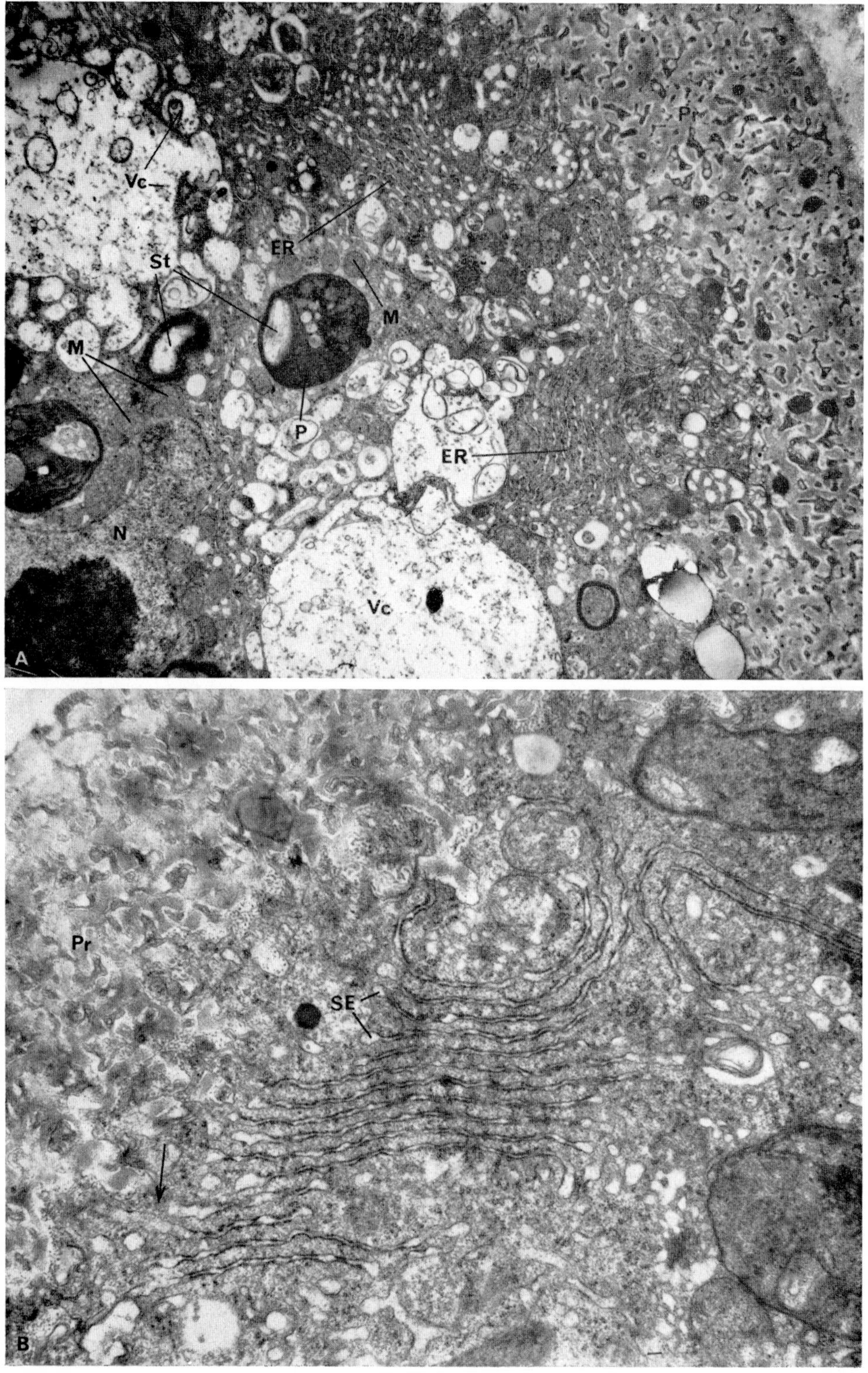

A. FAHN AND T. RACHMILEVITZ

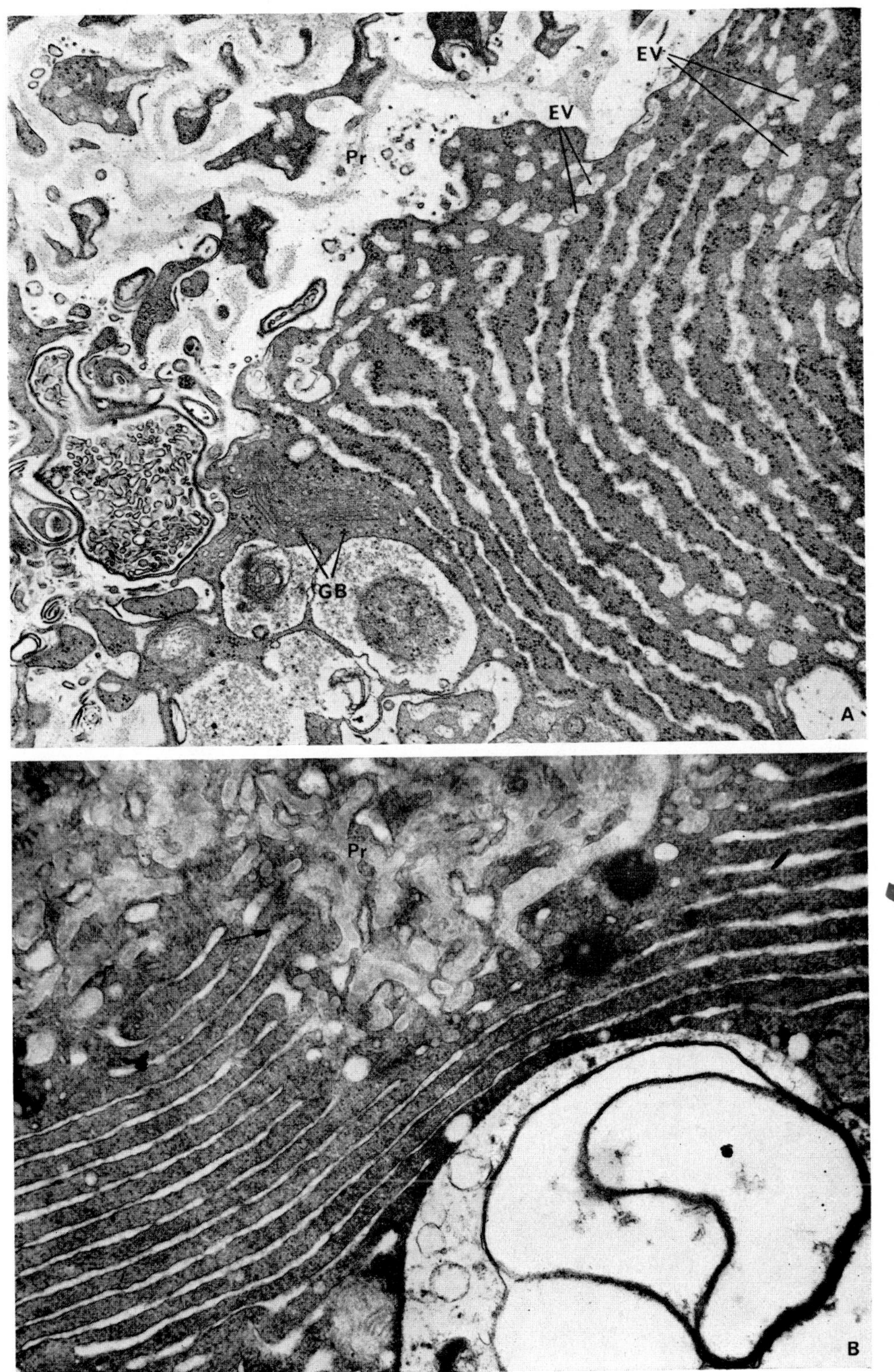

A. FAHN AND T. RACHMILEVITZ

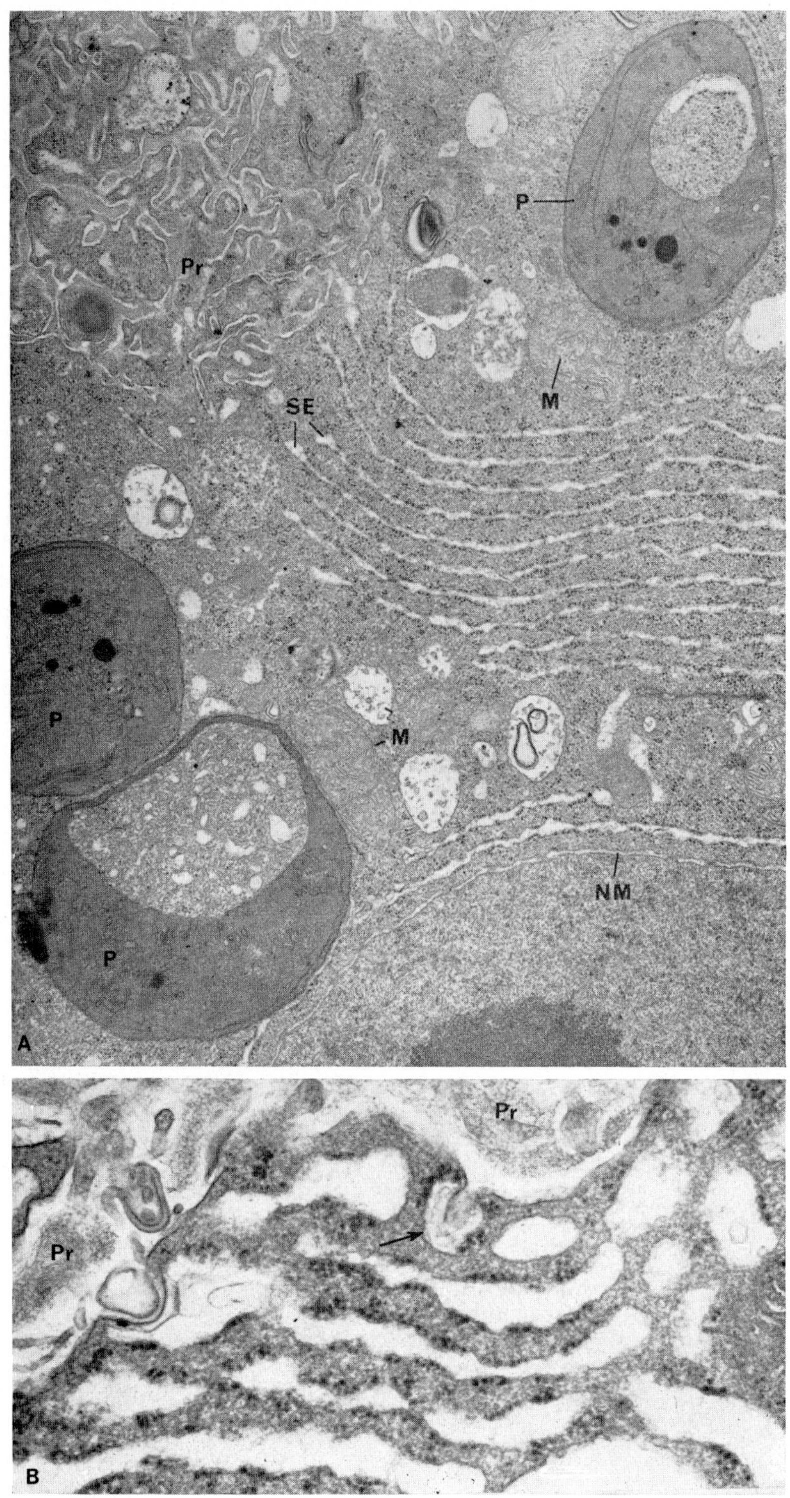

A. FAHN AND T. RACHMILEVITZ

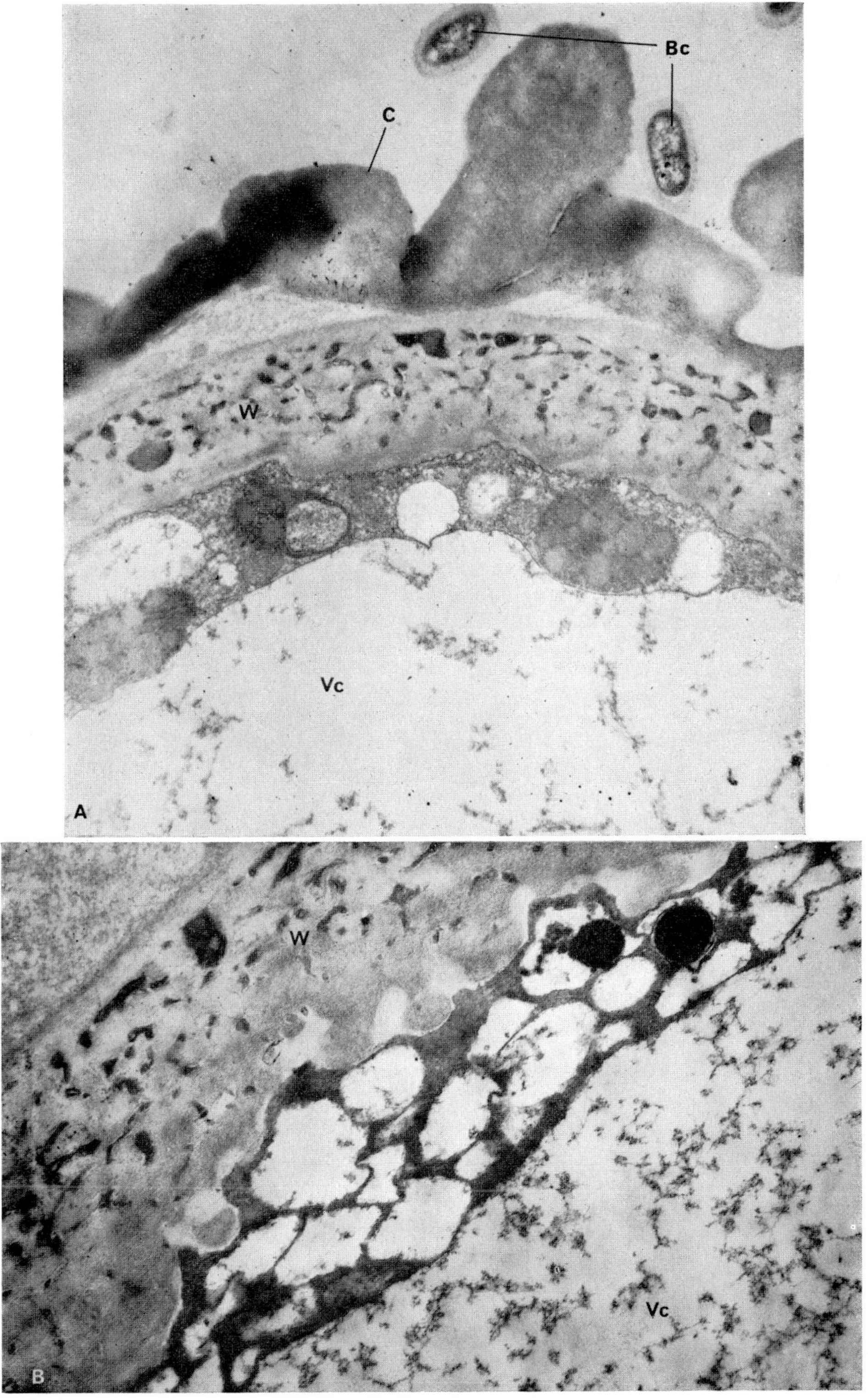

A. FAHN AND T. RACHMILEVITZ

A preliminary survey of the stem and leaf anatomy of *Thysanotus* R.Br. (Liliaceae)

N. H. BRITTAN

Botany Department, University of Western Australia, Nedlands, Western Australia

A preliminary account of the anatomy of 27 species of the genus *Thysanotus* is given. An attempt is made to correlate anatomical characteristics with morphological groupings. The occurrence of raphide canals is reported and the question posed as to whether this could indicate possible relationships between *Thysanotus* and the Commelinaceae. In *T. patersonii* anatomical differences found between plants from different localities appear to support a proposal to establish subspecific taxa based on exo-morphological differences.

CONTENTS

INTRODUCTION

The genus *Thysanotus* contained 19 species according to Bentham (1878), the present author (Brittan, 1960) described a further eight species and, arising from the revisional work at present being completed, at least a further ten taxa will be recognized. The genus therefore comprises a total of at least 37 species. All the species are Australian, two extend their range to New Guinea and one of these extends further into Thailand, Hong Kong and mainland China, some of the islands of the Malayan archipelago and the Philippines.

The genus is normally included in the Asphodeleae—Asphodelinae, which typically possess a rhizomatous rootstock and basal leaves. *Thysanotus* includes, in addition to rhizomatous species, species with fibrous roots and also ones in which the roots become tuberous either close to or distant from the rootstock. There are also species which produce leaves only in the early stages of the plant's development; these soon wither and the adult plant is leafless. Some species replace their leaves annually (*T. tuberosus*), while in other species (*T. triandrus*) the leaves persist for several seasons.

On the basis of these gross morphological features, taken in conjunction with the presence or absence of a separate scape, the four following groups can be recognized: Group I—those with persistent (i.e. more than one season) basal leaves and separate scapes; Group II—those with basal leaves replaced annually, also with separate scapes; Group III—those with leafless perennial vegetative parts on which flowers are borne; Group IV—those with leafless annual vegetative parts on which flowers are borne.

Table 1

Species	Distribution	Morphological group	Leaf bundles	
			Distribution	Type
T. multiflorus†				
60/58, 52/35	W.A.	I	II	I
53/9	W.A.	I	II	II
T. triandrus	W.A.	I	II	I
T. glaucus	W.A.	I	I	I
T. chinensis	Tropical Australia, New Guinea, Hong Kong, China, Thailand, Philippines, Malay Archipelago	I	I	I
T. baueri	N.S.W., Vic., S.A.	II	n.e.	n.e.
T. tenuis	W.A.	II	n.e.	n.e.
T. cymosus	W.A.	II	I	III
T. scaber	W.A.	II	I	III
T. tenellus	W.A., S.A.	II	I	III
T. tuberosus	Qld., N.S.W., Vic.	II	I	III
T. formosus	W.A.	II	I	IV
T. gageoides	W.A.	II	I	IV
T. thyrsoideus	W.A.	II	I	IV
T. vernalis	W.A.	II	I	IV
T. arenarius	W.A.	III	I	I
T. juncifolius	N.S.W., Vic., S.A.	III	I	I
**T. sabulosus*	W.A.	III	I	I
T. arbuscula	W.A.	III	n.a.	n.a.
T. anceps	W.A.	III	n.a.	n.a.
**T. heterocladus*	W.A.	III	n.a.	n.a.
T. dichotomus	W.A.	III	n.e.	n.e.
T. pseudojunceus	W.A.	III	n.a.	n.a.
T. sparteus	W.A.	III	n.a.	n.a.
T. spiniger	W.A.	III	n.a.	n.a.
**T. hirsutus*	S.A.	III	n.a.	n.a.
**T. virgatus*	N.S.W.	III	n.a.	n.a.
T. patersonii	N.S.W., Vic., Tas., S.A., W.A.	IV	n.e.	n.e.

* These four manuscript names are used informally prior to their valid publication.
† For authorities see Appendix.
n.a., Not applicable, leafless plants; n.e., not examined.

Group I includes four Western Australian species; *T. chinensis* is doubtfully included. The author has not seen this taxon in the field. It is, however, tropical in distribution and it is suggested that its leaves may well be perennial. All five species possess fibrous roots arising from a very small rootstock, and the leaves vary from ± terete (*T. glaucus*) to ± dorsiventrally flattened (*T. multiflorus*).

Group II species are characterized by the possession of tuberous roots, leaves which are produced annually, each season's growth withering before the new ones are produced, and separate scapes. The leaves vary in cross-section from terete (*T. formosus*), terete with a channel on the adaxial side (*T. tuberosus*) to markedly V-shaped (*T. vernalis*).

Group III have roots arising from rhizomes; leaves, if produced, soon wither, although they may persist in *T. arenarius* until flowering time. All bear flowers on the persistent vegetative axis.

Group IV contains the single species *T. patersonii*, which is the only scandent *Thysanotus*; it has tuberous roots, normally produces no leaves, but has a leafless annual axis on which flowers are borne. In some years the stem is not produced but is replaced by a few ± terete leaves.

MATERIALS

The material used in this investigation has mostly been the personal collection of the author; voucher herbarium specimens are deposited in the Herbarium, Botany Department, University of W.A. (UWA) or Herbarium, State Department of Agriculture, Perth (PERTH). Herbarium material from the Kew collection was used for the Australian and extra-Australian localities of *T. chinensis*. Details of the material used will be found in the Appendix.

ANATOMY

Leaf

The leaves are either more or less flattened dorsiventrally or more or less terete, becoming expanded in lower parts into sheathing bases of which the marginal parts are sometimes from one to a few cells thick and more or less dry and membranous. In flattened leaves, the adaxial surface is flat to concave, usually smooth, occasionally with slight ribs, some leaves are slightly asymmetrical in T.S. The abaxial surface is convex, smooth or markedly ribbed. A few species are triangular or tetragonal in T.S. Figure 1 shows representative leaf sections.

The chlorenchyma occurs in two to four rows in both stem and leaf. The individual cells vary in shape in T.S. from isodiametric to rectangular (elongated in a radial direction). Where the cells are elongated the long axis may be normal to the epidermal surface or as in the stem of *T. arenarius* at an angle of between 30 and 45 degrees to the normal (Plate 1**B**). In tangential L.S. the cells appear more or less circular to quadrangular. In *T. triandrus* and *T. multiflorus* adjacent chlorenchyma cells appear to be in contact only over a series of small more or less circular areas, and when seen in L.S. these cells have a somewhat papillate appearance.

There are variations in the disposition of the vascular bundles in the leaf, and two patterns can be distinguished:

(i) Bundles in a single line, more or less straight in dorsiventral, flattened leaves

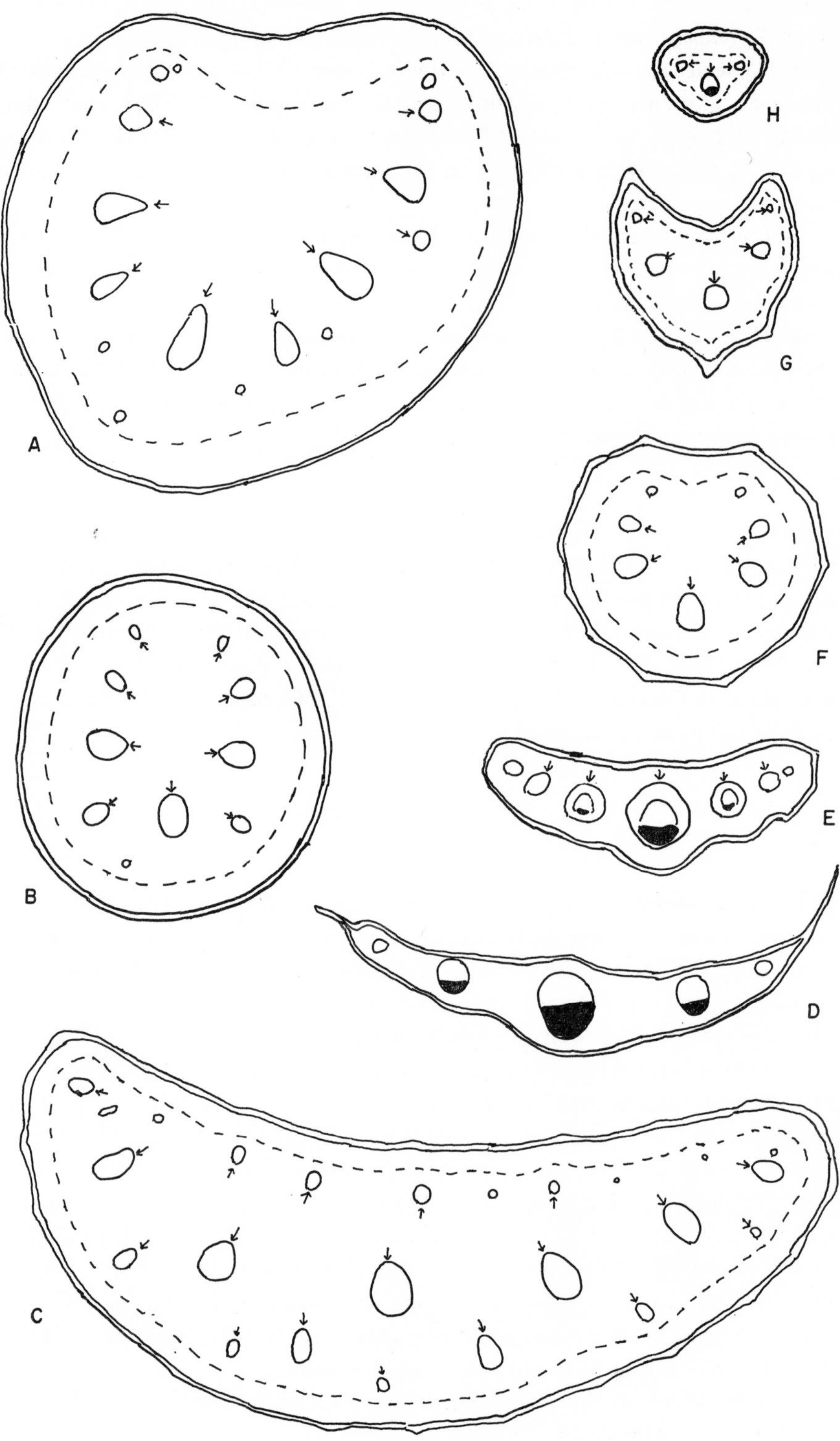

FIGURE 1

(*sabulosus*), deeply V-shaped in semi-terete leaves (*vernalis*, *thyrsoideus*) or forming almost a complete circle (*formosus*). Typically one bundle is larger than the rest; the others either show a gradual reduction in size towards the margins or to the abaxial surface, or else there may be the interposition of one to several much smaller bundles (*thyrsoideus*, *scaber*). To this group belong also *arenarius*, *cymosus*, *tenellus*, *juncifolius* and *gageoides*.

(ii) Larger bundles in a more or less straight line surrounded by an almost complete ring of smaller bundles. The larger bundles have adaxial xylem, except for the marginal bundles which tend to have xylem directed away from the margin. The outer bundles all have xylem directed towards the centre of the leaf irrespective of whether they are adaxial or abaxial in position. This situation is so far found only in the closely related species, *T. triandrus* and *T. multiflorus*. Metcalfe (1961) refers to the occurrence in the Liliaceae (*Narthecium*, *Nietneria*) of the vascular bundle arrangement commonly found in the Iridaceae, a row of vascular bundles just below each of the surfaces of the leaf, with the xylem of the opposed bundles facing each other. This arrangement differs from that in the second type reported from *Thysanotus* in that there is no line of vascular bundles along the middle of the leaf as seen in T.S. The only occurrence of an arrangement similar to the *Thysanotus* examples is in the midrib region of *Hanguana* (Flagellariaceae) reported by Tomlinson (1969, fig. 14A). It may be postulated that the present day leaf may in fact be the midrib remnant from a former leaf with an expanded lamina. The two species of *Thysanotus* with this distinctive leaf anatomy are, together with *T. glaucus*, the only species that are distinguished by the possession of three rather than six stamens.

Attention has been drawn by Metcalfe (1961, 1963) to variation in the size and disposition of the bundle components. Cheadle & Uhl (1948*a*, *b*) had previously made a detailed study of the position and arrangement of cell types in the xylem and phloem of monocotyledon bundles. In the leaf of *Thysanotus* it is possible on the basis of overall shape of the bundle, presence or absence of fibres and shape of the phloem mass to distinguish four types (Fig. 2): I, an oval bundle, well developed sclerenchymatous fibres, phloem in a narrow line, two groups of moderately sized xylem elements; II, oval-shaped bundle, some fibres above and below, phloem in a compact more or less semicircular mass and xylem V-shaped, not closely abutting on the phloem; III, triangular bundle, one or few fibres, phloem in triangular or crescent-shaped mass, xylem tending to V-shaped, closely abutting on the phloem; IV, elongated bundle, little fibre development, phloem tending to encircle a collenchymatous group, xylem in two lines with a roughly V-shaped disposition.

Most species fall into only one of the above categories; of the three specimens of *T. multiflorus*, however, two can be referred to type I, to which also the specimens of *T. triandrus* belong, but specimen 53/9 is referred to type II. Although it is not

---

FIGURE 1. T.S. leaves. **A.** *T. thyrsoideus*: asymmetric, almost terete with shallow adaxial groove. **B.** *T. formosus*: terete. **C.** *T. multiflorus*: asymmetric, flattened. **D.** *T. sabulosus*: basal region; margins one cell thick. **E.** *T. sabulosus*: upper median region, flattened, ribbed abaxial surface. **F.** *T. tuberosus*: ribbed, more or less terete. **G.** *T. vernalis*: V-shaped with marked ribs. **H.** *T. glaucus*: acicular leaf somewhat triangular in section.

Arrows, position of xylem; ---, boundary of chlorenchyma; solid black, sclerenchyma. All ×27.

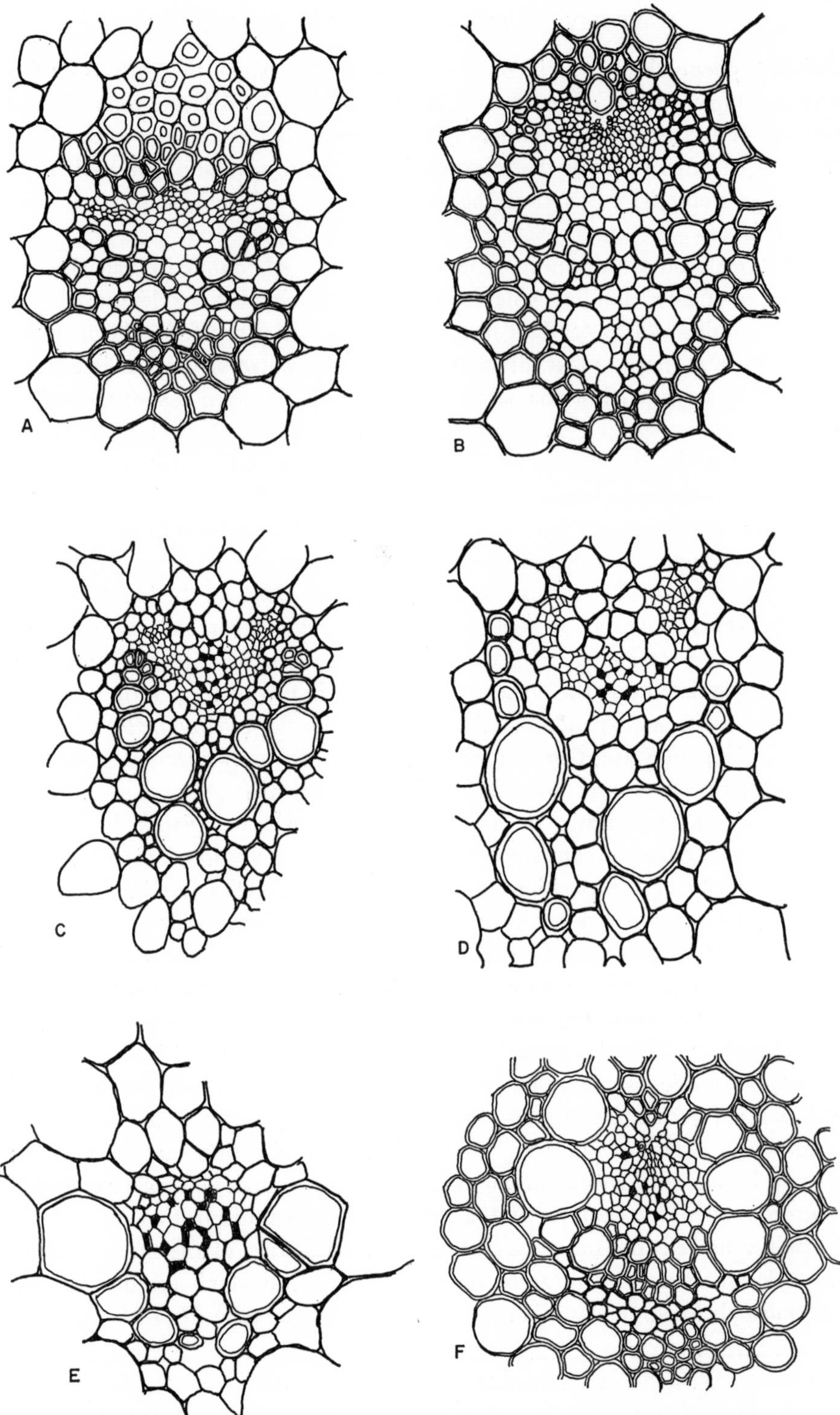

FIGURE 2. Vascular bundles of leaf, stem and scape in T.S. **A.** Leaf bundle *T. arenarius* 60/72, ×175. **B.** Leaf bundle of *T. multiflorus* 53/9, ×175. **C.** Leaf bundle of *T. tuberosus* 59/66, ×175. **D.** Leaf bundle of *T. formosus*, ×300. **E.** Scape bundle of *T. vernalis* 55/27, ×275. **F.** Stem bundle of *T. arbuscula* 54/25, ×200.

possible to separate specimen 53/9 from the others on morphological characters, it has been found to differ cytologically in that it possesses a tetraploid chromosome number as compared with the diploid condition in both *T. triandrus* and *T. multiflorus*. It has been tentatively suggested that specimen 53/9 may in fact be an amphidiploid arising from a *T. triandrus* × *multiflorus* hybrid. It may be that the anatomical difference may also be due to such a hybrid origin.

An attempted correlation of the four *Thysanotus* bundle types with those of Cheadle & Uhl (1948*a*, *b*) suggests that the author's type I belongs to Cheadle & Uhl's I, while the other three would all fit within their type II. Cheadle & Uhl, on the basis of their examination of 212 species of monocotyledons and after correlation of primitive characters shown by the species with their bundle organization, suggested that their type I was the primitive type and that probably type II was derived from it. In *Thysanotus* this would at least correlate with the rhizomatous rootstock habit, since three species so far found to have type I bundle organization have rhizomes bearing roots, and also possess six stamens. This may then be taken as an indication that these may possibly be primitive characters in *Thysanotus*. This is in accord with current evolutionary thinking which suggests that tuberous-rooted monocotyledons are intermediate stages in the development of the more highly evolved bulbs and corms. More evidence would be needed before one could confidently propose that, as the plants possessing primitive type bundles do not have scapes, the development of the scapose habit from non-scapose is also an evolutionary advance, which is in accord with one of Hutchinson's (1959) dicta.

Epidermis of leaf and stem

The epidermal cells are rectangular to narrowly spindle-shaped with blunt ends, arranged in longitudinal files. The length-width ratio varies from species to species (Fig. 3 **E–H**); the walls are always non-sinuous. In T.S. of some species the cells of the epidermis abutting on to a stoma may be larger in a radial direction than the rest of the epidermal cells. When ribs are found on leaves they result from increased height of the epidermal cells. Marked variation is found in the thickness of the epidermal cell walls.

In the majority of species the anticlinal epidermal walls of adjacent cells remain in contact until the cuticle is reached; in a few cases (in stems) the walls have interposed, for perhaps half their length, a wedge of cuticular material which gives the epidermis a somewhat crenellated appearance. Priestley (1943) attributed this to radial growth in the plant separating the outer walls of adjacent cells, the cuticular material then being deposited in the space thus created. It is in accord with this theory that the examples of this type of epidermis are found in older stems of *Thysanotus* species with persistent leafless stems.

The cuticle varies in thickness (up to 9 μm); the thicker deposits are found in plants from more arid environments. In some species the cuticular layer is not uniform but appears in T.S. to have ridges, which frequently occur at the edge of a stomatal aperture and give an indication of the presence of some included material between the cuticle and the cell wall (Fig. 3**D**). In stripped epidermal preparations it can be

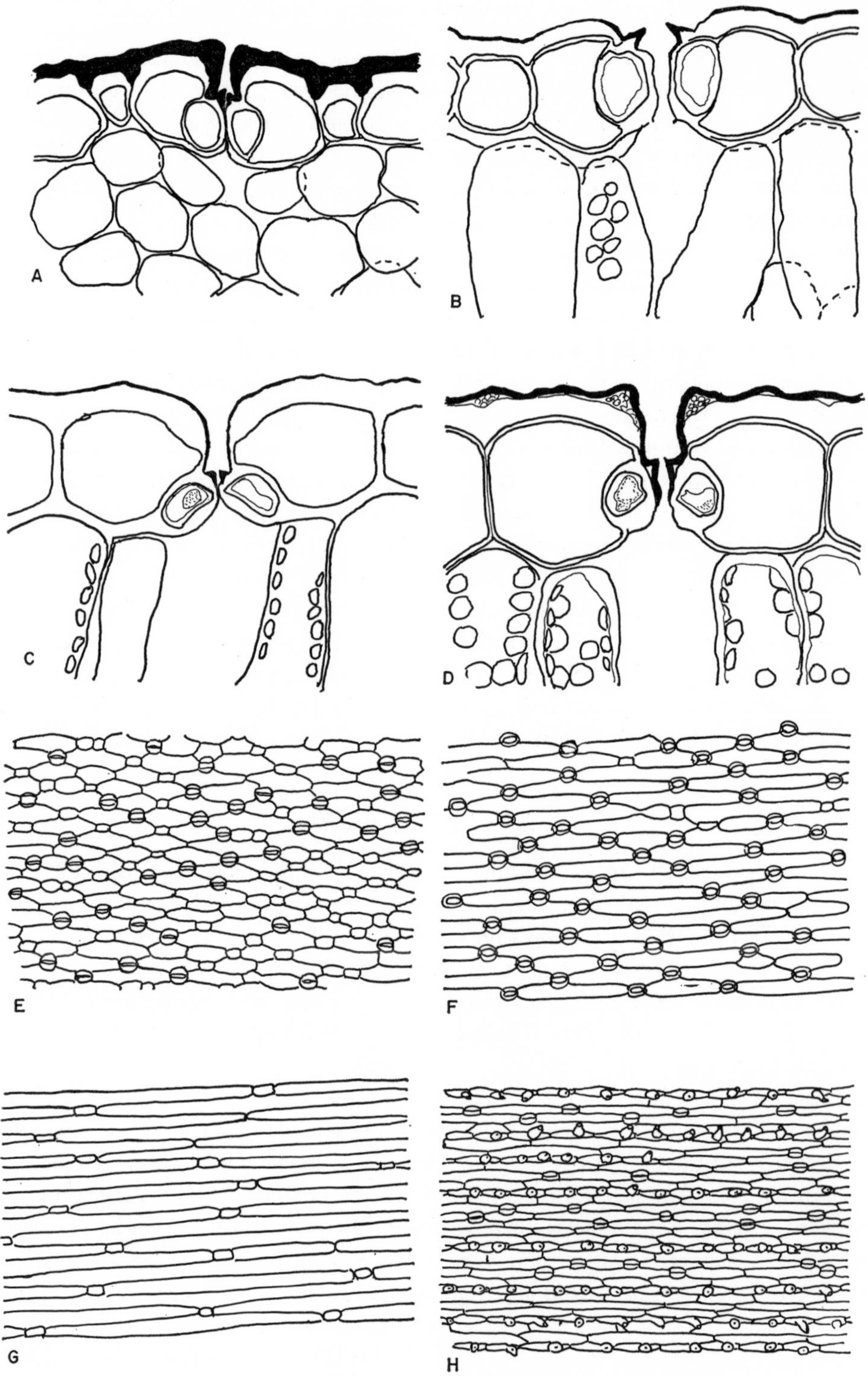

FIGURE 3. **A–D.** Stomata in T.S.: **A,** *T. heterocladus,* ×500; **B,** *T. tenuis* 58/20, ×550; **C,** *T. cymosus* 58/23, ×500; **D,** *T. formosus,* ×600.

E–H. Epidermis in surface view: **E,** *T. sabulosus* 60/136, leaf; **F,** *T. gageoides,* leaf; **G,** *T. chinensis* Specht 1001, leaf; **H,** *T. arbuscula* 54/25, stem. All ×75.

Cuticle shown in solid black.

seen that irregular patterns of papillae, sometimes aggregated into irregular ridges, are formed by these inclusions. Further investigation into their nature is being carried out.

Trichomes occur on both stems and leaves, either randomly distributed or, more commonly, in longitudinal files associated with ridges in both stems and leaves. Hairs are always simple, occurring either singly or sometimes (*T. triandrus*, *T. hirsutus*) two or three together. They may each arise from a conspicuously enlarged epidermal cell (*T. scaber*) and consist of cell wall material with an obvious tapering lumen, or else arise from a normal-sized epidermal cell and appear to consist entirely of cuticular material with a very narrow parallel-sided lumen. Present indications are that there appears to be a correlation between this latter type and the crenellated epidermis described above. Variation in length and density of hairs occurs within species; further work needs to be done to attempt to correlate this variation with the ecological habit of the plants.

In all species the stomata possess no subsidiary cells, the type called anomocytic by Metcalfe (1961). This type was shown by Stebbins & Khush (1961) to be characteristic of the Liliales. Stomata are found on both upper and lower surfaces of dorsiventral leaves, the entire surface of terete leaves and stems and are apparently absent only from the adaxial surface of the sheathing leaf bases. When a leaf or stem is ridged, stomata are found only in the intercostal regions. As seen in T.S. there are variations in the symmetry of the guard cells (cf. Tomlinson, 1969), and in the position of the stoma relative to the epidermal surface. Some of the more xerophytic species have very sunken stomata, often with a narrow pore formed by the overgrowing of the cuticle (Fig. 3**A, C, D**). Cuticular outgrowths are frequently found close to the junctions of the guard cell wall with the epidermal cell and these are particularly well developed on the outer side of the guard cell (Fig. 3**B–D**).

Crystals

Only calcium oxalate in the form of raphides has been observed. It is not quite a universal feature of the genus, not having been reported from *T. spiniger*, *T. dichotomus*, *T. virgatus* and *T. formosus*. These four do not provide sufficient basis to attempt to correlate crystal presence with type of growth, since the first three are rhizomatous, leafless plants, and the fourth a tuberous-rooted, annual-leaved plant. The raphides are found in sacs in articulated files (Plate 1**A, B**) recalling those observed by Tomlinson (1969) in Commelinaceae. The raphide canals are found most frequently in the chlorenchyma of the leaves, near the boundary of the chlorenchyma and parenchyma, or rarely in the parenchyma itself. In the stems they always occur in the chlorenchyma, occasionally quite closely associated with the sclerenchymatous sheath. The fact that these raphide canals possess very thin walls supports Tomlinson's suggestion that, once differentiated, raphide-sac cells undergo no further division and are stretched as the tissue in which they occur elongates. There is some evidence to contradict Tomlinson's statement that cross walls of adjacent raphide sacs do not break down naturally. Some considerable lengths of raphide canals have been seen with no indication of cross walls or of remnants of cross walls which may have been broken in the course of sectioning. Raphides have been observed lying across the boundary between two adjacent sacs. The raphides show variation in size between species over the range 45–92 μm. In two instances there is indication of intraplant variation, *T. anceps* 69–80 μm, *T. arenarius* 46–80 μm. In one instance, *T. triandrus*, the raphides (92 μm) are seen to be encased in distinct, presumably mucilaginous, sheaths (160 μm long).

Stem

The stem is taken here to include scapes. There are variations in the external characteristics of the stem. Some stems are smooth and terete, others more or less circular, but with varying degrees of development of ribs, others again markedly either pentagonal, hexagonal or polygonal (Plate 1**D–F**). All stems are photosynthetic and possess a chlorenchymatous layer, the details of which have been mentioned under the 'leaf' heading.

The vascular arrangement is in general uniform and consists of an outer ring of small bundles occurring outside, but in close association with, a ring of either collenchyma or sclerenchyma and enclosed within a more or less well marked ring of endodermal cells. Another series of small bundles, disposed in an irregular circle, occurs within the sheath. In the centre of the stem there occur the larger bundles and an outer ring of somewhat smaller bundles which tend to alternate with the inner ones. The relationships which exist between the various bundles have not yet been worked out, but on reading Tomlinson's (1969) account of stem vascularization in the Commelinaceae one cannot help being struck by the apparent similarity. There is, of course, the difference that whereas the Commelinaceae possess a series of cauline leaves at fairly close intervals, in *Thysanotus* the stem is either leafless or at the most possesses bracts at distant intervals. It can be foreseen that investigations of the sort that Scott & Priestley (1925) carried out would be difficult in *Thysanotus* because of the fact that, where leaves do occur, i.e. at the base of the plant, they are very closely packed, so that the internodes are very small.

In *T. patersonii* and *T. tenuis* the vascular structure differs from that described above in that the number of vascular bundles is reduced, to as low as two in one specimen of *T. patersonii*.

The main bundles of the stems examined are more or less circular in outline and show a U-shaped xylem mass with the larger elements occurring at the extremities of the arms (Fig. 2**E, F**). The xylem encloses on three sides a compact circular to more or less square mass of phloem. An exception to this occurs in the Western Australian collection of *T. patersonii*, where the xylem occurs in a more or less straight line (Fig. 4**C**).

Root

Roots have so far been sectioned only from two species, *T. gageoides* and *T. tuberosus*. Two large centrally located lacunae characterize the stele of *T. gageoides* (Plate 1**C**); these are surrounded by five xylem groups. There is a well marked endodermis with thickened radial and inner tangential walls. In *T. tuberosus* in place of the lacunae there is a central pith surrounded by five or six groups of xylem vessels and a well

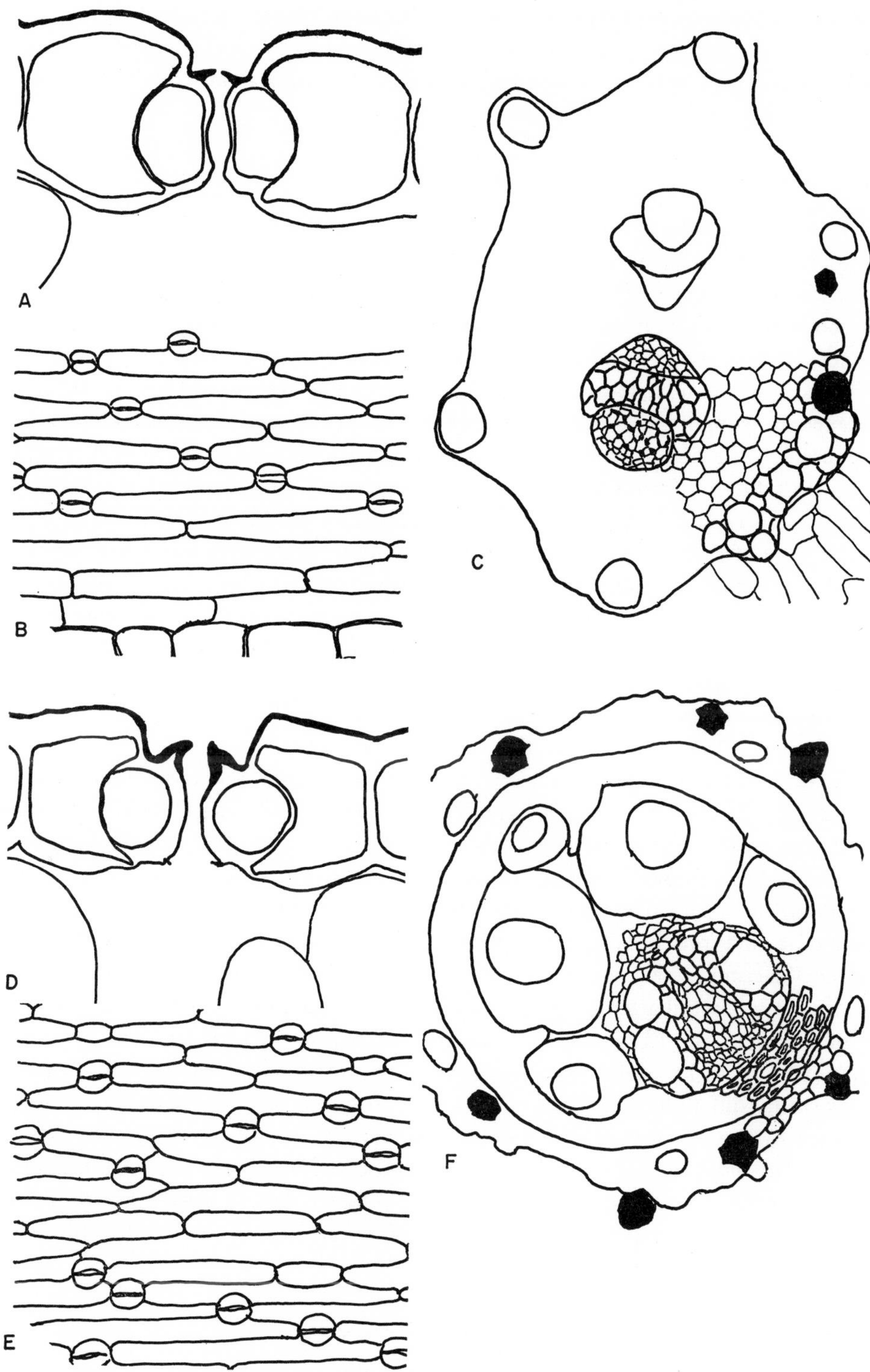

FIGURE 4. *T. patersonii.* **A, D.** T.S. stomata, ×600. **B, E.** Epidermis in surface view, ×100. **C, F.** T.S. stele ×110. **A–C.** Specimen 53/63. **D–F.** Specimen 59/23.
Solid black (in T.S. stele), raphide canals.

marked endodermis. Sections have been cut through one of the tuberous regions of the root of *T. tuberosus* and it is found that the vascular strand continues relatively unchanged except that the thickening of the endodermis is absent. The cortex is much enlarged and consists of parenchymatous tissue.

DISCUSSION

The anatomical detail reported here has in general been in accord with taxonomic groupings in the genus in the sense that anatomical differences occur between species and not within species. The results so far from *T. patersonii* are, however, not quite in accord with this pattern. It had been proposed (Brittan, 1962) to subdivide the species into two subspecies based on morphological characters from flower and fruit. Evidence from stem anatomy and cuticular morphology appear to provide additional evidence to support this subdivision (Fig. 4). A Western Australian collection attributed to the one subspecies shows a stelar structure made up of two bundles, each with rather uniform sized xylem elements arranged in a more or less rectangular pattern. The two collections of the other subspecies so far examined anatomically show a stele of three major bundles and three minor ones, with the xylem elements very dissimilar in size and arranged in a U-shaped pattern enclosing the phloem. In this latter case the bundles occupy most of the space inside the sclerenchyma sheath, whereas in the former there is a wide parenchymatous band surrounding the vascular tissue and also little development of sclerenchyma. The length of the epidermal cells, and the presence in one case of cuticular ridges and papillae (mentioned above) and their absence in the other, are additional pieces of evidence.

The occurrence of raphide canals is the other interesting point which arises from this survey. Of the Commelinales and Zingiberales (Tomlinson, 1969), only Commelinaceae were found to possess these; in the literature available to the author no record has been found of these occurring in other Liliaceae. If one looks at the taxonomic characters separating these families in Hutchinson (1959), they are placed in separate divisions based on the presence of similar (Liliaceae) versus dissimilar (Commelinaceae) calyx and corolla whorls. A comparison of the family characteristics shows a major dissimilarity in the form of the seed, which in Commelinaceae possesses an embryotega, in *Thysanotus* an arillate outgrowth. *Thysanotus*, which is normally included in Liliaceae, does, however, possess calyx and corolla whorls which are markedly different in size, the inner ones broad and bearing the fringed margins indicated by the generic name, the outer ones narrow. This suggestion of an affinity between *Thysanotus* and Commelinaceae on the basis of possession of similar organization of raphide canals may, of course, be far-fetched, but at least it points to the desirability of an investigation being made of other supposedly related liliaceous genera with regard to this particular feature.

ACKNOWLEDGEMENTS

The author is indebted to the Director of the Royal Botanic Gardens, Kew (Sir George Taylor) and to Dr Metcalfe (at that time Keeper of the Jodrell Laboratory) for the provision of laboratory facilities to enable the investigations reported here to be carried out during the author's absence from Western Australia on a period of study leave. It is a pleasure to acknowledge the helpful discussions afforded me by Dr Metcalfe and his staff and their co-operation throughout my stay.

REFERENCES

BENTHAM, G., 1878. *Flora Australiensis*, **7**: 36. London: Reeve.

BRITTAN, N. H., 1960. New Western Australian species of *Thysanotus* R. Br. (Liliaceae). *J. R. Soc. West. Aust.*, **43**: 10–29.

BRITTAN, N. H., 1962. Variation, classification and evolution in flowering plants—with particular reference to *Thysanotus*. *J. R. Soc. West. Aust.*, **45**: 1–11.

CHEADLE, V. I. & UHL, N. W., 1948*a*. Types of vascular bundles in the Monocotyledoneae and their relation to the late metaxylem conducting elements. *Am. J. Bot.*, **35**: 486–496.

CHEADLE, V. I. & UHL, N. W., 1948*b*. The relation of metaphloem to the types of vascular bundles in the Monocotyledoneae. *Am. J. Bot.*, **35**: 578–583.

HUTCHINSON, J., 1959. *Families of flowering plants. II. Monocotyledons*, 2nd ed. Oxford: Clarendon Press.

METCALFE, C. R., 1961. The anatomical approach to systematics. General introduction with special reference to recent work on monocotyledons. In *Recent advances in botany*, **1**: 146–150. Toronto: University of Toronto Press.

METCALFE, C. R., 1963. Comparative anatomy as a modern botanical discipline, with special reference to recent advances in the systematic anatomy of monocotyledons. In *Advances in botanical research* (ed. R. D. Preston), **1**: 101–148. London & New York: Academic Press.

PRIESTLEY, J. H., 1943. The cuticle in angiosperms. *Bot. Rev.*, **9**: 593–616.

SCOTT, L. I. & PRIESTLEY, J. H., 1925. Leaf and stem anatomy of *Tradescantia fluminensis* Vell. *J. Linn. Soc. (Bot.)*, **47**: 1–28.

STEBBINS, G. L. & KHUSH, G. S., 1961. Variation in the organization of the stomatal complex in the leaf epidermis of monocotyledons and its bearing on their phylogeny. *Am. J. Bot.*, **48**: 51–59.

TOMLINSON, P. B., 1969. *Anatomy of the monocotyledons* (ed. C. R. Metcalfe), **III**. *Commelinales—Zingiberales*. Oxford: Clarendon Press.

APPENDIX

Material examined by the author

(Voucher specimens in UWA unless otherwise indicated)

Species	Collection	Locality
Thysanotus anceps Lindl.	N.H.B. 58/40	National Park, nr Perth, W.A. 28.xi.1958
T. arbuscula Baker	N.H.B. 54/25	Balcatta Rd, nr Perth, W.A. 14.xii.1954
T. arenarius N. H. Brittan	N.H.B. 58/2	Stake Hill, nr Rockingham, W.A. 19.ix.1958
	N.H.B. 60/72	nr Qualup, W.A. 28.xi.1960
T. baueri R. Br.	N.H.B. 59/79	nr Swan Hill, Vic. 3.xii.1959
T. chinensis Benth.	Chung 1713	S. Fukien, China 1923 (K)
	Kerr 19044	Bangsak Trang, Siam. 20.iv.1930 (K)
	Merrill 4451	Suyoc, Luzon. Nov. 1905 (K)
	Specht 1001	Gove, N.T. 24.viii.1948 (K)
T. cymosus N. H. Brittan	N.H.B. 58/22	Kulin—L. Grace, W.A. 28.x.1958
	N.H.B. 58/23	Ongerup—Borden, W.A. 28.x.1958
T. dichotomus (Labill.)	N.H.B. A.3	Chittering. 31.x.1956 (anat. mat. only)
R. Br.	N.H.B. s.n.	Toodyay Rd, W.A. 9.xi.1951
	N.H.B. 60/66	Borden—Bremer Bay, W.A. 27.xi.1960
T. formosus N. H. Brittan	N.H.B. s.n.	Nannup—Augusta, W.A. 26.i.1953 (PERTH)
T. gageoides Diels	N.H.B. 53/24	nr Cranbrook, W.A. 30.x.1953
T. glaucus Endl.	N.H.B. s.n.	nr W. Mt Barren, W.A. 10.xii.1951
	N.H.B. 60/43	Forrestfield, nr Perth, W.A. 5.iii.1960
T. heterocladus N. H. Brittan ms.	N.H.B. A.14	South Coast, W.A. (anat. mat. only)

T. hirsutus N. H. Brittan ms.	N.H.B. 60/12, 60/15	Kangaroo Is., S.A. 20.i.1960
T. juncifolius (Salisb.) Willis & Court	N.H.B. 52/59	Tintinara, S.A. 27.xi.1952
	N.H.B. 59/105	Sutherland, N.S.W. 29.xii.1959
	N.H.B. 60/1	nr Nowra, N.S.W. 6.i.1960
	N.H.B. 60/2	nr Jervis Bay, N.S.W. 6.i.1960
T. multiflorus R. Br.	N.H.B. 52/35	Hill R., W.A. 24.ix.1952
	N.H.B. 53/9	nr Pearce, W.A. 6.x.1953
	N.H.B. 60/58	nr Mt Manypeaks, W.A. 10.xi.1960
T. patersonii R. Br.	N.H.B. 59/20	Keith—Bordertown, S.A. 8.x.1959
	N.H.B. 59/23	Risdon Hill, Tas. 12.x.1959
	N.H.B. 53/63	Balladonia—Israelite Bay, W.A. 30.x.1954
T. pseudojunceus N. H. Brittan	N.H.B. A.4	Nannup—Augusta, W.A. (anat. mat. only)
T. sabulosus N. H. Brittan ms.	N.H.B. 60/136	Newdegate—L. Grace, W.A. 15.xii.1960
T. scaber Endl.	N.H.B. A.8	Redhill, W.A. (anat. mat. only)
T. sparteus R. Br.	N.H.B. 60/57	nr Mt Manypeaks, W.A. 10.xi.1960
	N.H.B. 60/105	East of Esperance, W.A. 11.xii.1960
T. spiniger N. H. Brittan	N.H.B. 52/39	Hill R., W.A. 24.ix.1952 (PERTH)
T. tenellus Endl.	N.H.B. 58/4	Toodyay—Goomalling, W.A. 10.x.1958
T. tenuis Lindl.	N.H.B. 58/20	Tinkurrin, nr Wickepin, W.A. 26.x.1958
T. thyrsoideus Baker	N.H.B. 58/37	Kojonup—Mudiarrup, W.A. 4.xi.1958
T. triandrus (Labill.) R. Br.	N.H.B. 54/20	L. King—L. Grace, W.A. 5.xi.1954
	N.H.B. 59/2	Donnybrook—Jarrahwood, W.A. 7.i.1959
	N.H.B. 60/60	North of Cranbrook, W.A. 11.xi.1960
T. tuberosus R. Br.	N.H.B. 59/38, 59/39	South of Grafton, N.S.W. 26.x.1959
	N.H.B. 59/42	nr Brisbane, Qld. 31.x.1959
	N.H.B. 59/47	nr Jolly's Falls, Qld. 4.xi.1959
	N.H.B. 59/32	nr Caloundra, Qld. 6.xi.1959
	N.H.B. 59/53	Redcliffe—Petrie Rd, Qld. 8.xi.1959
	N.H.B. 59/66	nr Noosa Heads, Qld. 13.xi.1959
	N.H.B. 59/93	Nowa-Nowa—Orbost, Vic. 20.xii.1959
T. vernalis N. H. Brittan	N.H.B. 55/27	nr Mt Lesueur, W.A. 2.x.1955
T. virgatus N. H. Brittan ms.	N.H.B. 59/106, 61/02, 61/03	National Park nr Sydney, N.S.W. 29.xii.1959 and 3.vi.1961

EXPLANATION OF PLATE

Plate 1

A. Raphide canals, *T. anceps*, L.S. stem, ×110. **B.** Raphide canals, *T. arenarius*, L.S. stem, ×130. **C.** T.S. root, *T. gageoides*, ×70. **D.** T.S. scape, *T. tenellus*, ×42. **E.** T.S. stem, *T. virgatus*, ×34. **F.** T.S. stem, *T. sparteus* 60/105, ×34.

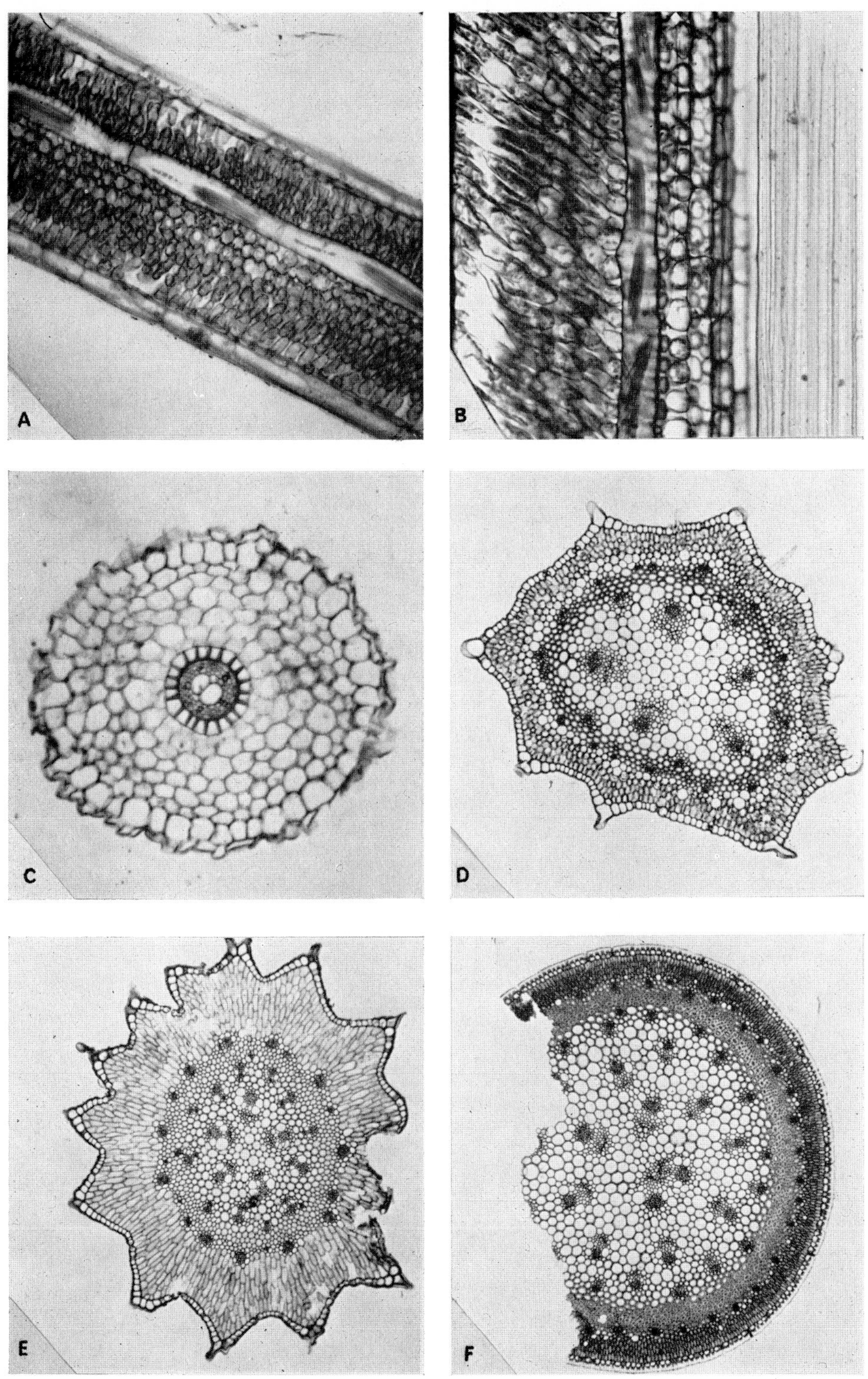

N. H. BRITTAN

(*Facing* p. 70)

Wood anatomy in three dimensions

G. W. D. FINDLAY

AND

J. F. LEVY

Department of Botany, Imperial College, London, England

Increasing use will be made of the scanning electron microscope for the examination of wood. The present paper illustrates the potential of this instrument, with reference to the wood structure of *Fitzroya cupressoides*, and also indicates some sources of artefacts.

CONTENTS

INTRODUCTION

The merits of the scanning reflection electron microscope as a tool in the study of wood anatomy have been outlined before, for example, by Resch & Blaschke (1968), Wagenfuhr & Zimmer (1968) and Findlay & Levy (1969). Findlay & Levy (1969) showed photomicrographs of a number of different woods, both sound and decayed, in order to show the breadth of application of the scanning electron microscope. The present paper attempts to show the application of the scanning electron microscope in greater detail to a single species.

MATERIALS AND METHODS

The specimens were prepared by first boiling the wood in water to soften it and then cutting out small blocks, with the desired face uppermost, using a new razor blade. A stereo-optical microscope was usually employed in this work when the precise orientation of the block was critical. In preparing longitudinal faces for examination, the wood was both split and cut. It was found that this gave two different aspects; splitting the wood caused the fibres to separate intact, at the middle lamella region, but cutting the wood is more likely to cut through cell walls and so expose the lumina of the fibres and tracheids. The material is left to air dry. The final blocks are mounted on a small metal stub with an adhesive such as 'Durofix' and provided with an electron-conducting surface by vacuum coating with gold/palladium alloy. No other preparation is necessary to obtain the differentiation of structure seen in the accompanying figures. Research is being carried out to investigate the possibility of obtaining better differentiation under the scanning electron microscope, but it is not possible as yet to comment on the results.

When examining a comparatively fragile material, such as wood, at high magnifications, it is most important to be aware of the possibility that the electron beam may damage or alter the specimen. Examples of such damage are shown in Plate 5**B, C, D,** which are taken from other investigations. Greatest damage has been found to occur with fine suspended structures such as pit membranes. If the accelerating voltage of the beam is lowered, such damage may be reduced; but, at the same time, this will involve a loss in the resolving power of the instrument. Insufficient coating of the specimen in preparation will also tend to increase the chances of beam damage. Care has been taken in the selection of the present photographs to avoid all such possible artefacts.

The following figures, with the exception of Plate 5**B, C, D,** are all of a South American softwood Alerce, *Fitzroya cupressoides* F. M. Johnston (*see* Phillips, 1957, for general wood anatomy). They were taken in the Botany Department of Imperial College on a Cambridge Instrument Company "Stereoscan" Mark II.

OBSERVATIONS

Plate 1**A** shows a general view of the wood cut to expose the transverse face and the tangential and radial longitudinal faces. The basic structure of the wood can be appreciated; one complete growth ring is visible and rays can be seen in both longitudinal and transverse sections.

Plate 1**B** and **C** are low magnification pictures of the tangential and radial faces respectively, and also demonstrate the nature of the rays. The 'rippled' effect along the walls of the tracheids in Plate 1**B** is due to bordered pits on the radial walls.

Plates 1**D** and 2**A** are both enlarged views of Plate 1**C**. Plate 1**D** shows how the innermost 'warty' layer of the tracheids is intact where the cells have been cut medianly (in the centre of the picture), but has been stripped off to reveal the pit membranes and primary wall area, to the left and bottom of the picture. Plate 2**A** shows the warty layer in greater detail and the cut, frayed edges of the tracheid walls. Plate 2**B**, a transverse face at a similar magnification, also demonstrates the warty layer. Plates 2**C**, **D** and 3**A** are examples of the effect shown in Plate 1**D**. Plate 2**C** shows the warty layer at the top of the picture which has split off, presumably along the line of microfibril orientation of the S_2 layer, to reveal the microfibril orientation of the S_1/primary layer beneath it. Plate 2**D** shows two tracheids; in the one to the left the warty layer is intact, but on the right this layer has been lost and the microfibril orientation of the S_2 layer is visible.

It should be pointed out that at this level of magnification it is not possible to observe the individual microfibrils, but their orientation can be deduced from the angle of the microfibril bundles. In Plate 3**A** all three wall layers are visible. In the middle of the central, displaced tracheid, a small area of the warty layer can be seen, beneath this the S_2 layer, and on either side the S_1/primary wall complex. In Plate 3**B**, at a higher magnification, most of the S_2 wall layer has been lost, but the remaining small area clearly demonstrates the difference in microfibril angle. Plates 3**C, D** and 4**A** show bordered pits in relation to the wall layers. In Plate 3**C** the warty layer and S_2 layer have been lost altogether, but a clear differentiation is visible between the S_1 layer (to the bottom left of the picture) and the primary wall supporting the pits. The warty layer is visible on the right, in Plate 3**D**, but has been lifted up together with the pit borders, and the pits corresponding to the borders can be seen on the far right of the picture. The structure of the pit membrane and border can be seen in more detail in Plate 4**A**. The margo and torus of the membrane are clearly visible in the pits to the left, whereas the pit to the right still has the border intact. Plate 4**B–D** are all views of the same pit, showing the warty interior of the border, but taken with the specimen tilted at different angles. Plate 4**B** was taken at a very low angle, Plate 4**C** was taken at 45° and Plate 4**D** at 90°.

The various wall layers can be seen clearly differentiated in Plate 5**A**, this is a transverse view of the late, summer wood. The presence of the warty layer can be observed from the lumen, next to a thick S_2 layer, and the middle lamella is also readily distinguishable. Plate 5**B–D** are examples of damage caused by the electron beam. Plate 5**B** is a macerated tracheid of Scots pine, *Pinus sylvestris* L., and dark horizontal lines have been caused by examining those areas at much higher magnifications.

Plate 5**C** and **D**, of bordered pits in Spruce, *Picea* sp., were taken within two minutes of each other. In Plate 5**C** the membrane of the upper pit can be seen to be intact whereas in Plate 5**D** the membrane has virtually disappeared and cracks are appearing in the lower membrane.

Although these pictures are in no way expected to provide a definitive account of the structure of this species, they are presented as a pioneer effort with higher magnifications in the field of wood anatomy. The field of plant anatomy is, of course, one in which Dr Metcalfe has achieved so much and where his kindly stimulation and enthusiastic encouragement has been given to so many students and colleagues. The authors take great pleasure in dedicating this paper to Dr Metcalfe with respect and affection.

ACKNOWLEDGEMENTS

The authors wish to acknowledge their gratitude to Dr J. Gay and Miss M. Martin for their help and advice on the use of the scanning electron microscope, and to Mr A. D. Greenwood for the use of photographic and other facilities in the Cytology Section. G. W. D. Findlay is at present supported by a fellowship from Hickson & Welch (Holdings) Limited for research at Imperial College.

REFERENCES

FINDLAY, G. W. D. & LEVY, J. F., 1969. Scanning electron microscopy as an aid to the study of wood anatomy and decay. *J. Inst. Wood Sci.*, **23**: 57–63.

PHILLIPS, E. W. J., 1957. *A Handbook of Softwoods.* Forest Products Research Laboratory: H.M.S.O.

Resch, A. & Blaschke, R., 1968. Über die Anwendung des Raster-Elektronenmikroskopes in der Holzanatomie. *Planta*, **78**: 85–88.

Wagenfuhr, R. & Zimmer, F., 1968. The surface of wood under the electron microscope. *Holztechnologie*, **9** (3): 151–2.

EXPLANATION OF PLATES

Plates 1 to 5

Scanning electron microscope photographs of wood structure.

Plates 1 to 5**A**. *Fitzroya cupressoides*.
Plate 5**B**. *Pinus sylvestris*, macerated tracheid.
Plate 5**C, D**. *Picea* sp., bordered pits.
For details see text.

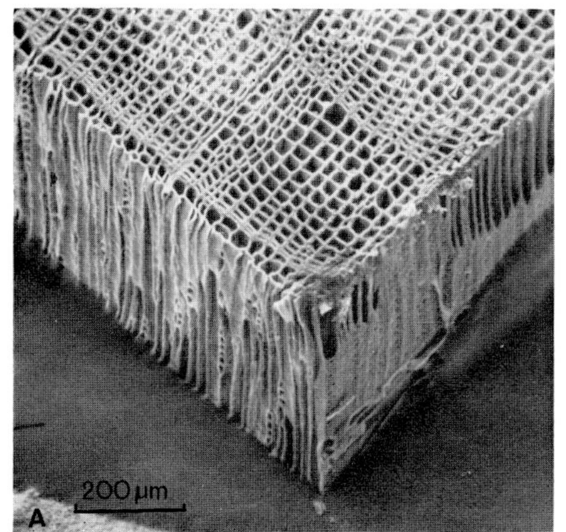
200 µm
A

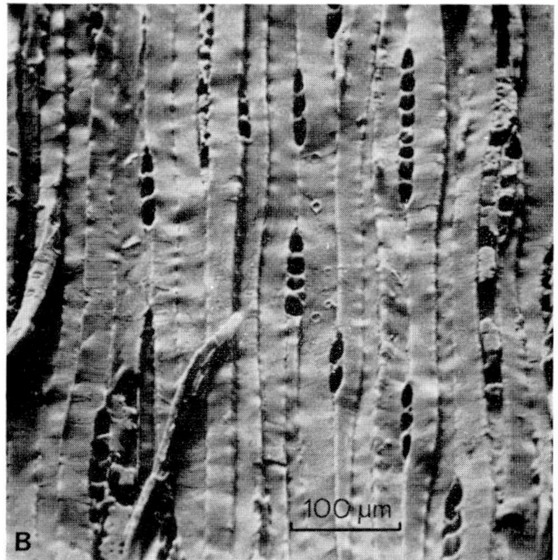
100 µm
B

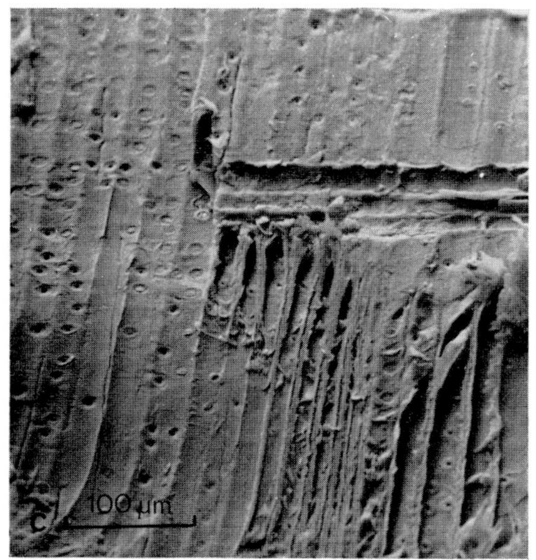
100 µm
C

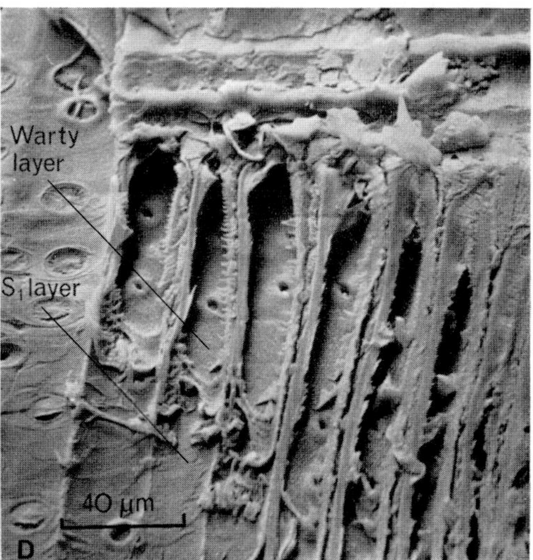
Warty
layer
S1 layer
40 µm
D

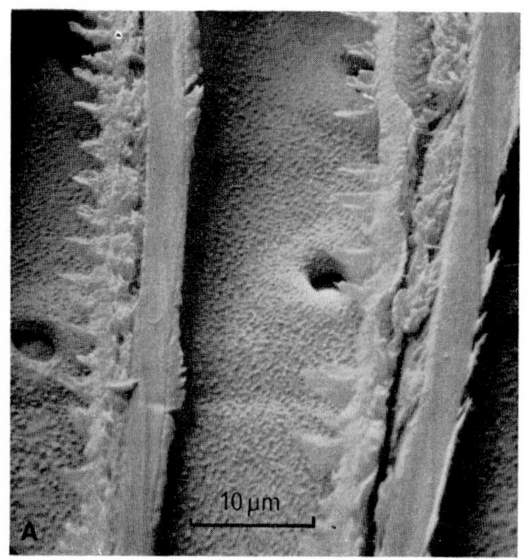

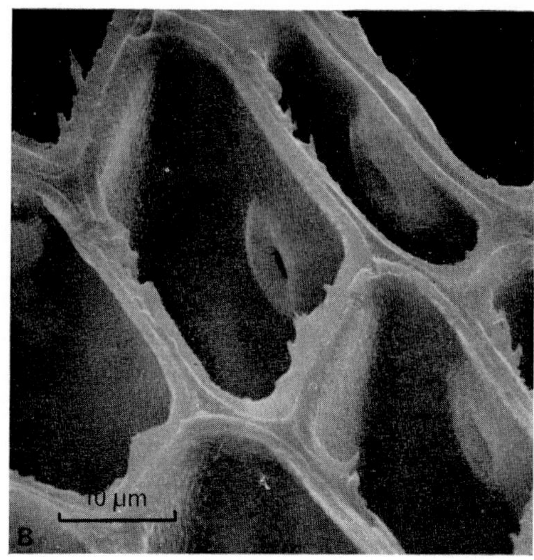

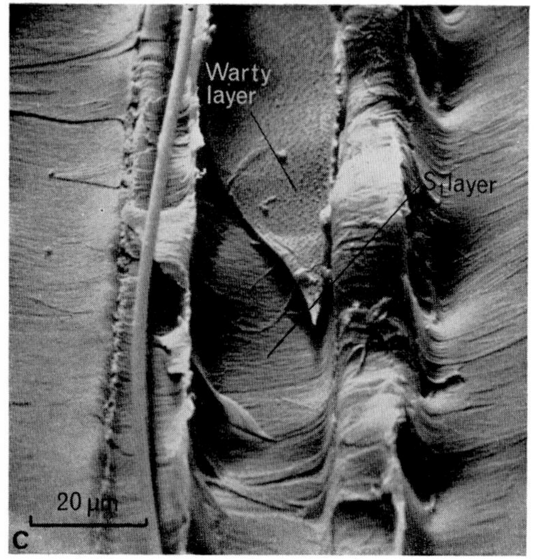

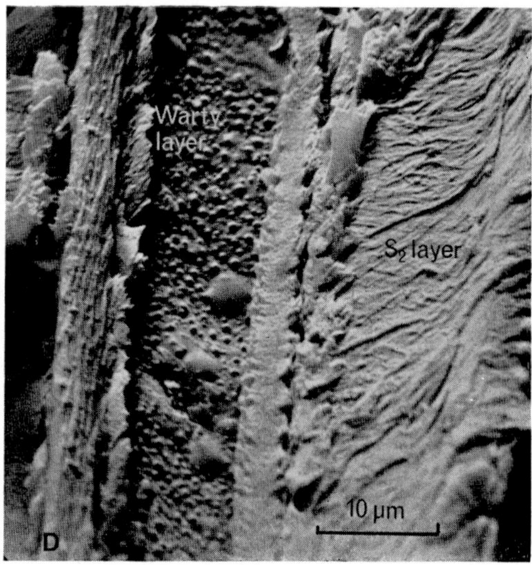

G. W. D. FINDLAY AND J. F. LEVY

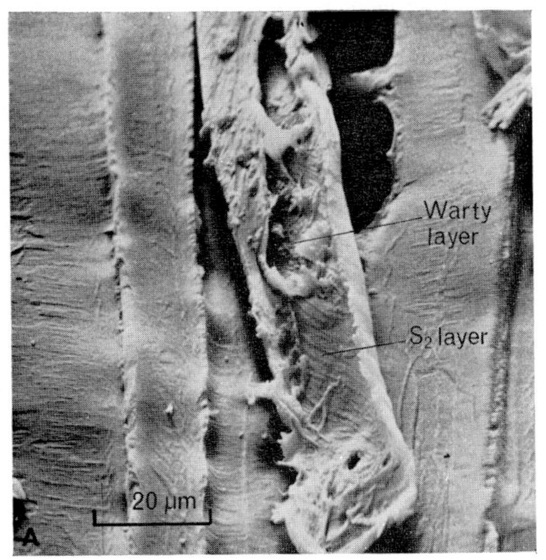

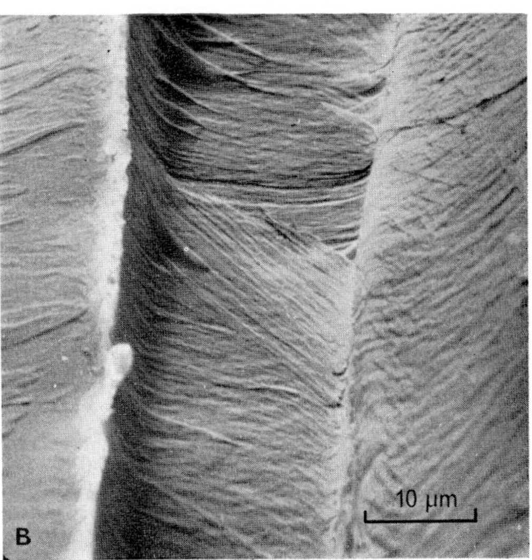

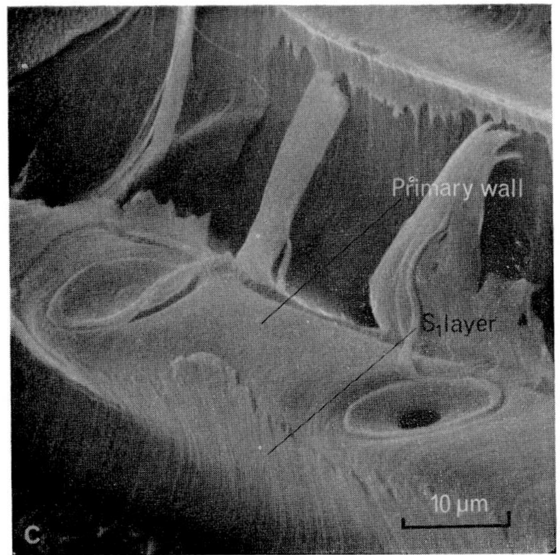

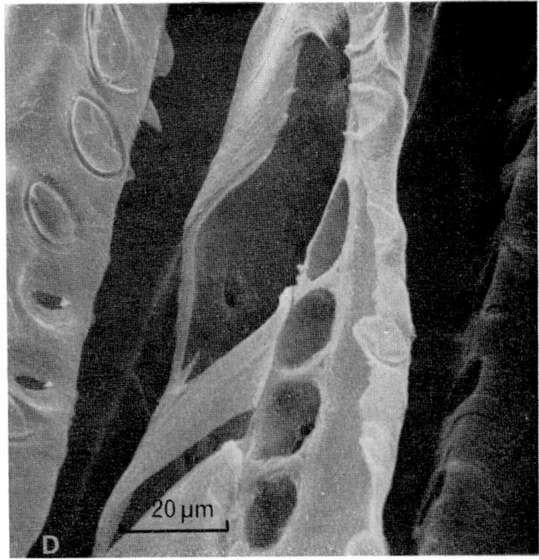

G. W. D. FINDLAY AND J. F. LEVY

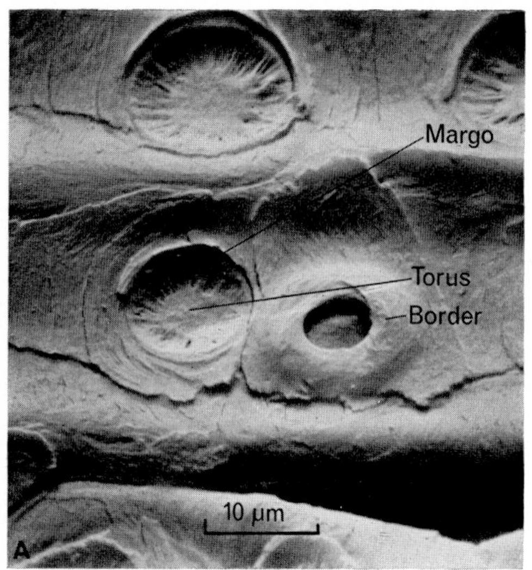

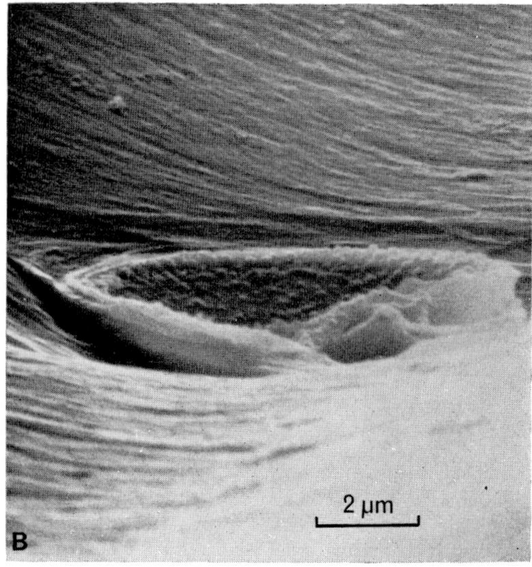

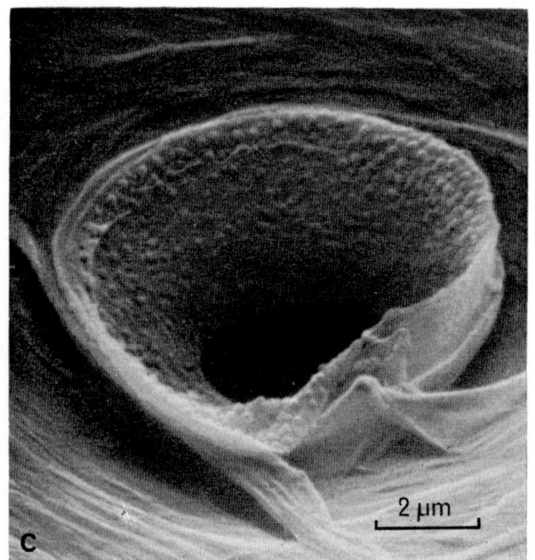

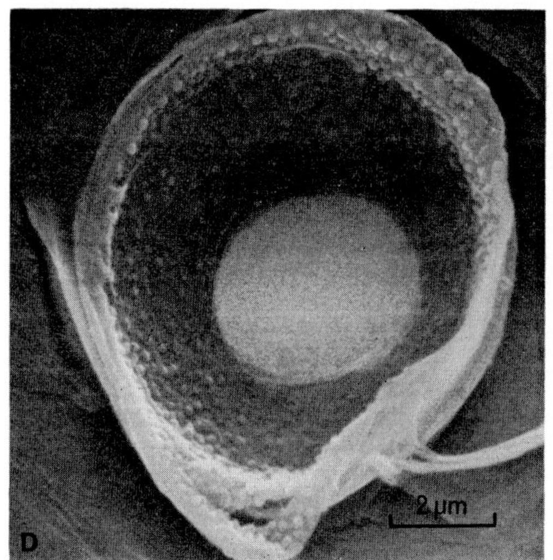

G. W. D. FINDLAY AND J. F. LEVY

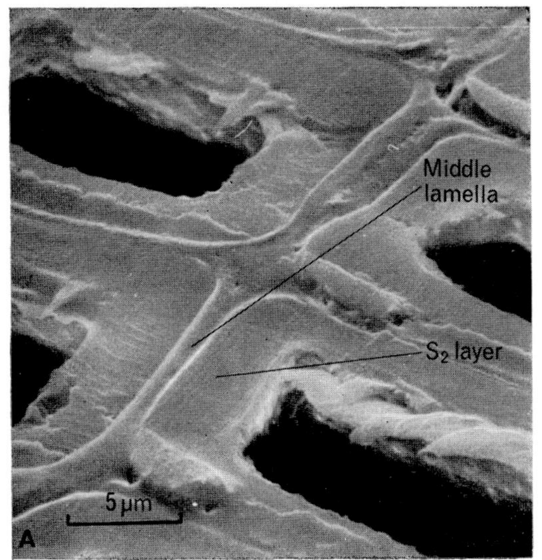

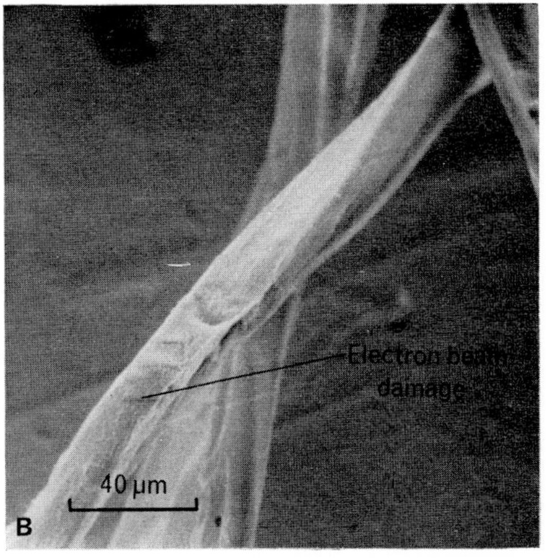

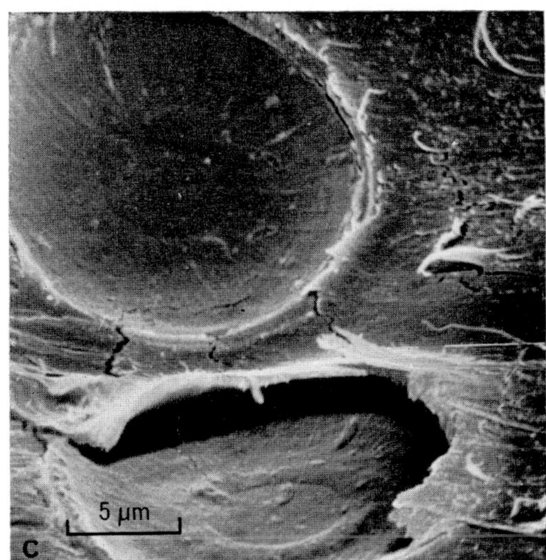

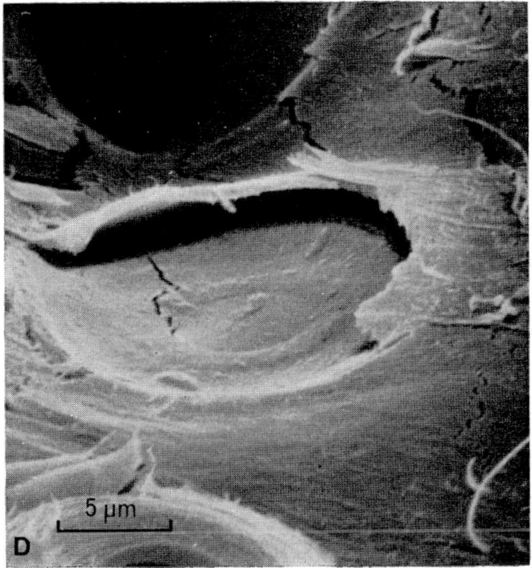

G. W. D. FINDLAY AND J. F. LEVY

Anatomy and taxonomy in *Juncus* subgenus *Genuini*

C. A. STACE

Department of Botany, The University, Manchester, England

Early classification of *Juncus* species was based on a wide range of characters, both morphological and anatomical. In the subgenus *Genuini* it is demonstrated that anatomical characters of stem and foliar bract can be used to distinguish most British species. Application of the anatomical data to the identification of possible parents of several hybrids is described.

CONTENTS

INTRODUCTION

The Bignoniaceae are widely cited as the family in which anatomical characters were first used to an appreciable extent in the diagnosis of taxonomic categories. Bureau (1864) found features of the anatomy of these plants useful aids to their precise delimitation at various taxonomic levels, and thus introduced the former wherever necessary. In *Juncus*, on the other hand, one cannot say that anatomical characters provide *supplementary* taxonomic evidence; from the early days of intensive study of this genus the vegetative anatomy has featured prominently in the assemblage of taxonomic characters. This is no doubt a measure of the extent of floral reduction in rushes, and also adds weight to the dictum that xeromorphic plants provide more anatomical characters of diagnostic value than do mesomorphic plants.

Buchenau (1875, 1890, 1906), in a long series of studies in the genus, divided *Juncus* into eight subgenera. Although many alternative schemes have been forwarded since that time, Buchenau's classification is still the most widely used, and it is significant that from the first his subgenera were partly characterized by leaf and stem anatomy. The plants to be discussed in this paper belong to what Buchenau called the subgenus *Junci genuini* or what is nowadays more often termed section *Genuini*. (This unlikely name is, in fact, the correct one, for this is not the type section or subgenus of the genus.) Cutler (1969) has produced a well-documented and up-to-date account of the anatomical characters of each of the subgenera in the fourth volume of 'Anatomy of the Monocotyledons'.

Subgenus *Genuini*, or the 'leafless rushes' as they have been called, has many species in the north and south parts of the Old and New World. In Britain there are five species: *J. effusus* L., *L. conglomeratus* L., *J. inflexus* L., *J. balticus* Willd. and *J. filiformis* L. Elsewhere in Europe two additional species are found: one (*J. arcticus* Willd.) extremely similar to *J. balticus* and often considered only a subspecies of it; the other (*J. jacquinii* L.) an alpine species of very distinct facies.

MORPHOLOGY

All the plants are rhizomatous perennials. The rhizomes bear scale-like leaves and from the axils of a number of these arise cylindrical erect aerial stems. The number of aerial stems produced in one growing season varies, largely according to the state of health of the specimen concerned, but there are always several. The length of rhizome internode is very variable, depending partly on environmental factors, which will not be gone into here, and partly upon the species. The three common species (*J. effusus*, *J. inflexus* and *J. conglomeratus*) and *J. jacquinii* usually possess rhizomes with very short internodes, so that in general the plants occur in dense clumps. *J. balticus*, *J. arcticus* and *J. filiformis* have longer internodes and are described in floras by such terms as 'far-creeping' or 'mat-forming' rather than tufted.

The aerial stems bear a number of leaves, usually very low down, but these are generally brownish and scale-like, representing only the sheath of the typical sheath-cum-blade type *Juncus* leaf. In *J. filiformis* and *J. jacquinii*, however, some of these leaf-sheaths bear small brownish or greenish vestiges of leaf-blades, and in the latter species there are in addition one or less often two leaves borne much further up the stem and comprising a sheath and a well-developed green blade which actually overtops the stem and inflorescence. Although, according to Buchenau, there are other species of this subgenus which do possess leaves with long green blades (albeit not European species), the presence of such leaves on the upper part of the stem and the overtopping of the inflorescence by them occurs nowhere else in the *Genuini*, and may in fact call to doubt the inclusion of *J. jacquinii* in that subgenus.

A similar morphological pattern to that of the *Genuini* is also found in the subgenera *Subulati* (with the single European species *J. subulatus* Forsk.) and *Thalassii* (more properly subgenus *Juncus*, including the well-known species *J. maritimus* Lam. and *J. acutus* L.). In these subgenera, however, some of the lower leaves always possess long green blades, and there are also anatomical differences.

In all the species except *J. jacquinii* the inflorescences are single and apparently sessile and lateral on the aerial stems. This is a unique feature of subgenus *Genuini*. Although opinion is not unanimous it is widely believed that the part of the aerial shoot which extends beyond the inflorescence is in fact a bract of the terminal inflorescence. Comparison of these rushes with those in subgenera *Subulati* and *Juncus* and with *J. jacquinii*, all of which have a much shorter and more obviously lateral main bract, makes it difficult to escape from that conclusion, despite the fact that the morphology and anatomy of the stem and bract are virtually identical in subgenus *Genuini*.

During development the inflorescence first becomes visible as a small swelling on the side of the aerial shoot, this position thus marking the juncture of the stem and presumed bract. Apart from this it is often not possible to distinguish these two organs, either internally or externally. A certain proportion of the aerial shoots, mostly those produced first and last during the season, never develop inflorescences, but they arise from the rhizome in exactly the same way as flowering shoots and there seems to be no reason for considering them as other than stems plus sterile bracts.

ANATOMY

Subgenus *Genuini* is equally distinctive in its anatomy. In this discussion only the structure of the aerial stem will be considered, and *J. jacquinii* will be ignored. In section the stems are basically circular, but in *J. inflexus* and *J. conglomeratus* they are rather regularly ridged. The tissues are in the main concentrically arranged. The ground tissue is divisible into three major zones: a broad band of chlorenchyma situated under the stoma-bearing epidermis; a very wide pith in the centre; and between these two a zone of rounded parenchyma cells usually extending roughly from the outer edges of the vascular bundles to just inside their inner edges. At various positions in the last zone, usually just outside the zone of vascular bundles, are to be found roundish patches of thin-walled large cells which break down eventually to form ill-defined longitudinal canals, probably containing air in nature. The vascular bundles are discrete regions, with very well developed xylem and phloem, surrounded by bundle sheaths. The latter typically consist of an inner sclerenchymatous layer and an outer parenchymatous zone, each developed to varying degrees. The main ring of vascular bundles is usually supplemented by one or two extra rings of smaller bundles lying nearer the epidermis and variously alternating with the main bundles. In the chlorenchyma zone, abutting upon the epidermis and extending some way towards the vascular bundles, are in most species conspicuous girders of heavily lignified fibres. In those species with ridged stems the ridges mainly coincide with the position of such sclerenchyma girders. The chlorenchyma cells are generally somewhat palisade-shaped.

The pith may be a continuous tissue or possess a smaller or larger pith cavity in the centre, especially in the lower parts of the stems. In *J. inflexus* there is always a large central cavity which usually occupies the whole of the pith zone. This cavity is, however, traversed at intervals by transverse diaphragms of normal pith tissue. The pith is in most species composed of conspicuously stellate cells with long arms and large intercellular spaces. Such tissue is familiar to most sixth-form biologists, and formed part of the experimental material of Houwink & Roelofsen's (1954) classical work on primary cell-wall growth.

The five British species of subgenus *Genuini* are largely distinguishable on the basis of stem anatomy. *Juncus arcticus* probably does not differ anatomically from *J. balticus*. The latter species is undoubtedly the most distinctive of the five. Three conspicuous diagnostic features may be mentioned. Firstly, the outer cortical sclerenchyma girders are completely lacking, and the chlorenchyma is particularly extensive in the radial direction. This lack of sclerenchyma accounts for the fact that the stem of *J. balticus* is always completely smooth whether fresh or dry, and Buchenau placed such species in a group called *Junci genuini laeves*. They are further characterized by the fact that the stomata occur all over the surface of the stem, whereas in those species with sclerenchyma girders the stomata are confined to bands alternating with the girders. The other four British species, which possess sclerenchyma girders, fall into the *Junci genuini valleculati*. (It is of interest that *J. jacquinii* was the only species of this subgenus which Buchenau did not place in one or other of these two groups.) Secondly, the bundle sheaths are very much thicker in *J. balticus* than in the other four species, due to the multi-layered inner sclerenchyma sheath. Thirdly, the typical stellate pith cells are absent, being replaced by angular cells leaving relatively small intercellular spaces. It is very easy, however, to envisage the development of these cells into the stellate type, and the stellate cells of other species do indeed pass through such stages during their growth. According to Buchenau some *Junci genuini laeves* do possess stellate pith cells.

Juncus inflexus is distinguished from the remaining species, apart from its interrupted pith, by the fact that the stem is very markedly ridged, each ridge coinciding with a large main vascular bundle, and on the same radius, occupying much of the ridge, a very well-developed, broad and deep sclerenchyma girder. In the other three species the sclerenchyma girders are much less strongly developed. In addition, the epidermal cells upon each ridge in *J. inflexus* are conspicuously larger, particularly in the radial direction, than those in the furrows between the ridges.

Juncus filiformis differs from *J. effusus* and *J. conglomeratus* in its slenderer stems with fewer vascular bundles, in its sparse pith with irregularly stellate cells, and in its shallow sclerenchyma girders which extend only a short way across the chlorenchyma zone. In the other two species the vascular bundles are more numerous and the sclerenchyma girders are longer and also more numerous.

Juncus effusus and *J. conglomeratus* have been confused from Linnean times to the present day. Much of this confusion has stemmed from the fact that the inflorescences, usually compact in the latter and spreading in the former, may actually be compact or spreading in either species. But even when the inflorescence character is ignored, much difficulty may be encountered in distinguishing the two species. These difficulties are usually under-estimated in modern floras, probably due to a failure to observe a sufficient range of material. The usual situation is that one can quite easily distinguish between the two species in any one locality, but that if one takes the whole range encountered in different areas then the characters of the two species overlap. This state of affairs is exactly paralleled by the anatomical features.

Juncus effusus and *J. conglomeratus* are usually separable macroscopically by a good number of characters of the inflorescence, stems and scale-leaves, but, as mentioned above, none of these is absolute. When fresh the stems of *J. effusus* are shining and smooth with numerous longitudinal striations representing the sclerenchyma girders. When the stems dry these striations become slight ridges. The fresh stems of *J. conglomeratus* are dull and markedly ridged, the ridges being further apart and thus fewer in number than in *J. effusus*. Such a situation would seem ideal for anatomical investigation, but strangely enough there seems no anatomical basis for the morphological appearance. There are on average fewer bundles and fewer sclerenchyma girders in *J. conglomeratus*, but the ranges of the two species overlap considerably.

Moreover there is no close correlation between the numbers of bundles, girders or, indeed, ridges. Girders are opposite most of the main bundles, but they are unaccountably absent from opposite some of the larger bundles, and well-developed girders lie opposite many of the very minor bundles. The ridges mostly coincide with the main bundles, but this is not always the case; on occasions a very small bundle lies opposite a ridge, and main bundles may lie in a groove. What causes the difference between the two species in the number and elevation of the ridges is thus so far unknown. Difficulties are added to by the fact that the number of ridges decreases from stem base to apex, and in *J. conglomeratus* the ridges are more prominent just below the inflorescence than elsewhere. Moreover this species frequently produces a second crop of flowering stems late in the year, and these possess much less markedly ridged stems than are typical for the species. In *J. conglomeratus* it has been noticed that the epidermal cells overlying the ridges are frequently somewhat larger than the other epidermal cells. This is, however, by no means always the case, and the cells are never as enlarged as they are in *J. inflexus*. Such cells are never present in *J. effusus*.

The two subgenera most similar to the *Genuini* on morphological features also resemble it quite closely in their anatomy. Subgenus *Subulati* has stellate pith cells but no sclerenchyma girders; it is distinguished by its sunken stomata overarched by papillate epidermal cells. Subgenus *Juncus* has sclerenchyma girders and non-stellate pith cells; there are extra bundles and sclerenchyma strands (most probably vestigial bundles) in the pith, which at once separates it from the *Genuini*.

HYBRIDS

By far the most widespread hybrid is that between *J. effusus* and *J. inflexus* (*Juncus* × *diffusus* Hoppe), which is found over much of the British Isles and the rest of Europe. It is, however, rather uncommon considering the frequency with which its parents are found growing together. There is little agreement as to the occurrence of *J. effusus* × *J. conglomeratus* either in Britain or elsewhere. This is perhaps not surprising in view of the difficulties encountered in distinguishing between the would-be parents. Some of the intermediate specimens are more or less sterile, and these are the plants usually described as hybrids. But some workers hold the view that hybrids are frequent and fertile, and backcross (see Agnew, 1968), while the writer has seen no specimens to convince him of the occurrence of hybrids at all. The combination *J. conglomeratus* × *J. inflexus* has been recorded but is of much more uncertain occurrence. It would presumably closely resemble *J.* × *diffusus*. In Scandinavia and the Baltic region the hybrid *J. balticus* × *J. filiformis* is not infrequent wherever the parents meet. The parents are never found together in Britain. There is good reason to believe that in Scandinavia the hybrid *J. arcticus* × *J. filiformis* also occurs, evidence resting mainly on the presence of *J. arcticus* rather than *J. balticus* in the area concerned. Intermediates between the latter two species occur but are as yet of uncertain significance. The other two hybrids are extremely rare, being endemic to Lancashire (Stace, 1970); they are *J. balticus* × *J. effusus* and *J. balticus* × *J. inflexus*.

Anatomical characteristics have undoubtedly aided the study of hybrids as much as that of the parent species, for they show features which are often of great help in pinpointing parentage. Four cases will be mentioned. *Juncus* × *diffusus* is intermediate between *J. effusus* and *J. inflexus* in most features of morphology and anatomy. The size and shape of the sclerenchyma girders, the degree of elevation of the stem ridges, and the degree of enlargement of the epidermal cells overlying the ridges, are the best diagnostic characters. The pith, continuous in the former parent but interrupted in the latter, is continuous, although the cells may be less densely packed than in *J. effusus*.

The other three examples are the hybrids of *J. balticus* with *J. filiformis*, *J. inflexus* and *J. effusus*, which all bear three important points in common. In the first place sclerenchyma girders are completely lacking, an obvious mark of *J. balticus* as one parent. This view is strengthened by the anatomy of the pith cells, which are intermediate between the angular *J. balticus* type and the stellate type of the other species. Thirdly, the bundle sheaths are always better developed in the hybrids than in *J. inflexus*, *J. effusus* or *J. filiformis*, and may sometimes approach the size of those in *J. balticus*. On the whole, therefore, one concludes that the characters of *J. balticus* tend to be dominant over those of the other species, but that the pith cells are unmistakable signs of a hybrid between a species with stellate cells and one with angular cells. The identity of all three hybrids would be extremely dubious without the use of anatomical studies.

These three hybrids are extensively rhizomatous and form large patches mainly in coastal dune-slacks. *J. inflexus* × *J. balticus* differs from the other two in that the stems are slightly ridged, the ridging being entirely due to enlarged epidermal cells opposite the main vascular bundles, often almost to the same extent as in *J. inflexus* itself. *J. filiformis* × *J. balticus* and *J. effusus* × *J. balticus* are very similar to each other. The latter, so far as is known, is now present only in cultivation. It was discovered in 1933 on the Lancashire coast, and in 1968, when its site was built over, it formed a large clone perhaps 500 m^2 in extent. It was recorded from Prussia in 1893 and described as *J. scalovicus* Aschers. & Graebn., but I am informed by Dr S. Snogerup of Lund that the type specimen is in fact *J. balticus* × *J. filiformis*. This hybrid is not uncommon on the Swedish coast and elsewhere. It may be distinguished anatomically from the English hybrid by the slenderer stems, the smaller number of vascular bundles, and the very sparse and irregular-shaped pith cells.

As mentioned previously, *J. inflexus* × *J. balticus* is also endemic to Lancashire. Two large patches, both discovered in 1952 and still thriving, occur within a few kilometres of each other. In general appearance they do not very closely resemble either parent, but exhibit a good deal of hybrid vigour. The rhizome internodes are much longer, the inflorescence larger, and the aerial stems much thicker and taller than in either parent, the stems sometimes attaining 2 m in height. The stem anatomy, however, makes the identity quite certain. Unlike the case with *J.* × *diffusus*, which has the continuous pith of *J. effusus* rather than the discontinuous pith of *J. inflexus*, these colonies of *J. balticus* × *J. inflexus* have an interrupted pith. In 1966 a third patch of this hybrid was discovered in another area of the Lancashire coast. This clone is much smaller and less robust in all respects than the other colonies, and moreover it has a continuous pith. In all other anatomical aspects it is, however, identical with the two first-discovered patches.

CONCLUSION

Anatomical features of the Juncaceae are clearly of immense taxonomic value. In *Juncus* they help distinguish the subgenera or sections, and are important aids in deciding the systematic position of such dubiously placed species as *J. jacquinii*. They are also frequently of equal use at the specific level, and in the study of inter-specific hybrids. In such a genus it is obvious that one cannot and should not separate anatomical from morphological characters, nor reproductive from vegetative ones; and it is important that all monographers should in future incorporate anatomical data into their formal taxonomic descriptions.

But it is, of course, equally clear that anatomical investigations will still leave a great many questions, particularly concerning the hybrids, which only experimental work will answer. The Lancashire sites for *J. balticus* × *J. inflexus* represent the only English localities of *J. balticus*. But in Scotland and elsewhere in Europe the two species frequently come into contact, yet no hybrids have been found. The reason is unknown. All hybrids in the *Genuini* seem to be sterile, but it is not certain whether a degree of fertility and hence of backcrossing may occur. There are some indications of this in *J.* × *diffusus*, and *J. balticus* × *J. inflexus* appears to possess perfectly good pollen. Problems such as these, and the origin of the differences between clones of the last hybrid (whether they be due to backcrossing or to hybridization between different strains), are at present under investigation, and wherever the characterization of a plant is required anatomical studies are deeply involved.

REFERENCES

AGNEW, A. D. Q., 1968. The interspecific relationships of *Juncus effusus* and *J. conglomeratus* in Britain. *Watsonia*, **6**: 377–388.
BUCHENAU, F., 1875. Monographie der Juncaceen vom Cap. *Abh. naturw. Ver. Bremen*, **4**: 393–512.
BUCHENAU, F., 1890. Monographia Juncacearum. *Bot. Jb.*, **12**: 1–495.
BUCHENAU, F., 1906. Juncaceae. In Engler & Gilg, *Das Pflanzenreich*, **4** (36): 1–284.
BUREAU, E., 1864. *Monographie des Bignoniacées*. Paris: Baillière.
CUTLER, D. F., 1969. Juncaceae. In *Anatomy of the monocotyledons*. **IV**. *Juncales*, pp. 17–77. Oxford: Clarendon Press.
HOUWINK, A. L. & ROELOFSEN, P. A., 1954. Fibrillar architecture of growing plant cell walls. *Acta bot. neerl.*, **3**: 385–395.
STACE, C. A., 1970. Unique *Juncus* hybrids in Lancashire. *Nature, Lond.*, **226**: 180.

Heterogeneous medullary rays in Araucariaceae

PÁL GREGUSS

The University, Szeged, Hungary

The author distinguishes five medullary ray types in secondary wood of gymnosperms; the types can be considered as constituting an evolutionary series. The Araucariaceae have rays of the first type, with thin-walled parenchymatous cells having small, simple pits. In *Agathis macrophylla* and *A. rhomboidalis* the author has demonstrated for the first time occasional thick-walled cells with simple pits or pits intermediate between simple and bordered; consequently these rays are heterogeneous. *A. macrophylla* also exhibits thin-walled, cambial-like cells at the upper and lower margins of the ray. It is suggested that these cells divide to produce ray cells proper.

CONTENTS

INTRODUCTION

The structure of rays plays a very important role in recognizing woods anatomically. These present the greatest variety of diagnostic characters for separating not only single genera but also single species. Closer investigation of the medullary ray patterns of about 100 species of *Podocarpus*, 90 of *Pinus*, about 50 of *Juniperus* or nearly 40 species of Araucariae, shows that they still differ from one another in small details and, therefore, the determination of single species is possible on the basis of the detailed structure of their rays. There are some families, e.g. Araucariaceae, Podocarpaceae, of which the ray pattern is composed of quite simple parenchyma cells with thin walls having, it is true, pits different to a certain degree; but in others, however, such as *Pinus*, the rays are composed of many cell types and the pits are also much more varied. This greater diversity may probably be explained by their evolutionary history.

From among the gymnosperms, the author (1955) has selected a probable series demonstrating the evolutionary history of medullary rays, according to which, for example, the highly varied ray structure of Pinaceae, composed of many cell types, is to be considered more developed and representing a more recent state than the ray structure of *Cycas*, *Ginkgo* and *Araucaria*, which is of a more ancient type, composed of parenchyma cells with simple thin walls. The finer structure of the ray cells of the latter is still very similar to that of the thin-walled parenchyma cells occurring around the centre, thus denoting that in these woods the origin of rays can in general be traced back to the parenchyma cells of the central medulla. When, however, these thin-walled parenchyma cells began to carry out other functions, for instance conducting food, or storage, they began to change slowly; their walls became thickened and lignified to various degrees, and conspicuous pits developed.

The author has determined (Greguss, 1955) five stages in the evolutionary history of rays. In the first stage, every wall of all the ray cells remains thin, possibly with small simple pits similar to those of the medullary parenchyma cells. This state is generally characteristic of the following gymnosperms, the Cycadaceae, *Ginkgo*, Araucariaceae and most Podocarpaceae. In the second stage, the horizontal walls of ray cells are thicker; the tangential walls, however, remain thin. That is characteristic of the Taxaceae, Taxodiaceae and, to some extent, of Cupressaceae. In the next stage, apart from the thin-walled parenchyma cells, there also occur parenchyma cells with rather thick walls. This stage is found in some Pinaceae and Taxodiaceae, although the pits are of different kinds. In the following stage, beside the thin- and thick-walled parenchyma cells, there develop ray tracheids with walls of differing thickness, e.g. in Cupressaceae, Taxodiaceae and Pinaceae, while in the fifth stage, apart from the cells of ray types mentioned above, there develop diverse ducts and parenchyma cells showing a wide range of morphology. Varied medullary rays of this type occur in the Pinaceae.

A single type of ray structure is more or less characteristic of a single gymnosperm family or genus. In this way, for instance, only the thin-walled ray parenchyma cells are characteristic of the Araucariaceae. In the development of medullary rays the first stage is represented by these.

The author has investigated the xylotomy of about 40 species of Araucariaceae and found that in most species only so-called homogeneous thin-walled parenchyma cells are present. In a few species, however, there is evidence for the presence of stage two rays in that there occur here and there occasional ray cells with thick walls and simple pits as well as the thin-walled ray cells. Consequently, the rays exhibit a heterogeneous structure. A heterogeneous medullary ray structure like this has not so far been noted in the literature for either recent or fossil members of the Araucariaceae. Therefore it is worth mentioning that in the wood of *Agathis macrophylla* Mast. and *Agathis rhomboidalis* Warburg heterogeneous rays also occur in exceptional cases.

OBSERVATIONS

Agathis rhomboidalis

Plate 1**A, B** shows the medullary rays of *Agathis rhomboidalis*. Both photos show a ray five cells in height. The cross-field areas of the thin-walled ray parenchyma cells have the araucaroid pits characteristic of the Araucariaceae. The walls of the ray cell in the second layer down have, however, grown very thick; in the communicating walls between the adjacent thin-walled parenchyma cells and the narrow lumen of the thick-walled idioblast-like ray cell there are crowded simple pits. This can be well observed in Plate 1**A**. In the cross-fields the bordered pits occur, in fact, in the wall of the longitudinal tracheids. In the thin-walled ray parenchyma cells there is only a small pit corresponding to the aperture of the bordered pits, the field itself being indistinct. That is verified also through the pittings of the thick-walled parenchyma cell. In it only the aperture of the simple pit can be observed, but not the field. The long (radial) walls of the idioblast-like ray cell decrease in thickness at one end of the cell and that end (tangential) wall is thin over its entire surface and is in contact with a thin-walled parenchyma cell.

In connection with the ray structure of this species another interesting phenomenon is also worth mentioning. There runs on both, i.e. upper and lower, margins of the medullary ray the so-called marginal parenchyma layer of thin-walled cells first described by the author (Greguss, 1955, 1957). This marginal parenchyma of thin-walled cells (denoted by white triangles in Plate 1**A, B**) is characteristic of Araucariaceae. Since their walls are thin, they are probably capable of dividing. In the opinion of the author, these cells of probable cambial or more nearly monopleuric cambial nature produce by division the so-called medullary ray cells proper. There is no doubt that these marginal cells occur first in the Araucariaceae and Podocarpaceae.

Agathis macrophylla

The second species in which this special type of thick-walled ray cell occurs is *Agathis macrophylla*. Such cells are represented in Plate 1**C–E**. Their common characteristic is that their walls have grown very thick, and some simple pits occur in every wall. If these cells are adjacent to similar thick-walled cells, the simple pit-pairs are opposite (Plate 1**D**). In contrast to *Agathis rhomboidalis*, pits occur in all thick walls including the tangential walls, whether or not they are in contact with a thin-walled parenchyma cell (Plate 1**C**).

It is evident from the simple nature of the pitting in the radial walls of the thick-walled ray cells that these cells are not ray tracheids, which would exhibit bordered pits. Though it is true that some of the simple pits are very similar in structure to bordered pits, forming a transitional type between a simple pit and a bordered one (Plate 1**C** at the arrow), this phenomenon suggests that the bordered pits are, in fact, transformed from the simple pits of the thick-walled cells, as already mentioned by the author in his monograph on the transfusion cells of Cycadaceae (Greguss, 1968, 1969).

Plate 1**F, G** show rays in tangential section. In Plate 1**F**, we see the arrangement of thick-walled ray cells one above another in a high medullary ray. The same is shown, at a higher magnification, in Plate 1**G**. The lumen of these thick-walled ray cells is narrow, nearly point-like, demonstrating clearly that they differ essentially from the thin-walled ray cells. A medullary ray composed of such thin-walled cells is shown by the upper ray on the left in Plate 1**F**.

SUMMARY

Within the Araucariaceae in some *Agathis* species ray cells with thick walls sometimes occur among the thin-walled ray parenchyma cells. The rays are, therefore, heterogeneous. However, they represent a lower degree of development than those heterogeneous rays which have occasional ray tracheids in addition to the thin- and thick-walled parenchyma cells. These occur, anyway, only in the Cupressaceae, Taxodiaceae and Pinaceae. To the best of my knowledge this is a new observation in the Araucariaceae.

DEDICATION

The author dedicates this paper with a deep sense of appreciation to Dr C. R. Metcalfe on the occasion of his retirement.

REFERENCES

GREGUSS, P., 1955. A phylogenetic system of the gymnosperms in the light of xylotomy. In *Identification of living gymnosperms on the basis of xylotomy*, pp. 33–38. Budapest: Akadémiai Kiadó.

GREGUSS, P., 1957. Marginal ray parenchyma in Araucariaceae and in Podocarpaceae. *Acta biol. Szeged*, (*N.S.*), **3**: 15–17.

GREGUSS, P., 1968. *Xylotomy of living Cycads*, 226 pp. Budapest: Akadémiai Kiadó.

GREGUSS, P., 1969. Transfusion tissue in the stems of Cycads. *Phytomorphology*, **19**: 34–43.

EXPLANATION OF PLATE

PLATE 1

A. Radial section of rays of *Agathis rhomboidalis*. Above, the walls of the ray cell in the second layer down are much thickened, except one of the tangential walls touching a parenchyma cell, which is thin. In the thick-walled cell there are simple araucaroid pits. In the cross-fields of the thin-walled cells there are 1–6 araucaroid pits. Above and below, at the arrows, thin-walled marginal parenchyma cells are present. All the walls of the other ray cells are thin (× 500).

B. A five-cell high medullary ray from *A. rhomboidalis*. In the second layer down is a long, thick-walled parenchyma cell, with simple pits in the thick walls. At the upper and lower triangles there is a marginal parenchyma cell (× 260).

C. An isolated thick-walled ray parenchyma cell from *Agathis macrophylla*. In the tangential and horizontal walls there are simple pits (× 500).

D. Two layers of thick-walled medullary ray cells in *A. macrophylla*, with simple pits in the thick walls (× 300).

E. An isolated thick-walled medullary ray cell with simple pits in the thick walls; above and below it are thin-walled ray cells (× 260).

F. Tangential section of the same. In the high medullary ray of a single layer there are thin- and thick-walled parenchyma cells, the lumina of the thick-walled parenchyma cells being narrow (× 160).

G. A detail from **F.** The thick-walled ray cells are well separated (× 300).

Photos by Greguss.

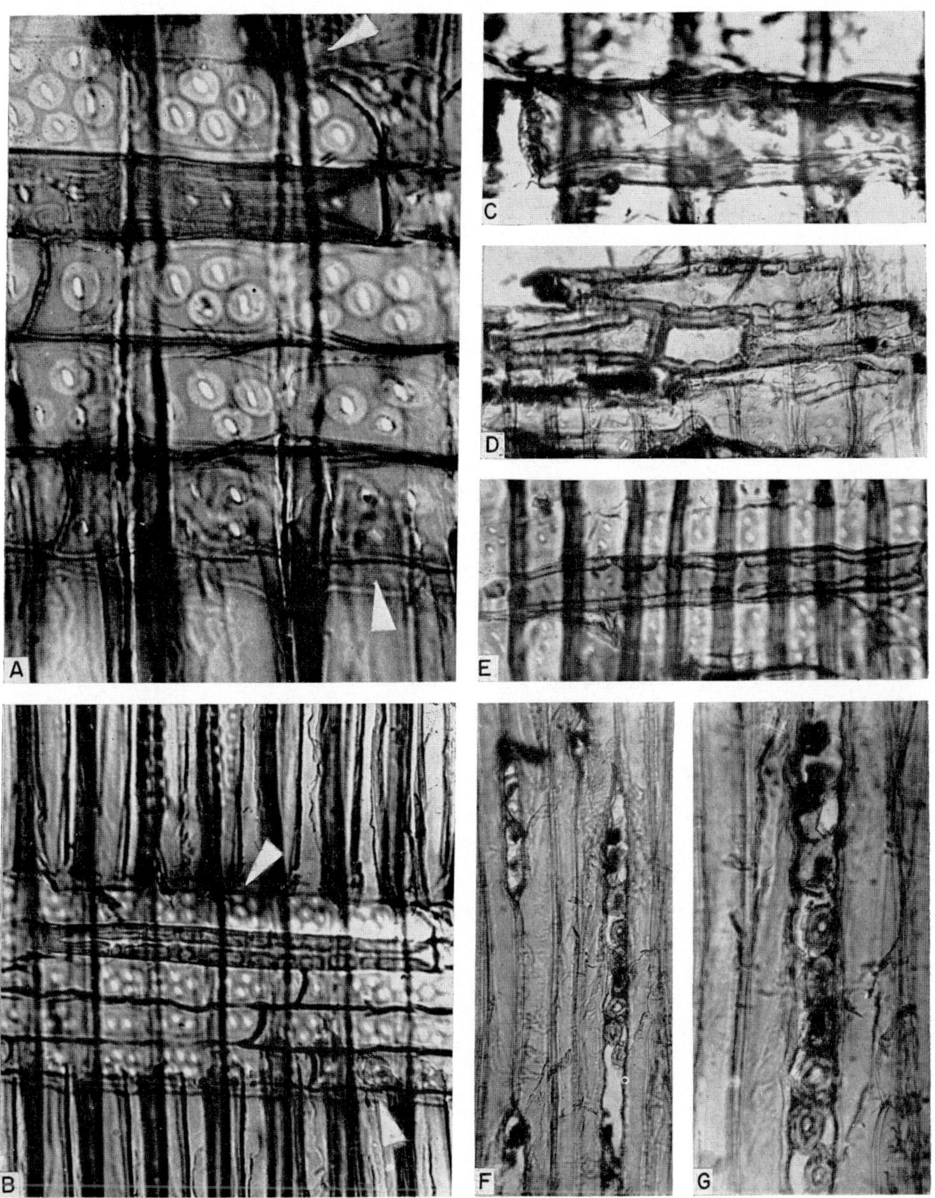

P. GREGUSS

(*Facing p.* 86)

Constant and variable features of the Araliaceae

W. R. PHILIPSON

Department of Botany, University of Canterbury, New Zealand

The range of vegetative form found among members of the Araliaceae is described and some features of their floral morphology are discussed, with particular reference to unilocular gynoecia. It is concluded that truly monocarpellary flowers occur in the family. The constant presence of a second reduced ovule in each loculus emphasizes the close relationship with the Umbelliferae, forming an order isolated from other families. The identity of the vascular pattern of such genera as *Harmsiopanax* with that general in the Umbelliferae also confirms the unity of the Umbellales. The incidence of pleiomery of the androecium is recorded and the opposed views that this character is primitive or advanced are examined in the light of the vascular supply to the stamens and the correlation of this character with a racemose arrangement of the flowers. It is concluded that pleiomery is a primitive feature.

CONTENTS

INTRODUCTION

The Araliaceae and the Umbelliferae taken together form a natural group. As currently understood, these two families contain no genera which are doubtfully placed in them, and only *Mastixia*, and possibly also *Helwingia*, remain as satellite genera with some claims to inclusion. While the concept of the group is thus clear, there are few invariable characters that can be used to define it absolutely. The Umbelliferae are comparatively uniform with regard to habit, inflorescence, flower and fruit in contrast to the more diverse Araliaceae. Some account is given here of the range of form found in the latter family, and the primitive or advanced nature of some characters is discussed.

VEGETATIVE MORPHOLOGY

The Araliaceae are predominantly trees, usually of moderate size, though sometimes attaining the stature of large forest trees. The largest araliad is the tree currently known as *Peekeliopanax spectabilis* Harms, which attains a height of 130 feet with a diameter of the bole at breast height of six feet. This tree occurs in the rain forests of New Guinea, the Bismark Archipelago and the Solomon Islands. Other growth forms frequent within the family are shrubs, lianes and woody epiphytes. The herbaceous habit is less frequent, but occurs in all species of *Panax* and *Stilbocarpa* and in some species of *Aralia* and *Boerlagiodendron.*

Most araliads are large-leaved and thick-stemmed (pachycaul), though trees with small leaves and fine twigs (leptocaul) do occur rarely, an example being *Pseudopanax simplex* (Forst. f.) Philipson, of temperate rain forests of New Zealand. The extreme pachycaul habit is seen in *Harmsiopanax*, in which there is a simple erect trunk attaining heights of up to 40 feet, with a crown of huge palmately lobed leaves, with petioles in excess of six feet and blades four feet in diameter. This palm-like tree is monocarpic, the crown of foliage leaves dying as the apex is transformed into a richly branched panicle 15 feet high and 20 feet in diameter.

The vegetative shoots of most woody araliads are of one type, though in *Evodiopanax* long and short shoots are reported. The habit of the shrub *Pseudopanax anomalum* (Hook.) Philipson is unique in the family since the long shoots frequently bear scale leaves with no (or only a few) true foliage leaves. All, or most, of the foliage is borne on specialized short shoots (Philipson, 1970). Sucker shoots arising from the root system occur in species of *Aralia* and *Tetrapanax*. The herbaceous araliads are mostly rhizomatous, as in *Panax* and *Stilbocarpa*. Their underground shoots may be tuberous, as in species of *Aralia* and *Panax*, some of which provide the ginseng of commerce.

The habit of branching may change with the age of the plant. The best known example of this is the English Ivy, *Hedera helix* L., which clings tightly to its support by adventitious roots during its vegetative phase, but whose branches grow freely into the air when about to flower. More striking examples of habit heteroblastism are found in the New Zealand 'lancewoods' (*Pseudopanax ferox* Kirk and its allies). These grow for many years as erect unbranched shrubs before developing crowns and becoming branched trees. This change in habit is accompanied by complex changes in leaf form.

The leaves are borne in spiral phyllotaxis or rarely stand in opposite pairs (*Cheirodendron*, *Eremopanax*) or in whorls (*Panax*). The blade of the foliage leaves is usually compound, either pinnately or digitately, to the first, second or even higher degrees. In digitate leaves which branch to the second degree, this feature may be limited to the central leaflet(s) (*Mackinlaya*, *Boerlagiodendron*). The bases of the petiolules of digitate leaves may be connected by a fan-like webbing (*Trevesia* and *Boerlagiodendron*). Simple leaves may be entire, palmately lobed, or more rarely pinnately lobed (*Aralidium*). Peltate leaves occur in some species of *Harmsiopanax*. The leaf is usually large and may be very large (ten feet in *Aralia*) and is rarely small and simple (e.g. *Pseudopanax simplex*, adult) and only most exceptionally very small (about 1–2 cm in *Pseudopanax anomalum*). Variation in leaf shape on the same plant is frequent and heteroblastic changes in leaf-shape are common. The leaf-base is usually clasping and may ensheath the stem (*Mackinlaya*). The node in the clasping leaves is multilacunar but in small leaves the node may be trilacunar. The sheathing leaf base often extends upwards as a ligule-like intrapetiolar stipule, which may be single or consist of two lanceolate structures (*Tetrapanax*). In several genera the stipular extension is reduced to a minute margin or is quite absent (*Gastonia* and allied genera). The petiolar base is unique in *Boerlagiodendron* in bearing a series of transverse flaps of tissue in addition to the normal stipular appendage.

In many Araliaceae all the leaves, except the prophylls at the base of lateral branches, develop as foliage leaves. In others some leaves may be differentiated into cataphylls of various types which enclose terminal buds. Cataphylls are unusual even in species growing in temperate climates (Philipson, 1970), but are known to occur in tropical genera (Borchert, 1969).

The surface of the leaves and stems may be glabrous, but is often covered with a tomentum, sometimes thick, of stellate or simple hairs. Several of the Araliaceae are inordinately spiny (*Aralia*, *Harmsiopanax*).

REPRODUCTIVE MORPHOLOGY

General

The inflorescences usually terminate the principal shoots, so that growth is sympodial. Lateral shoots may rapidly overtop the terminal inflorescences, which then appear lateral but which can be seen to be leaf-opposed (*Mackinlaya*). In the 'umbrella tree', *Peekeliopanax*, two or three rapidly growing and thick branches arise below each inflorescence bud. The inflorescences develop very slowly, so that at anthesis the inflorescences stand several crotches below the terminal tufts of leaves. Successively younger inflorescence buds occur in the higher crotches. When the inflorescences terminate short lateral branches they may appear to form axillary fascicles (*Pseudopanax anomalum*). Truly lateral inflorescences occur in *Aralia*, *Stilbocarpa* and some species of *Schefflera*, and are reported in *Woodburnia*. The form of the inflorescence is varied and will be discussed later in this account. The pedicels in many genera (see Harms, 1898) show an articulation immediately below the ovary.

The flowers are actinomorphic with the ovary inferior or partially superior, except for two species of *Tetraplasandra* with completely superior ovaries (Eyde & Tseng, 1969). The flowers are mostly hermaphrodite, though in some genera male flowers occur grouped with hermaphrodite (e.g. *Plerandra stahliana* Warb. and other species, *Mackinlaya*), and monoecious and dioecious species are known to occur, especially in New Zealand. In *Pseudopanax ferox* the male inflorescences are racemose while the female are umbellate.

The calyx, when present, consists of a whorl of small lobes (elongated in *Woodburnia*) equal in number to the petals, or it forms a more or less entire rim. Rarely it is quite lacking (*Meryta*). In most genera there are five calyx lobes alternating with five petals. However, the perianth members may vary in number from three to many, though the petals may not always be completely divided from one another. One flower of *Munroidendron*, for example, had 13 corolla segments, of which ten were clearly single, one double and two corresponded to three lobes. In *Peekeliopanax spectabilis* the numerous petals are indicated by clefts that extend only halfway through the corolla tissue (Plate 1**D**). Not infrequently the corolla is completely calyptrate with no indication of individual petals (*Tupidanthus*). The corolla of some species of *Boerlagiodendron* is most aberrant, being united into a tube below and opening above by spreading lobes. The petals typically are rather fleshy and valvate, and are inserted around the whole circumference of the upper part of the ovary by broad bases. However, in many genera (forming the tribe Aralieae) the petals are imbricate, and in the small group of genera forming the Mackinlayeae the petals are inserted by a narrow base, or claw, as in the Umbelliferae.

The stamens are usually equal in number to the petals and alternate with them, but they may be more numerous. The occurrence of anisomerous androecia will be discussed fully later in this account. The filament is usually fleshy and short, but is filamentous and long in *Tupidanthus* and *Plerandra*, in which genera the most numerous stamens occur. The filaments curve inwards in the bud, often strongly, with the dorsifixed anthers fitting into more or less well defined cavities formed by the recurved apices of the petals and also exceptionally by projections from the lower part of the petals (*Arthrophyllum diversifolium* Bl.). The pollen-sacs are four or rarely eight (*Dizygotheca*, *Octotheca*) and open longitudinally by introrse slits. In *Tetraplasandra oahuensis* (A. Gray) Harms, the anthers appear to have only two pollen-sacs as pairs fuse at an early stage of development.

The gynoecium

The ovary normally contains two or more loculi, each with a single functional ovule. The loculi are usually few in number (five or fewer) but may be more numerous (about 20 in some species of *Reynoldsia*, *Plerandra* and *Gastonia*), reaching a maximum of about 100 in *Tupidanthus*. Genera with single-seeded fruits have attracted much interest. In *Aralidium* three or four loculi are present at anthesis, but only one seed normally matures, the other ovules and loculi aborting, though traces of them can still be discerned in the fruit. There are four genera described as having only one loculus in the flower, namely *Arthrophyllum*, *Eremopanax*, *Diplopanax* and *Wardenia*. *Wardenia* should almost certainly be eliminated from this list. Its flower is described as having a single loculus with two pendulous ovules. These are said to become separated by a septum which develops in the fruit. I have sectioned a young fruit from the type collection (kindly sent me by Mr J. F. M. Cannon of the British Museum, Natural History) and find that the ovules are separated by a septum up the centre of which the ventral vascular system runs (Fig. 1). Since the ovules receive their traces from the ventrals in the normal way, it is difficult to visualize how the septum could form at a late stage of floral development. I have sectioned flowers of two of the three remaining genera with a single loculus, namely *Arthrophyllum* and *Diplopanax*. In a former account I interpreted the loculus of *Arthrophyllum* as bounded by tissue derived from several carpels and the style as also compound (Philipson, 1967). Since then, better material has become available and it is evident that the carpel wall contains a dorsal and ventral bundle between which is a ring of carpellary bundles (Plate 1**C**). That is to say, the loculus is bounded by the tissue of a single carpel, and the style also appears simple. However, a vascular bundle which accompanies the ventral bundle appears unrelated to either the peripheral or carpellary systems. This bundle runs parallel to the ventral over its entire course and joins with it at the base of the ovary. Traced upwards, it can be seen to branch and enter the disk. This bundle and the ventral are related to each other exactly as are the two ventrals of *Harmsiopanax* (see below). It is suggested that it may be the only remaining evidence of a second carpel. The difference between *Aralidium* and *Arthrophyllum* is, therefore, that carpel development fails at anthesis in the former, but at an early stage

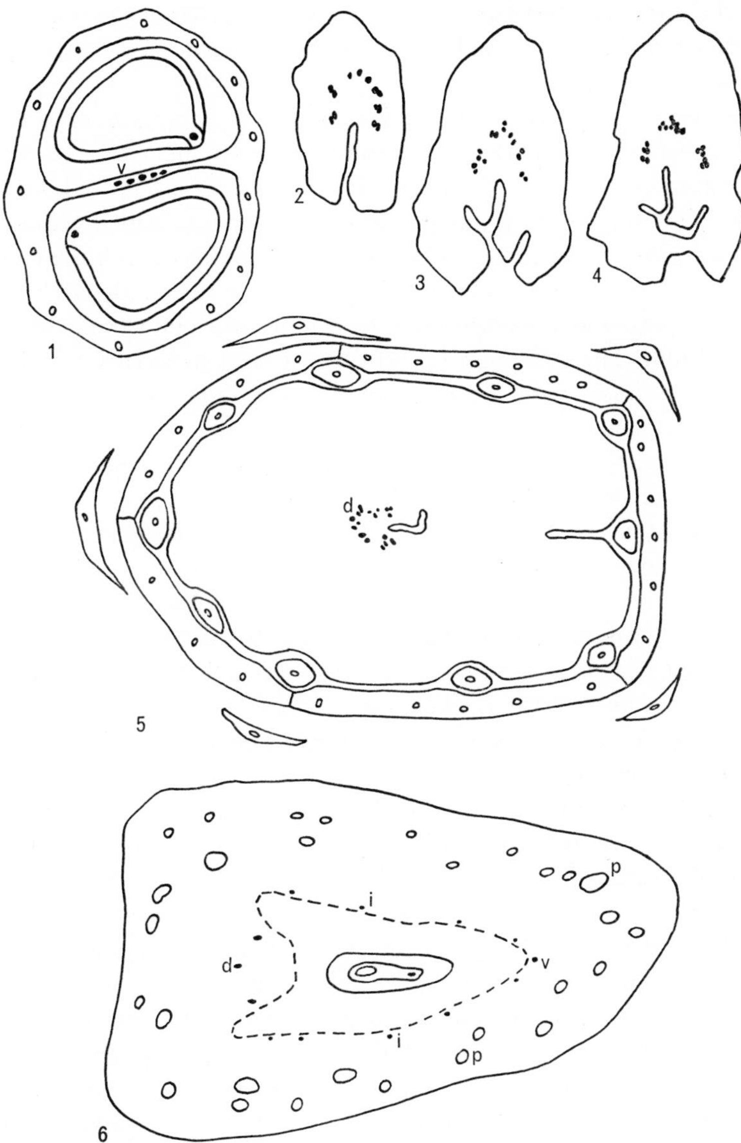

FIGURES 1 to 6. 1. *Wardenia simplex*, T.S. of fruit (× 16). Each of the two loculi contains a seed; ventral bundles occur in the septum. 2 to 6. *Diplopanax stachyanthus*, T.S. of flower bud (× 30): 2 to 4, style at successively lower levels; 5, above insertion of corolla and stamens; 6, ovary; bundles of peripheral system indicated by open circles and of carpellary system in solid black.

d, Dorsal bundle; i, intermediate carpellary bundles; p, peripheral bundle; v, ventral bundle.

of floral development in the latter. If this sequence were taken further, no primordium of a second carpel would ever appear. In that event the gynoecium would be truly monocarpic. Since flowers of the Araliaceae are known with carpels ranging from 100 to two, there appears to be no difficulty in admitting the possibility of monocarpic flowers. No evidence of a second carpel was detected in *Diplopanax* (Figs 2 to 6), so that it may well represent the final stage of reduction. Baumann-Bodenheim (1955) interpreted the carpel of *Eremopanax* as pseudomonocarpellary by extrapolation from ovaries like those of *Arthrophyllum*, but records that no trace of a second carpel occurs in *Eremopanax*.

Other araliaceous gynoecia are to be interpreted as composed of a number of carpels, the radiating walls between the loculi being common to adjacent carpels. Towards the apex of the carpels the ventral margins are closely applied, but each has a discernible epidermis and the loculi may be in open communication with each other. The carpels are prolonged upwards into a stylar portion into which the loculus projects a greater or less distance, and which is capped by a frequently double stigmatic crest. The stylar portion of the carpels may be free or united into a solid or sometimes hollow stylopodium. The loculi may open freely into this stylar cavity, which in turn may communicate freely with the exterior. A nectariferous disk lies between the stamens and the stylopodium.

In each loculus one half of the axile placenta bears a single pendulous anatropous ovule, of which the funiculus and well developed ovular vascular bundle are axial, the micropyle facing outwards (Fig. 7). The morphology of the ovule of comparatively few genera has been investigated in detail, but the usual condition is for a long fleshy funicle to be extended on the micropylar side as a distinct obturator with surface cells differentiated as a continuation of the conducting tissue. There is a single thick integument and a long straight micropyle. The nucellus is thin and disintegrates in all genera observed by the writer, but crassinucellate ovules are recorded by others (see Davis, 1966). A well defined endothelium develops in *Harmsiopanax* and other genera, but is not general.

The other half of the placenta also receives a vascular strand similar in origin to the ovular supply. This half-placenta is prolonged into a tapering organ that curves upwards into the cavity of the stylar portion of the carpel (Fig. 8). This organ has been noted in all the genera of the Araliaceae investigated, about 30 in number, that is nearly half of the family. It has also been observed in several umbelliferous genera. The organ has frequently been reported in these two families (e.g. Pigott, 1914; Singh, 1954), and its ovular nature has been confirmed by the presence in it of archesporial cells (Beghtel, 1925) and by the occasional development of two normal ovules in one loculus. (Jurica (1922) encountered a plant of *Eryngium* in which all flowers had two ovules in each loculus). It has been regarded as a vestigial ovule by most authors and often its presence has been considered exceptional. However, its constancy, and the presence within it of a vascular supply, suggest that this organ is not merely vestigial but has some other function. The conducting tissue of the Umbellales lines the stylar arms and continues over the placentae and down the outer side of the funiculus (obturator) (Borthwick, 1931). The organ formed by the second ovule projects into this system and its apex is composed of cells with densely cytoplasmic contents. It is suggested, therefore, that an ovule has been transformed into an organ associated with the conduction of pollen tubes into the loculus in the vicinity of the microphyll.

As the presence of this second ovule appears to be constant in the two families it may be accepted as one of their diagnostic characters, and may be added to those which already serve to unite the two families into a natural group. The allocation of the Araliaceae and Umbelliferae into widely separated parts of his system by Hutchinson (1959) cannot be accepted. It is also interesting to note that the single-ovulate carpels of the Cornaceae show no trace of such an organ as a normal feature

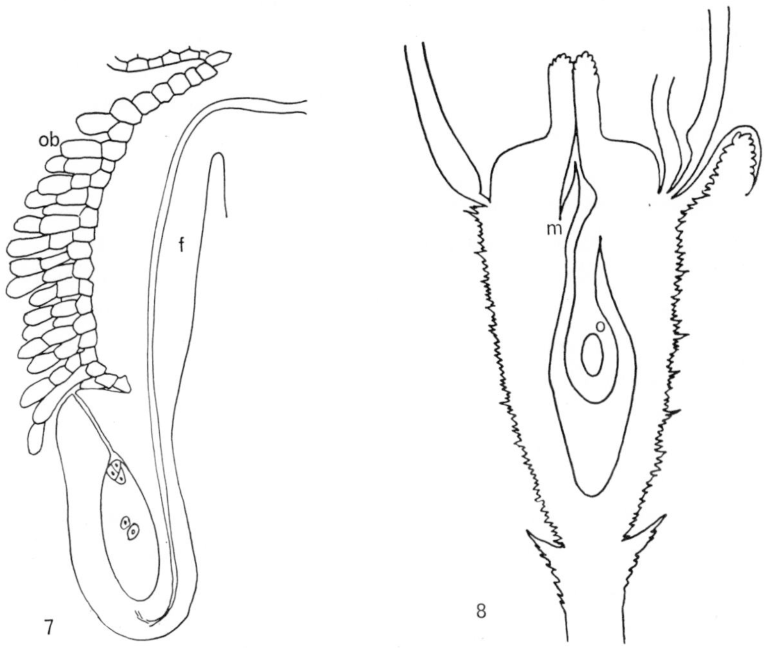

FIGURES 7 and 8. 7. *Boerlagiodendron novoguineense*, L.S. of ovule (×200). 8. *Harmsiopanax aculeata*, L.S. ovary (×75) to one side of the septum.
f, Funicle; m, modified ovule; o, ovule; ob, obturator.

of their morphology. For this reason, and because of the orientation of the ovule in the Cornaceae, and also for chemotaxonomic reasons (Hegnauer, 1964), I should now be less inclined than formerly (Philipson, 1967) to regard the Umbellales as part of a large complex of families to which the Cornaceae also belong. This view, with the two families making a well defined and isolated order, the Umbellales, is in accord with the treatment of Cronquist (1968).

The outer layers of the ovary wall develop in the fruit into a succulent or leathery covering around the hardened inner layers which surround the seed as a pyrene. The fruit is indehiscent, but in *Myodocarpus* each of the two carpels expands into broad wings and tends to split into two mericarps (Baumann, 1946). The fruit of *Harmsiopanax* also splits into two one-seeded mericarps, a condition approaching that of the Umbelliferae. Grushvitsky, Tikhomirov, Aksenov & Shibakina (1969) have drawn attention to histological resemblances between the fruits of *Stilbocarpa* and the

Umbelliferae. The sessile female flowers of *Meryta* either develop into globular multiple fruits or the individual fruits may remain separate. In *Boerlagiodendron* the flowers of the outer rays of the inflorescence develop into normal araliaceous fruits, whereas those of the central rays are sterile and develop directly into seedless, berry-like 'pseudofruits'.

Floral vasculature

The vascular system of the flower of the Araliaceae is simple, and, with minor variations, conforms to a uniform plan. For convenience of description this can be divided into three systems: (i) an outer system of peripheral bundles which supply the two perianth whorls and the stamens; (ii) a system supplying the carpels; and (iii) a system supplying the ovules (ventral system).

A description of the vasculature may begin with that found in a flower with the three outer whorls each with five parts (Plate 1**A**, *Pseudopanax anomalum*). In this species ten peripheral bundles provide the vascular supply to the sepals, petals and stamens (first system). Most bundles of the second system are united with the first up to the level of the top of the ovary. The exceptions are four bundles closely associated with the ventral system. Other bundles of the carpel wall, including the two dorsal bundles, depart from the common bundles at the top of the ovary and supply the disk and stylopodium. Eight of the ten peripheral bundles contribute in this way, the two exceptions lying symmetrically opposite to the septum between the two loculi. The third system comprises a pair of ventral bundles which originate from the stele of the pedicel and continue up the centre of the septum. At the top of the ovary these fuse momentarily before branching to supply bundles to the two ovules, and also to the two transformed ovules. They then unite with the four closely associated strands of the second system and enter the base of the style.

The principal variations on this pattern concern the different meristic variations of the whorls, and in particular the variations due to the presence of fewer or more carpels. These have been discussed in a previous account (Philipson, 1967). Other variations concern the degree of union between the second and first systems. For example, the dorsals are free for most of their length in *Hedera* (Plate 1**B**) and more intermediate carpellary bundles are present in that genus. In *Plerandra* (Fig. 9), *Gastonia* and *Indokingia* the numerous dorsal and intermediate bundles are free (or predominantly so), and in *Peekeliopanax* (Plate 1**E**) the first system of bundles forms two distinct rings, the outer serving the perianth and partly the androecium while the inner serves the greater part of the androecium. In *Harmsiopanax* (Fig. 10) the system is even simpler than in Plate 1**A** because no part of the second system is distinct from the peripheral bundles except in the disk. Another interesting difference is that the relative position of the two ventrals is at right-angles to those of *Pseudopanax*, though like them they divide into four bundles before supplying the four ovules. In *Pseudopanax anomalum* the ventral bundles represent bundles from adjacent carpels, which have united, whereas in *Harmsiopanax* they represent the fusion of bundles from the two halves of the same placenta. In this respect, as in the separation of its fruit into two mericarps, *Harmsiopanax* resembles the Umbelliferae. Although its mericarps are not held by a carpophore, the tissues are arranged very similarly. Both Grushvitsky *et al.* (1969) and Eyde & Tseng (1969) have considered united ventral bundles as an advanced character, and it is interesting to realize that in *Stilbocarpa*, which Grushvitsky describes, the union has been in the same manner as in *Harmsiopanax*, though more than two carpels may be involved. That is to say, the ventral bundles lie opposite the loculi, as is normal in the Umbelliferae. On the other hand, in *Tetraplasandra gymnocarpa* (Hillebr.) Sherff, described by Eyde & Tseng, and probably in

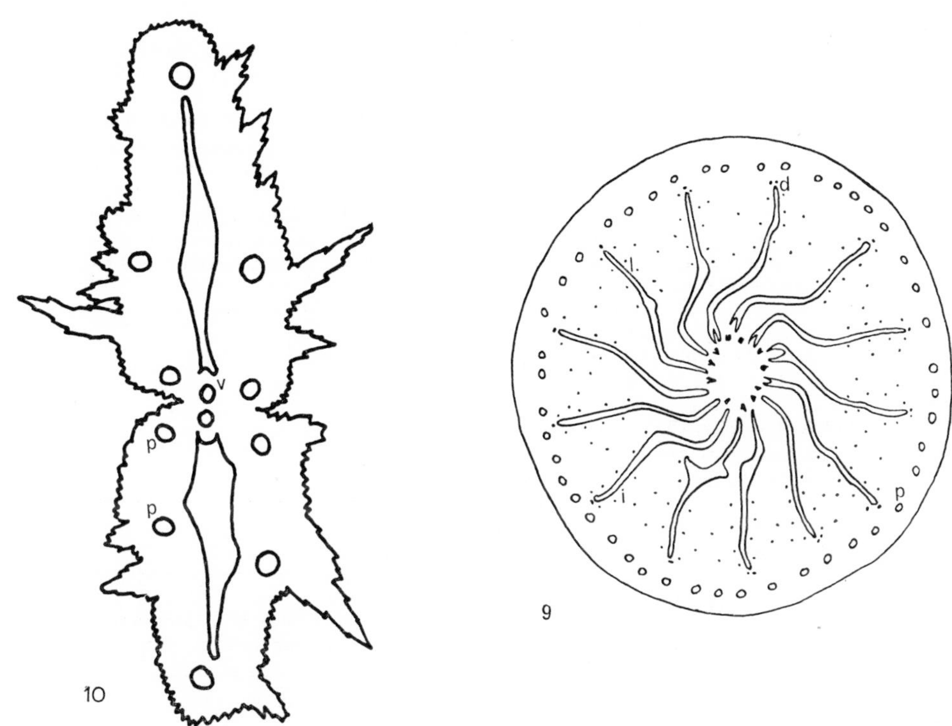

FIGURES 9 and 10. 9. *Plerandra stahliana*, T.S. ovary (×6). 10. *Harmsiopanax aculeata*, T.S. ovary (×55). d, Dorsal bundle; i, intermediate carpellary bundle; l, loculus; p, peripheral bundles; v, ventral bundle.

most Araliaceae in fact, the united ventrals stand between the loculi. In this respect, therefore, *Stilbocarpa* and especially *Harmsiopanax* approach the Umbelliferae. *Mackinlaya* and *Myodocarpus*, which approach the Umbelliferae very closely in other respects (Baumann, 1946; Philipson, 1951), also have a ventral vascular system similar to that of *Harmsiopanax*. In *Hedera* the bundles of the ventral system lie between the loculi for most of their course, but divide and reform opposite the locule before supplying the ovules (Plate 1**B**). In all genera investigated the ventrals divide to form two placental bundles, and usually also give off branches to the disk. This forking may take place some distance below the placentas, in which case there are approximately twice as many bundles in the ventral system as there are loculi, or sometimes several more. This feature is strongly developed in *Indokingia*.

PLEIOMERY OF THE ANDROECIUM

Two features of the Araliaceae which are variable, namely the number of stamens and inflorescence form, may now be considered more fully, especially in so far as they throw light on the nature of the flower in primitive members of this alliance.

The stamens are more numerous than the petals in the genera listed in Table 1.

Table 1. Genera with stamens more numerous than the petals

GROUP 1: leaves pinnately compound
(1) *Gastonia*; about six species from East Africa, Madagascar, the Seychelles, Mascarenes, Malaya Peninsula and Archipelago, Solomons. Most species with stamens equal to the petals, but others with up to six times as many
(2) *Tetraplasandra*; about 19 species, Hawaiian Islands (Malaysian records are considered to refer to *Gastonia*), 13 species have stamens more numerous than the petals
(3) *Peekeliopanax*; 1 species, New Guinea, Bismark Archipelago, Solomons. Stamens up to 70
(4) *Indokingia*; 1 species, Seychelles. Stamens very numerous (to 100)
(5) *Munroidendron*; 1 species, Hawaiian Islands. Doubtfully to be included. The corolla is partially calyptrate with at least as many ribs as stamens
Note. All the genera in group 1 are very closely allied. It is proposed to combine three genera (1, 3 and 4) into one, *Gastonia*
GROUP 2: leaves digitately compound or palmately lobed
(6) *Plerandra*; about 14 species, New Guinea, Solomons, Fiji. Stamens 15–500
(7) *Tupidanthus*; 1 species, Western Himalay, Yunnan, Upper Burma, Thailand. Stamens very numerous
(8) *Plerandropsis*; 1 species, Vietnam. Stamens numerous
(9) *Geopanax*; 1 species, Seychelles. Stamens 10, petals 5
(10) *Octotheca*; 1 species, New Caledonia. Stamens 15, petals 5
GROUP 3: leaf simple, elliptic, entire
(11) *Diplopanax*; 1 species, Kwangsi. Stamens 10, petals 5
GROUP 4: leaf generally palmate, inflorescence specialized
(12) *Boerlagiodendron*; about 40 species, Formosa, Philippines, Marianas and Carolines, Malaysia, Solomons, New Hebrides. Stamens equal to the petals or up to four times as many (30)

The sporadic occurrence of this character in so many genera of diverse affinity within the family has been thought to suggest the relict nature of the character, especially as so many of the genera are monotypic. However, the opposite conclusion is reached by Cronquist (1968) who regards the pentamerous flower, so characteristic of the whole order, as primitive. Eyde & Tseng (1969) regard pleiomery as a primitive character in the genus *Tetraplasandra*, in which the possibly advanced character of a superior ovary is associated with few and regular stamens. A choice between these diametrically opposed interpretations can be attempted with the aid of further evidence drawn from, (i) the vascular anatomy of the androecium, and (ii) the association of pleiomery of the androecium with other primitive or advanced features.

The proposition that flowers with numerous stamens are to be regarded as primitive is, of course, a very broad generalization which could not be held to apply to any particular group of plants without supporting evidence. Several examples of the multiplication of stamens during the course of evolution are generally accepted, so that the primitiveness of the pleiomerous araliaceous flower cannot be assumed. Flowers in which numerous stamens are a primitive feature might be expected to have the stamens arranged in a spiral or in a series of whorls. On the other hand, the arrangement of the vascular supply to the androecium might provide evidence for the secondary increase in the number of stamens.

The number of stamens present is exceedingly variable in some species. This can be seen in *Tetraplasandra* (Sherff, 1955) and *Plerandra* (Smith & Stone, 1968), and is striking in *Peekeliopanax spectabilis* (stamens 25–66) and *Gastonia papuana* Miq. (stamens 12–55). Some interesting comparisons of vascular systems can be made among the species of *Gastonia*. Flowers of *Gastonia cutispongia* with ten petals and ten stamens have approximately 20 peripheral bundles in the ovary wall, of which only ten contribute to the androecium. *G. seychellarum*, and small flowers of *G. papuana*, each with approximately 18 stamens and five or six petals, have about 20 peripheral bundles, and in these species all the peripheral bundles contribute to the androecium. That is to say there is an approximate correspondence between the number of stamens and the number of their traces in the ovary wall. In flowers of *G. papuana* with high pleiomery the number of peripheral bundles in the ovary wall is no greater, the numerous stamen-traces arising by branching at the top of the ovary. A similar relationship is found in flowers of *Peekeliopanax*. A small flower, with 37 stamens and ten petals, has 24 stamen traces and ten peripheral bundles (which also supply the androecium) that is, a total of 34 bundles to the stamens. However, in large flowers, a departure from this relationship is found. A flower with 57 stamens also has 24 stamen traces and 11 peripheral bundles, that is, a total of only 35 bundles supplied 57 stamens by branching at the summit of the ovary. A comparison of the vasculature of species of *Tetraplasandra* with few and many stamens also shows that the number of peripheral bundles does not equal the stamen traces once a moderate degree of pleiomery has been exceeded.

As might be expected the flowers of *Plerandra stahliana* Warb., with many more stamens, have a more complex vascular system. In a flower with approximately 270 stamens, the peripheral bundles more than double in number a short distance below the top of the ovary. But even then they are less than half the number of the stamens, and they also give rise to the traces of the calyx and corolla. Just below the level of insertion of the corolla and stamens, the peripheral bundles become joined into a complex ramifying collar surrounding the disk. Numerous small vascular bundles from the carpellary system enter the inner edge of this ring and therefore contribute to the vascular supply of the androecium. The presence of these bundles does not, however, alter the fact that the peripheral bundles initially are much fewer in number than the stamens and that much of the branching required to provide sufficient stamen traces occurs in the vascular collar.

There is also a difference in the arrangement of the stamens between *Peekeliopanax* and *Gastonia*, on the one hand, and *Plerandra* on the other. All species of the first pair of genera examined have their stamens inserted in a single ring (Plate 1**D**), or, in flowers with very high numbers of stamens, two adjacent filaments may overlap but their traces form a single ring. In *Plerandra stahliana*, however, the stamens are inserted in three, four or five series (Plate 1**F**). The figure of *P. grayi* Seem. given by Smith & Stone (1968) also shows the stamens in three series.

Any interpretation based on the above evidence can only be tentative especially as the evidence of vascular pattern and of the insertion of the filaments in series tend to contradict one another. The vasculature of the genera investigated, which belong to groups 1 and 2 of Table 1, suggests an original pleiomerous condition with the stamens in one whorl but more numerous than the petals. Species with very high numbers of stamens may have arisen by multiplication of primordia. Since these are the only species in this alliance with stamens inserted in more than one series, it would follow that this is a derived condition. On the other hand, if a spiral arrangement of many stamens is accepted as primitive, then high pleiomery of stamens would also be a primitive character. We may now consider the relationship between stamen number and form of inflorescence in the hope of resolving this problem.

THE FORM OF THE INFLORESCENCE

A range of inflorescence types occurs within the family. The umbel is characteristic of the Araliaceae, but it is not so prevalent as in the Umbelliferae. That family is considered to be the more advanced on several grounds, one of the strongest of which is the nature of their secondary xylem (Rodriguez, 1957). Consequently there are grounds for regarding the umbel as the advanced condition of flower arrangement within the Araliaceae. Genera which lack an umbellate inflorescence, or which include other arrangements, may be suspected of retaining a primitive feature. This conclusion can only be very tentative because inflorescence morphology is well known to be subject to variation.

A new subdivision of the family into tribes has recently appeared (Hutchinson, 1967). In this system the interest of the racemose floral arrangement in some genera has been recognized and this character has been the principal one used for the definition of the tribes. This dependence on a single inflorescence character has some unfortunate results. These arise partly because the presence or absence of racemose inflorescences is not at all so clear cut as is required of a key character. At least two of the tribes which are defined as having only umbellate inflorescence include genera with some species regularly racemose. The evolution of the umbel from a raceme is vividly illustrated by Hutchinson with reference to *Cuphocarpus*, in which the lower flowers are male and racemose while the upper and hermaphrodite flowers are umbellate. An inflorescence of this type is also characteristic of the genus *Plerandra*, and racemosely arranged flowers also occur in *Tetraplasandra* (even without *Dipanax*), although both these genera are placed in the Plerandreae. In the tribe Panaceae the genus *Schefflera* has several species with racemose inflorescences, as have some *Pseudopanax* species, at least on male trees. The converse is also true since some species of *Reynoldsia* have strictly umbellate inflorescences, though this genus is placed in the tribe Cussonieae. Moreover, reliance on the inflorescence has separated plants belonging to the same genus into distinct tribes, and has also resulted in the juxtaposition of unrelated genera. Two examples will suffice. *Anomopanax* is revived and placed in a distinct tribe from *Mackinlaya*, although these two groups of species share several characters unusual in the family (Philipson, 1951). The only other genus placed in the Anomopanaceae, *Aralidium*, is very distinctive, much further removed from *Anomopanax* than *Mackinlaya*, but with an inflorescence somewhat similar to that of *Anomopanax*. Secondly, *Tetraplasandra* is placed in the Plerandreae, whereas the very closely related *Munroidendron* and *Dipanax* are placed in the Cussonieae and in different subtribes. The equally closely related *Gastonia* is placed in the Panaceae. *Gastonia*, *Tetraplasandra* and *Dipanax*, which in my opinion are very closely related, if not congeneric, thus appear in three tribes when inflorescence characters are considered.

It is not possible to state accurately the number of genera in which a racemose inflorescence occurs but it is not likely to greatly exceed ten. The total number of genera in the family is probably not much more than 70, if that. That is to say, between one-seventh and one-sixth of the genera are completely or partly racemose. The genera with pleiomerous stamens listed in Table 1 show a higher proportion of racemose inflorescences. Four of the 12 genera, namely *Munroidendron*, *Tetraplasandra*, *Plerandra* and *Diplopanax*, have some or all of their species with racemose flower arrangement. This alone suggests an association between pleiomery of the androecium and a racemose inflorescence, and this conclusion appears stronger when it is realized that *Gastonia*, *Peekeliopanax* and *Indokingia* are either congeneric with *Tetraplasandra* or very close to it, and thus the list of genera might be reduced to nine, with four racemose representatives. It is important also to note that of the four groups of genera exhibiting pleiomery, three include members with racemose inflorescences. The exception is *Boerlagiodendron*, one of the most distinctive genera in the whole family with its pseudofruits, its tubular corolla and fringed petiole.

Since a racemose inflorescence is probably primitive within the Order, its correlation with another unrelated character, namely pleiomery of the androecium, may be considered evidence for the primitiveness of that character. Finally, it should be noted that the association of a racemose inflorescence is with regular and slight pleiomery in Group 3 (*Diplopanax*) as well as with the first two groups, in which all conditions of pleiomery occur. To this extent the evidence of the inflorescence would favour the view that high pleiomery such as is found in *Tupidanthus* and *Plerandra* is a derived condition. Evolution in *Gastonia* (with *Peekeliopanax*) and *Tetraplasandra* might, therefore, have resulted in both multiplication and reduction of stamen number from an original condition of moderate pleiomery.

If an attempt were to be made to visualize the ancestral type of the order Umbellales, it would include the characters of a racemose inflorescence and possibly a limited pleiomery as seen in *Diplopanax*. It may also be significant that the only species with perigynous flowers occur in a genus in which racemose inflorescences and pleiomery of the stamens are both frequent.

REFERENCES

BAUMANN, M. G., 1946. *Myodocarpus* und die Phylogenie der Umbelliferen-Frucht. *Ber. schweiz. bot. Ges.*, **56**: 13–112.

BAUMANN-BODENHEIM, M. G., 1955. Ableitung und Bau bicarpellat-monospermer und pseudomonocarpellater Araliaceen- und Umbelliferen-Früchte. *Ber. schweiz. bot. Ges.*, **65**: 481–510.

BEGHTEL, F. E., 1925. The embryology of *Pastinaca sativa*. *Am. J. Bot.*, **12**: 327–337.

BORCHERT, R., 1969. Unusual shoot growth pattern in a tropical tree, *Oreopanax* (Araliaceae). *Am. J. Bot.*, **56**: 1033–1041.

BORTHWICK, H. A., 1931. Development of the macrogametophyte and embryo of *Daucus carota*. *Bot. Gaz.*, **92**: 23–44.

Cronquist, A., 1968. *The evolution and classification of flowering plants.* Boston: Houghton Mifflin.
Davis, G. L., 1966. *Systematic embryology of the angiosperms.* New York: Wiley.
Eyde, R. H. & Tseng, C. C., 1969. Flower of *Tetraplasandra gymnocarpa.* Hypogyny with epigynous ancestry. *Science, N.Y.*, **166**: 506–508.
Grushvitsky, I. V., Tikhomirov, V. N., Aksenov, E. S. & Shibakina, G. V., 1969. Succulent fruit with carpophore in species of the genus *Stilbocarpa* Decne et Planch. (Araliaceae). [In Russian.] *Bull. Moscow Soc. Exp. nat. Sci.* (*Biol*)., **74**: 64–76.
Harms, H., 1898. Araliaceae. In Engler & Prantl, *Die natürlichen Pflanzenfamilien,* **3** (8). Leipzig: Engelmann.
Hegnauer, R., 1964. *Chemotaxonomie der Pflanzen.* **III.** *Dicotyledoneae: Acanthaceae—Cyrillaceae.* Basel: Birkhäuser.
Hutchinson, J., 1959. *The families of flowering plants.* **I.** *Dicotyledons,* 2nd ed. Oxford: Clarendon Press.
Hutchinson, J., 1967. *The genera of flowering plants.* **II.** *Dicotyledons.* Oxford: Clarendon Press.
Jurica, H. S., 1922. A morphological study of Umbelliferae. *Bot. Gaz.*, **74**: 292–307.
Philipson, W. R., 1951. Contributions to our knowledge of Old World Araliaceae. *Bull. Br. Mus. nat. Hist.* (*Bot.*), **1**: 1–20.
Philipson, W. R., 1967. *Griselinia* Forst. fil.—anomaly or link. *N.Z. Jl Bot.*, **5**: 134–165.
Philipson, W. R., 1970. Shoot differentiation in the Araliaceae of New Zealand. *J. Indian bot. Soc.*, **48**: (in press).
Pigott, E. M., 1914 [1915]. Notes on *Nothopanax arboreum* with some reference to the development of the gametophyte. *Trans. Proc. N.Z. Inst.*, **47**: 599–612.
Rodriguez, R. L., 1957. Systematic anatomical studies on *Myrrhidendron* and other woody Umbellales. *Univ. Calif. Publs Bot.*, **29**: 145–318.
Sherff, E. E., 1955. Revision of the Hawaiian members of the genus *Tetraplasandra* A. Gray. *Fieldiana* (*Bot.*), **29**: 49–142.
Singh, D., 1954. Floral morphology and embryology of *Hedera nepalensis* K. Koch. *Agra Univ. J. Res., Sci.*, **3**: 289–299.
Smith, A. C. & Stone, B. C., 1968. Studies of Pacific Island Plants. **XIX**. The Araliaceae of the New Hebrides, Fiji, Samoa, and Tonga. *J. Arnold Arbor.*, **49**: 431–493.

EXPLANATION OF PLATE

Plate 1

A. *Pseudopanax anomalum*, T.S. ovary (×50). Note peripheral system of ten bundles; v, ventral bundle.
B. *Hedera helix*, T.S. ovary (×25). Note five dorsal bundles in addition to ten peripherals.
C. *Arthrophyllum diversifolium*, T.S. ovary (×100); v, the two ventral bundles.
D, E. *Peekeliopanax spectabilis*. **D,** T.S. flower (×15) at level of insertion of the stamens. t, calyptrate corolla. **E,** T.S. ovary (×10); p, outer peripheral bundles; s, inner peripheral (staminal) bundles.
F. *Plerandra stahliana*, T.S. flower (×8) at level of insertion of the stamens. The filaments are in several series.
c, Carpellary bundles; d, dorsal bundles; f, filaments; i, intermediate carpellary bundles.

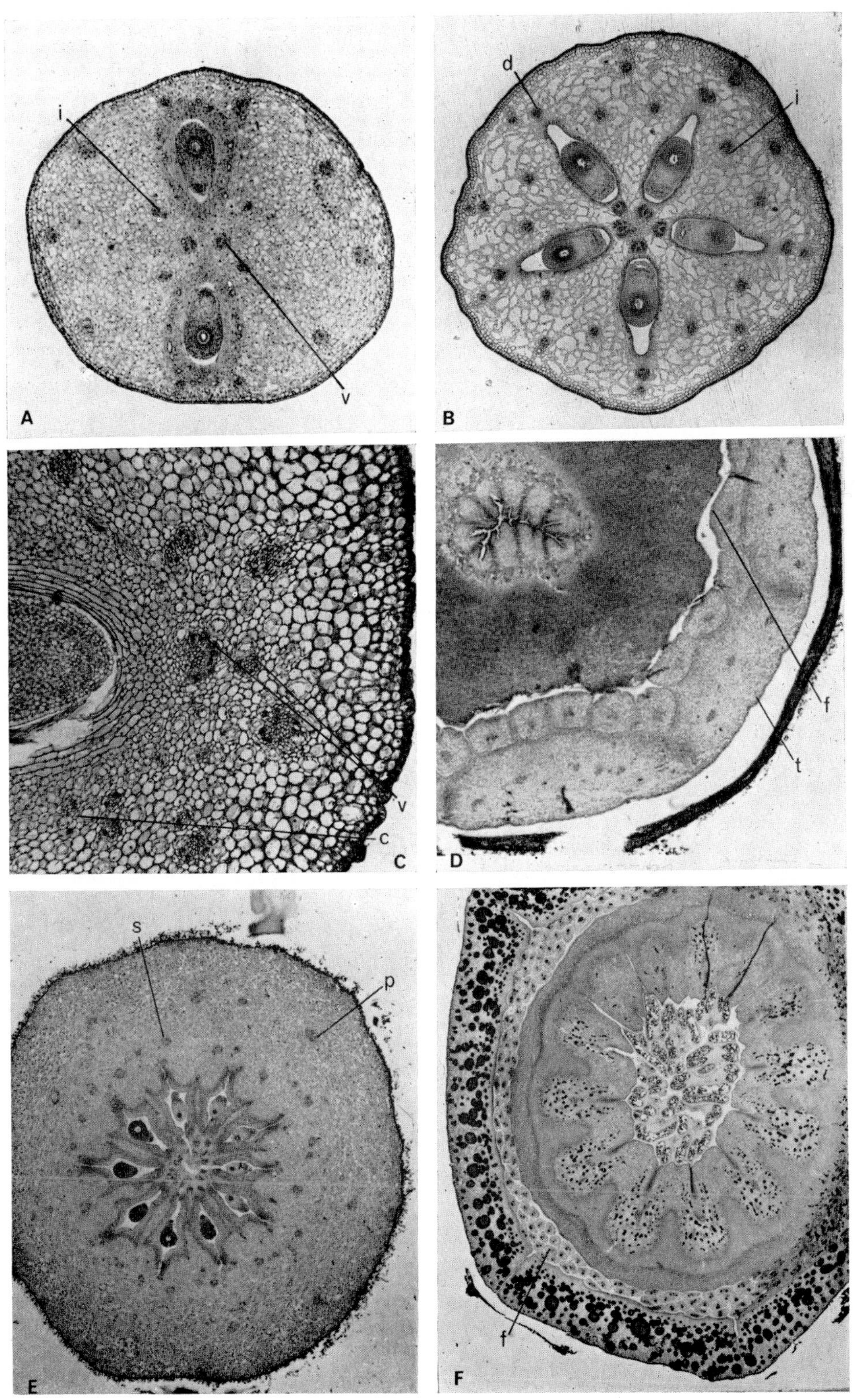

W. R. PHILIPSON

Comparative development and morphological interpretation of 'rachis-leaves' in Umbelliferae

DONALD R. KAPLAN

Department of Botany, University of California, Berkeley, California, U.S.A.

The foliage leaves of *Lilaeopsis occidentalis* and *Oxypolis greenmanii* are linear, unifacial appendages differentiated into a sheathing base and an upper leaf zone articulated into 'nodes' and 'internodes'. Leaves of both genera lack any obvious signs of pinnae at maturity, except for a series of reduced nodal appendages (hydathodes) inserted along the adaxial surface. The question of whether these leaves are equivalent to the petiole or rachis region hinges on the morphological interpretation of these nodal protuberances; therefore, a comparative histogenetic investigation was undertaken to elucidate their nature. Following initiation, leaves of both genera undergo a period of apical and intercalary growth in length and show an acropetal course of tissue maturation during this early growth phase. Although these leaves arise as dorsiventral primordia, they become unifacial through adaxial meristematic activity which leads to a rounding of the leaf axis. Comparison of histogenesis in these two genera with that of the pinnatifid leaf of *Carum carvi* indicates that the nodal appendages in these species of *Lilaeopsis* and *Oxypolis* arise in a manner identical to pinnae in *Carum* but soon become arrested and differentiated as hydathodes. The discovery of subunifacial to bifacial morphology in both juvenile and stem leaves of *Oxypolis* and the related lateral insertion of nodal appendages also supports their homology with pinnae. Furthermore, comparative developmental observations between adult and juvenile leaves of *Oxypolis* suggest that the median versus lateral position of their reduced leaflets is related to the timing of pinna initiation versus axis thickening: if leaflets are initiated before adaxial meristematic activity promotes axis thickening, then they are lateral in insertion and paired, whereas if they arise after rounding of the leaf axis has occurred, then the leaflets are single and median in insertion. It is concluded that the leaves of *Lilaeopsis* and *Oxypolis* are equivalent to the rachis of a pinnately compound leaf with the nodal appendages equivalent to reduced, transformed pinnae which function as hydathodes at maturity.

CONTENTS

INTRODUCTION

Within the vegetatively polymorphic family Umbelliferae, foliage leaves of *Lilaeopsis*, *Ottoa* and some species of *Oxypolis* are particularly distinctive and diverge morphologically from the pinnatifid appendages that characterize many genera in this family (Bitter, 1897; Drude, 1898; Domin, 1908; Troll, 1934*a*, *b*, 1939). Leaves in these three genera* are linear, terete appendages which are segmented into conspicuous 'nodes' and 'internodes' but without an obvious lamina at maturity. In their overall form and organization they resemble unifacial leaves characteristic of monocotyledons such as *Juncus*, *Allium* and *Ornithogalum*.

Survey of the literature of leaf morphology in Umbelliferae reveals a lack of agreement on the morphological interpretation and hence evolutionary origin of these divergent leaves in the family. As indicated by Briquet (1897), their proper interpretation has been inhibited by the lack of transition forms between these leaves and those of other members of the family. On the one hand, Goebel (1891), Bitter (1897), Drude (1898) and Rennert (1903) interpreted them as 'septate phyllodes', suggesting that they represented extensions of the petiolar region in which the lamina was reduced or even lost in evolution. On the other hand, Briquet (1897), Domin (1908), Gaisberg (1922) and Troll (1934*a*, *b*) suggested that they were equivalent to the entire rachis-axis of a compound leaf (hence the term 'Rhachisblätter') and supported this interpretation by demonstrating a correspondence between the positions of nodal septa and the points of pinna insertion in leaves of other Umbelliferae. They also indicated that because a nodal septum occurs at the juncture of the upper leaf axis and leaf base, a petiolar zone was absent. Goebel, in the various editions of his *Organographie der Pflanzen*, altered his original phyllode interpretation of the leaf of *Lilaeopsis* (Goebel, 1891). In the first edition he suggested that the adaxial nodal protuberances were aborted leaflets (Goebel, 1900). However, in the last edition he changed his mind and raised doubts as to their leaflet character, largely because of their aberrant median (adaxial) insertion (Goebel, 1933). Troll (1934*a*), however, was able to remove Goebel's objection by indicating the existence of median leaflets in some other species of Umbelliferae (e.g. *Cicuta virosa*) and correlating their adaxial position with the unifaciality of the subjacent rachis.

It is evident that despite the long history of morphological study, there have been no thorough developmental investigations of these distinctive foliage leaves. In this paper I present new comparative morphological and developmental evidence for leaves of *Lilaeopsis* and *Oxypolis*, which supports Troll's interpretation of the leaflet nature of these nodal appendages and hence of the rachis nature of these foliage leaves. Furthermore, these ontogenetic data are compared with similar information for the highly pinnatifid leaf of *Carum carvi*, in an effort to specify some of the developmental mechanisms responsible for the morphological divergence of 'Rhachisblätter' from compound leaves typical of members of this family.

Although Briquet (1897), Rennert (1903) and Gaisberg (1922) have described previously the anatomy of the leaves of *Lilaeopsis* and *Oxypolis*, a brief account of

* It should be noted that all the 20+ species of *Lilaeopsis* and the single species of *Ottoa* bear 'rachis-leaves'; in *Oxypolis*, on the other hand, of the seven component species, three have 'rachis-leaves' while the other four develop 'normal' pinnate leaf blades.

their mature morphology and anatomy will be given in this paper as background for the developmental information to be presented. However, since this paper describes only those anatomical features that are related to the organography of these leaves, the reader is referred to the above-cited studies for more detailed histological descriptions.

MATERIAL AND METHODS

Material of *Lilaeopsis occidentalis* Coult. & Rose, *Oxypolis greenmanii* Math. & Const. and *Carum carvi* L. was obtained from plants in cultivation in the University of California Botanical Garden, Berkeley. The sources of material and original collection sites are indicated in Table 1. Voucher specimens for each of these species have been deposited in the Herbarium of the University of California, Berkeley.

Table 1. Sources of material

Species	Locality	Collector
Lilaeopsis occidentalis Coult. & Rose	Bodega Bay, Sonoma Co., Calif., U.S.A.	L. Constance 3880
Oxypolis greenmanii Math. & Const.	Between Port St Joe and Wewahitchka, Gulf Co., Fla., U.S.A.	R. Godfrey and P. Redfearn 53756
Carum carvi L.	Botanical Garden, Gatersleben, W. Germany	Unknown

Buds of both lateral and terminal shoots of each species were dissected and fixed in CRAF III (Sass, 1958), dehydrated and embedded in paraffin by conventional procedures outlined elsewhere (Kaplan, 1970). Serial sections were cut 7–10 μm in thickness in both longitudinal and transverse planes and the sections stained with Heidenhain's iron-alum hematoxylin and counterstained with safranin and fast green. In addition, mature shoots and foliage leaves were fixed in FAA (50% ethanol) and cleared in 75% lactic acid for observation of leaf vasculature. In order to study the three-dimensional changes in leaf form during development, whole leaf primordia fixed in FAA were observed and photographed with a Leitz epi-illuminator.

In this study, leaf length is equated with developmental age and the length has been determined by measuring median longitudinal sections of primordia along a straight line parallel to the long axis of the leaf, from its axil to its apex. Measurements from transverse sections were made by the conventional method of counting the number of sections and multiplying by section thickness.

TERMINOLOGY

Because of the range of leaves exhibited by the shoot of *Oxypolis greenmanii*, it is useful, for purposes of description, to give specific designations to each type, depending upon its position within the total shoot system. It should be emphasized that, because of the gradual changes in morphology between serial organs of the shoot, delimitation of appendage types is somewhat arbitrary. The first series of reduced leaves of a lateral shoot, beginning with the prophyll, will be called *juvenile leaves*; the largest leaves of the shoot, which are the appendages of the vegetative rosette, will be designated *rosette* or *adult leaves*; and those borne on the elongated portion of the shoot will be referred to as *stem leaves*. Since the discussion of leaf morphology and development in *Lilaeopsis* will be restricted to leaves of the vegetative rosette and because there is far less variation in form among serial appendages of this genus, the above categories will refer only to leaves of the shoot of *Oxypolis*.

Use of the terms 'node' and 'internode' with reference to the leaves of *Lilaeopsis* and *Oxypolis* is purely descriptive and does not imply any comparison with stems. They refer only to the articulated character of these appendages and the use of quotation marks around these terms emphasizes their descriptive nature.

In the Observations section, the terms *nodal appendage*, *adaxial protuberance* and *adaxial* or *nodal hydathode* will be used interchangeably with reference to those median nodal protuberances of the leaves of *Lilaeopsis* and *Oxypolis*, which are interpreted in the Discussion section as aborted pinnae. It seemed best to use terms that were morphologically noncommittal in the Observations part of the text.

In all transverse sections of leaves illustrated, whether of primordia or of mature organs, the adaxial surface is orientated toward the bottom of the page.

OBSERVATIONS

Lilaeopsis occidentalis Coult. & Rose

General organography and morphology and anatomy of the mature leaf

Lilaeopsis occidentalis is a low, tufted, perennial herb which grows partially submerged in brackish waters and coastal mudflats in northern California and northwestern North America (Mathias & Constance, 1945). The genus is a member of tribe Ammineae of Drude (1898). Its shoot system consists of a series of plagiotropic runners with elongated internodes and scale leaves which, in turn, produce orthotropic, rosette shoots bearing linear, terete leaves in a distichous phyllotaxis.

The adult foliage leaves measure 11–17 cm in length and approximately 1·5 mm in diameter. Each is differentiated into a sheathing to stipular base (referred to subsequently as the lower leaf zone or 'Unterblatt' in the terminology of Eichler (1861)) which may comprise one-sixth of the total length, and a cylindrical upper leaf zone ('Oberblatt') segmented into 'nodes' and 'internodes' (Plate 1**A**). The 'nodes' appear as constrictions along the upper leaf axis (Plate 1**A, B**) and vary in number from 8 to 15 in the different leaves sampled, with 12 being the predominant value. The 'internodes' are hollow, and transverse nodal septa subdivide the upper leaf axis into a vertical series of internodal chambers.

Small, rounded protuberances occur at the adaxial surface of each of the eight to nine basal 'nodes' (in leaves with at least 12 nodes; arrow, Plate 1**B**). Histological sections indicate that these protuberances are hydathodes and possess characteristic epithem tissue and epidermal water pores, and are vasculated by numerous tracheary elements from the girdling commissural vein at the node (Plate 1**F**). A larger, but less protuberant hydathode is located toward the adaxial surface of the leaf apex and is vasculated by the conjunction of longitudinal bundles at the leaf tip.

Transverse sections taken at various levels along the length of the upper leaf axis reveal the radial symmetry and unifacial character of that sector, i.e. the longitudinal vascular strands are radial in transectional disposition with the xylem pole of each collateral vascular bundle oriented towards the centre of the leaf axis (Fig. 1**E**–**G**). The peripheral photosynthetic tissue is also radial in distribution and consists of hypodermal palisade tissue two to three cells in depth.

Transverse sections through the base of the upper leaf zone demonstrate the transition to the bifacial sheathing base and the gradual expression of the adaxial surface, beginning as a small furrow (Fig. 1**H**) and opening out as a broadened surface in the basal regions of the lower leaf zone (Fig. 1**I, J**). Related to expansion of the adaxial surface is a reorientation of the vasculature so that the xylem pole of each strand faces the adaxial or upper surface of the leaf (Fig. 1**J**).

In overall pattern, vasculature of the leaf is a longitudinally striate system, consisting of individual strands, some of which run the length of the leaf but which anastomose only in the region of the terminal hydathode. These longitudinal bundles are united laterally only at the 'nodes', where they are interconnected by a ring-like commissural bundle. This bundle has been called an 'anastomosing girdle' ('Anastomosengürtel') by Troll (1934*b*) and traverses the periphery of each nodal septum (Plates 1**F** and 4**A**).

A peculiar feature of leaf vasculature in this species is the change in number of vascular bundles along the length of the leaf axis. For example, transverse sections through the three or four distal 'internodes' reveal six bundles: a dorsal median bundle is located at the abaxial side of the leaf (DM, Fig. 1**E**), and what appears to be a ventral median bundle directly opposite to it on the adaxial side (VM, Fig. 1**E**), as well as a pair of lateral strands situated between these two median bundles to the left and right sides of the leaf (L1, L2 and L1′, L2′ respectively, Fig. 1**E**). In the fifth or sixth 'internodes' from the tip, seven bundles are evident with *two* adaxial, lateral bundles (L3–L3′, Fig. 1**F**) in place of the single ventral median strand. Since these two adaxial bundles show no sign of union to form the ventral median vein, but are merely separate strands which merge into the anastomosing bundle in the node above, it is not clear whether the ventral median bundle is a fusion product of strands L3–L3′ or simply a continuation of one of these strands into the distal 'internodes' of the leaf. Observations on cleared leaves suggest the latter is true; therefore the vasculature of this leaf does not constitute a 'closed system' characteristic of other unifacial leaf types (Troll, 1939).

Sections through the more basal 'internodes' (Fig. 1**G**) show a marked reduction in the number of vascular strands from seven to three. The dorsal median and lateral vein pair, L2–L2′, continue through the basal part of the 'Oberblatt' (Fig. 1**G**) and leaf base (Fig. 1**H**–**J**); however, the xylem and phloem components of bundle pairs L1–L1′ and L3–L3′ terminate at higher levels in the leaf.

It is interesting to note that each vascular bundle has a duct located adjacent to the phloem which seems to accompany the bundle throughout its length. These ducts appear to run the length of the leaf, and where vascular tissue is not differentiated in a given 'internode', its position is marked by the secretory duct. Therefore, while there are only three vascular bundles in the leaf base, there is a larger number of ducts

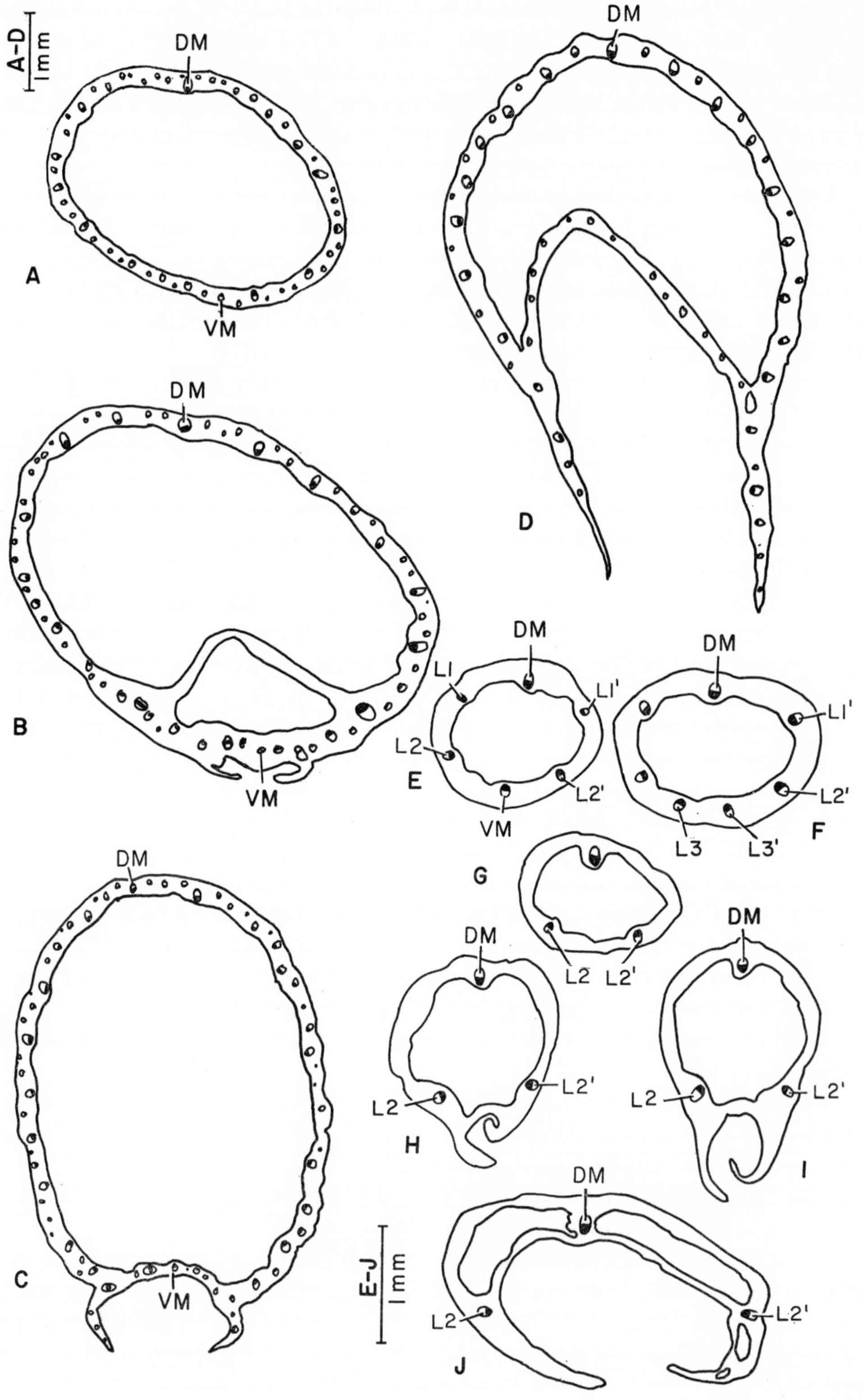
A-D
1mm
DM
A
VM
DM
D
DM
B
VM
DM
L1
L1'
L2
E
L2'
VM
DM
L1'
L2'
F
L3
L3'
DM
C
VM
G
L2
L2'
DM
L2
L2'
H
DM
L2
L2'
I
E-J
1mm
DM
L2
L2'
J

FIGURE 1

occupying the positions of veins present only in the distal region of the leaf (the duct positions are not illustrated in Fig. 1**E–J**).

I am not sure that the ducts are procambial in origin, nor have I studied the relative timing of differentiation of ducts and provascular tissue. All that can be said at present is that the ducts seem to arise from densely cytoplasmic tissue which is clearly demarcated from ground tissues but at the time of origin cannot be distinguished easily from procambium that will differentiate into conducting elements. Transverse sections of procambial strands along the length of leaf primordia show a reduction in diameter from apex to base undoubtedly related to their transition to the duct region (for example, follow bundle L1 from Fig. 3**F, H, I**), but in the basal regions of the procambium it is impossible to decide whether the strand is small because of its incipient character or its limitation to duct initiation. These observations underline the necessity for anatomical observations along the entire length of the leaf before morphological conclusions based on vascular organization are drawn.

Leaf development

Leaf primordia in *Lilaeopsis* are initiated in a distichous phyllotaxis, from nearly opposite positions at the shoot apical meristem. From the time of its initiation, growth of the leaf is erect (Plate 3**A, B**). Apical as well as intercalary meristematic activity are responsible for early longitudinal growth (Plate 3**A–C**). Tissue maturation begins on the abaxial side when the leaf is only 50 μm long (Plate 3**A**) and proceeds acropetally during subsequent development (Plate 3**B, C**). Coupled with precocious vacuolation of abaxial tissues is greater elongation of cells of that surface than those at the adaxial side (Plate 3**B**); however, abaxial elongation is balanced by more rapid division on the ventral side; and therefore, the erect growth of the primordium is maintained (Plate 3**B, C**). A similar but less rapid acropetal wave of tissue maturation becomes evident on the adaxial side when the leaf is approximately 150 μm in length (Plate 3**B**), but it does not extend to the tip until after the cessation of apical meristematic activity.

Apical meristematic activity is fairly prolonged, but comes to a close when the leaf is 1000–1500 μm long. Cessation of apical growth is marked by differentiation of subapical cells as the epithem of the terminal hydathode. From that point on, longitudinal growth is strictly intercalary (Plate 3**E, F**).

The sheathing base is initiated around the shoot apex and encircles the axis by the time the leaf is 400–500 μm long (Plate 2**A**). Meristematic activity in the wings of the sheath brings its margins into contact and then into an overlapping configuration (Plate 2**C**). Marginal meristematic activity responsible for extension of the wings (Fig. 2**E, N**) is identical with that described for the leaf of *Acorus calamus* (Kaplan, 1970): accentuated division of marginal initials initiates and extends the biseriate protodermal wing (MI, Plate 4**A**). Submarginal initial activity is reduced by comparison and builds up the central tissues of the wing, some distance from the margin.

FIGURE 1. Diagrams in T.S. taken along the length of mature leaves of *Oxypolis greenmanii* and *Lilaeopsis occidentalis*. **A–D,** T.S. of the leaf of *Oxypolis greenmanii*. **E–J,** T.S. of a leaf of *Lilaeopsis occidentalis*. Phloem—unshaded; xylem—black; DM, dorsal median bundle; L1–L3, L1′–L3′, lateral bundles at the left and right sides of the leaf, respectively; VM, ventral median bundle.

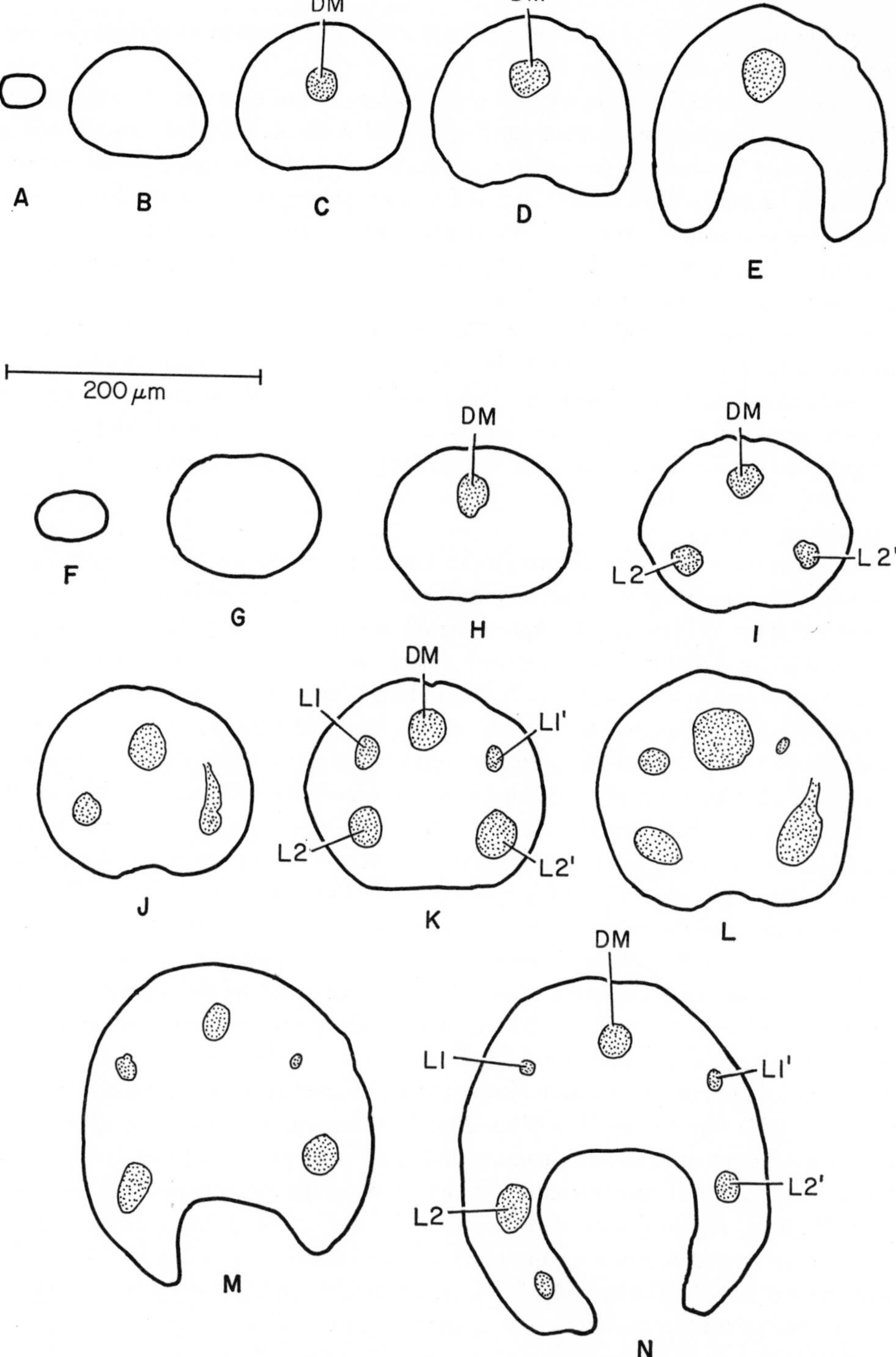

FIGURE 2. T.S. through leaf primordia of *Lilaeopsis occidentalis*. **A–E,** T.S. through a leaf primordium 190 μm in length at the following distances below the tip: **A,** 10 μm; **B,** 50 μm; **C,** 100 μm; **D,** 140 μm; **E,** 190 μm. **F–N,** T.S. through a leaf primordium 440 μm in length at the following distances below the tip: **F,** 10 μm; **G,** 50 μm; **H,** 100 μm; **I,** 150 μm; **J,** 200 μm; **K,** 250 μm; **L,** 300 μm; **M,** 350 μm; **N,** 400 μm. Procambial tissue—stippled; DM, dorsal median bundle; L1–L2, L1′–L2′, lateral bundles at left and right sides of the leaf, respectively.

The young primordium exhibits radial growth in thickness by means of an adaxial meristem. Repeated periclinal division of adaxial hypodermal lineages forms short, radially aligned cell files characteristic of this type of polarized meristematic activity (Plate 3**D**) and causes the lowermost region of the upper leaf zone to extend slightly in a radial direction (Plate 3**B**). Transectional series along the length of successively older primordia indicate that the leaf is adaxially flattened and dorsiventral in its earliest stages of development (Fig. 2**A**–**E**; Plate 3**D**). Its final terete outline is due in part to rapid division of cells of the adaxial meristem (AdM, Plate 3**D, G**) which leads to a rounding out of the adaxial surface of the leaf axis. Adaxial meristematic activity is reduced at each 'node' and accounts for the slight indentation above and subtending each nodal hydathode (Figs 2**J** and 3**D, E, G**; Plate 1**B**).

Plate 3**G** is a transverse section of a leaf 400 μm in length, cut through the boundary between the unifacial upper leaf and the bifacial sheathing base. Although extension of sheath wings is evident at this level, remnants of adaxial meristematic activity are still apparent in the form of radially-aligned cell files at the adaxial side (AdM, Plate 3**G**). In the sheathing base, the balance between adaxial thickening and marginal growth is shifted in favour of the latter, whereas in the unifacial upper leaf zone the reciprocal relationship occurs.

In addition to adaxial meristematic activity, the final terete form of the upper leaf axis is due also to division and enlargement of central 'pith' cells (Plate 4**A**). Cells at the periphery of the primordium accommodate to swelling of the 'pith' by dividing in an anticlinal plane and thereby maintaining their small, isodiametric form (Plate 4**A**). These peripheral hypodermal layers ultimately differentiate into the photosynthetic and procambial tissue of the leaf axis.

The course of procambial differentiation in the leaf is correlated with the patterns of apical and radial meristematic activity as well as the general course of tissue maturation. Differentiation of the major procambial strands (DM and L2–L2′ (Fig. 1**E**–**J**)) is continuous and acropetal. The first strand to differentiate is the dorsal median bundle at the abaxial side of the leaf. It becomes evident when the leaf is less than 100 μm in length and differentiates acropetally during the apical growth phase (Fig. 2**C**–**E**; Plate 3**B, C**). The major lateral vein pair (L2–L2′) appears next, followed in turn by pair L1–L1′ intercalated between L2–L2′ and the DM (Fig. 2**H**–**N**). Because of the aforementioned complications in determining the actual status of procambial and duct differentiation, it is not possible, without more detailed study, to determine the sequence or direction of vein inception for the other vascular strands. It appears that differentiation of the lateral vein pair L3–L3′ follows that of L1–L1′ and then any additional minor bundles are intercalated between the pre-existing veins as the intercostal panels are expanded (Fig. 3**A**–**I**), but this has not been followed in detail in this investigation.

When the leaf is approximately 200–250 μm long, localized periclinal divisions occur in hypodermal layers at the adaxial surface of the primordium (Plate 3**B**). With continued division these centres become prominent adaxial protuberances which resemble leaflet primordia in their histogenesis and appearance (NA, Plate 3**C, E**). Initiation of these emergences from derivatives of the adaxial meristem signifies the cessation of adaxial meristematic activity in the leaf axis.

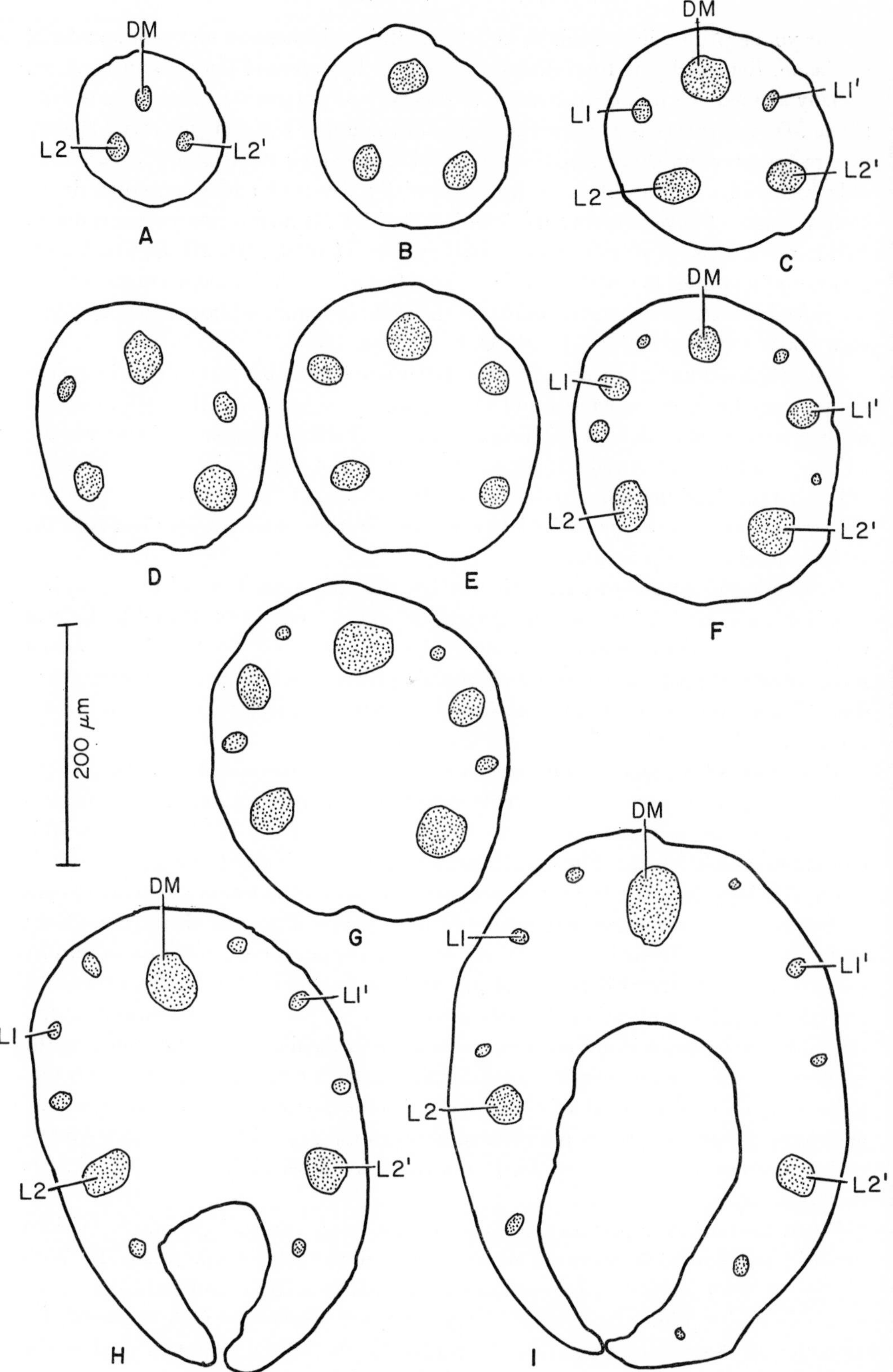

DM
L2
L2'
A
B
DM
L1
L1'
L2
L2'
C
D
E
DM
L1
L1'
L2
L2'
F
200 μm
G
DM
L1'
L1
L2
L2'
H
DM
L1
L1'
L2
L2'
I

FIGURE 3

As the leaf axis elongates these protuberances continue to be initiated in a close acropetal succession until a total of from six to nine have originated (Plate 3**C**, **E**). In whole primordia the appendages first appear as a series of transverse ridges (Plate 2**A**), which later extend upward as prominent appendages (Plate 2**B**, **C**) and ultimately mature as the nodal hydathodes of the adult leaf (Plate 1**B**).

Initiation and differentiation of nodal diaphragms occur at the same time as initiation of nodal appendages. These septa also arise in an acropetal sequence and are first evident when the leaf is 200–250 μm in length (NS, Plate 3**B**). Diaphragm initiation is largely through differential cell enlargement: cells that will form the internodes increase in diameter and become conspicuously vacuolate, whereas cells that will form a septum remain small in volume and may even become further subdivided during the later development of the diaphragm (Plate 3**B**, **C**, **E**). The most distal septa are initiated without hydathodes (Plate 3**E**, **F**), and this accounts for the lack of appendages at the most apical 'nodes' of the mature leaf.

Following enlargement and vacuolation, central 'pith' cells of each 'internode' elongate anticlinally and divide repeatedly forming anticlinal cell files resembling a rib meristem (Plate 3**C**, **E**). This division is responsible for intercalary growth of the primordium following differentiation of the apex. Cells which will form the peripheral chlorenchyma divide for a longer time than those of the pith; hence when internodal stretching commences, pith cells are torn apart, forming the internodal cavities of the mature leaf (Plate 3**F**).

Internodal extension, marking the final phases of leaf growth, begins at the apex and proceeds basipetally (Plate 3**F**). Because cell division ceases first in the distal regions of the leaf and is protracted at the base, the lower 'internodes' elongate more than those at the tip (Plate 1**A**).

Oxypolis greenmanii Math. & Const.

General organography of the shoot

This species is a member of tribe Peucedaneae (Drude, 1898) and, although not closely related systematically, morphologically it is essentially a gigantic form of *Lilaeopsis*. It grows in similar aquatic habitats but apparently is geographically restricted to cypress swamps in western Florida (Godfrey & Kral, 1958). Its shoot consists of a plagiotropic rhizome which produces erect lateral shoot rosettes bearing linear leaves in a distichous phyllotaxis. With the onset of flowering each orthotropic shoot exhibits considerable internodal elongation as well as a reduction in stem diameter. Flowering shoots of plants under cultivation at Berkeley reached a length of 200–230 cm and perhaps are even taller in nature. Correlated with axis elongation is a gradual change from an alternate to a helical phyllotaxis as well as a reduction in leaf length in transition to the flowering region. The orthotropic shoots terminate in a compound umbel and equivalent umbels may develop in the axils of the penultimate and even lower leaves of each axis.

FIGURE 3. T.S. through a leaf primordium of *Lilaeopsis occidentalis* 960 μm in length at the following distances below the tip: **A,** 100 μm; **B,** 200 μm; **C,** 300 μm; **D,** 400 μm; **E,** 500 μm; **F,** 600 μm; **G,** 700 μm; **H,** 800 μm; **I,** 900 μm. Procambial tissue—stippled; DM, dorsal median bundle; L1–L2, L1′–L2′, lateral bundles at left and right sides of the leaf, respectively.

Morphology and anatomy of the basal, rosette leaves

The adult rosette leaves are similar in form but considerably larger than those of *Lilaeopsis* (Plate 1**C**). They measure 36–44 cm in length and 0·9–1·3 cm in width at the region of maximal diameter (Plate 1**C**). The lower leaf zone measures 14–16 cm in length and may comprise up to one-third of the total length. Typically, the margins of the sheathing base are united above its point of insertion on the leaf axis and, therefore, the sheath has the form of a median stipule (Plate 1**C**). I have observed numerous examples where the margins of the sheath were not united in stipular form; nonetheless, the vascular pattern in the sheath and the occurrence of nodal septa below the upper limit of the leaf base suggest the existence of median stipule formation in the leaf of *Oxypolis*.

Like the leaf of *Lilaeopsis*, the upper leaf of *Oxypolis* is segmented into 'nodes' and 'internodes' (Plate 1**C**). Because of the abundance of hypodermal sclerenchyma and hence greater rigidity of the leaf axis, the nodal regions are evident only as slight indentations, without the markedly constricted character of the 'nodes' of *Lilaeopsis* (Plate 1, cf. **C, D** with **A, B**). As in *Lilaeopsis*, small appendages are located on the adaxial surface of each node. They are triangular in shape and are subtended by a deep, pit-like depression (Plate 1**D, E**). Histological examination indicates that these appendages also are hydathodes (Plate 1**E**) and are vasculated by a strand from the anastomosing vein at each node. Except for the septa below the level of the sheath, hydathodes occur at every nodal position along the upper leaf axis, including that at the juncture of the upper and lower leaf zones (Plate 1**C**).

Transverse sections taken at different levels of the leaf of *Oxypolis* demonstrate the unifacial symmetry of the upper leaf (Fig. 1**A**) and the bifacial character of the leaf base (Fig. 1**B–D**). The vasculature of this leaf is similarly striate in pattern, but in contrast to the more diminutive leaf of *Lilaeopsis*, exhibits a far greater number of vascular bundles in a given section (Fig. 1, cf. **A** with **E**). Despite its greater complexity, the vasculature can be resolved into the same components described for the leaf of *Lilaeopsis*. A dorsal median vein is evident at the abaxial side of the leaf and runs the length of the leaf (DM, Fig. 1**A–D**). In addition, there is a ventral median strand opposite to it (VM, Fig. 1**A–C**), which arises by union of the two most marginal minor veins at the adaxial surface of the leaf. Between these two median strands is a series of 14 or more major lateral bundles, and between these an even greater number of secondary and tertiary lateral bundles traverse the leaf axis (Fig. 1**A–D**). In spite of the greater number of vascular strands, lateral union of these bundles occurs only at the nodal septa.

Transverse sections through the tip of the median stipule reveal its structural complexity. The sheathing base has its own series of vascular strands (Fig. 1**B, C**), which eventually anastomose with those of the dorsal half of the leaf axis in the most basal part of the leaf. This means that a ring of vascular bundles is evident in the lower leaf zone as well (Fig. 1**D**). Only in the most basal regions of the leaf does the adaxial stipular portion approximate the abaxial side and close off the central cavity with a concomitant union of vascular strands of the abaxial and adaxial sides of the leaf. With closure of the central 'pith' cavity, the leaf base exhibits a more conventional bifacial vascular configuration.

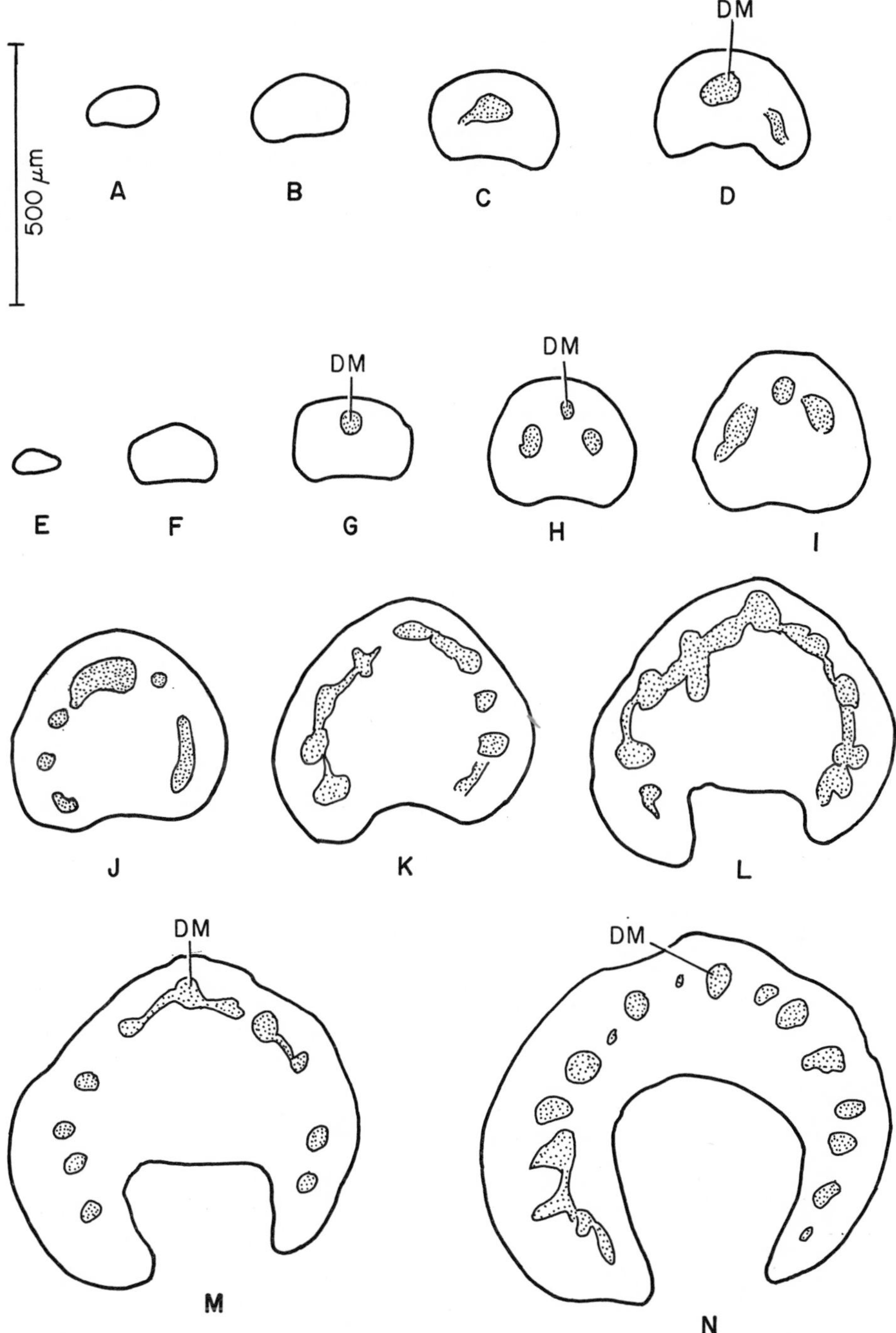

FIGURE 4. T.S. through leaf primordia of *Oxypolis greenmanii*. **A–D,** T.S. through a leaf primordium 140 μm in length at the following distances below the tip: **A,** 10 μm; **B,** 40 μm; **C,** 80 μm; **D,** 120 μm. **E–N,** T.S. through a leaf primordium 450 μm in length at the following distances below the tip: **E,** 10 μm; **F,** 50 μm; **G,** 100 μm; **H,** 150 μm; **I,** 200 μm; **J,** 250 μm; **K,** 300 μm; **L,** 350 μm; **M,** 400 μm; **N,** 450 μm. Procambial tissue—stippled; DM, dorsal median bundle.

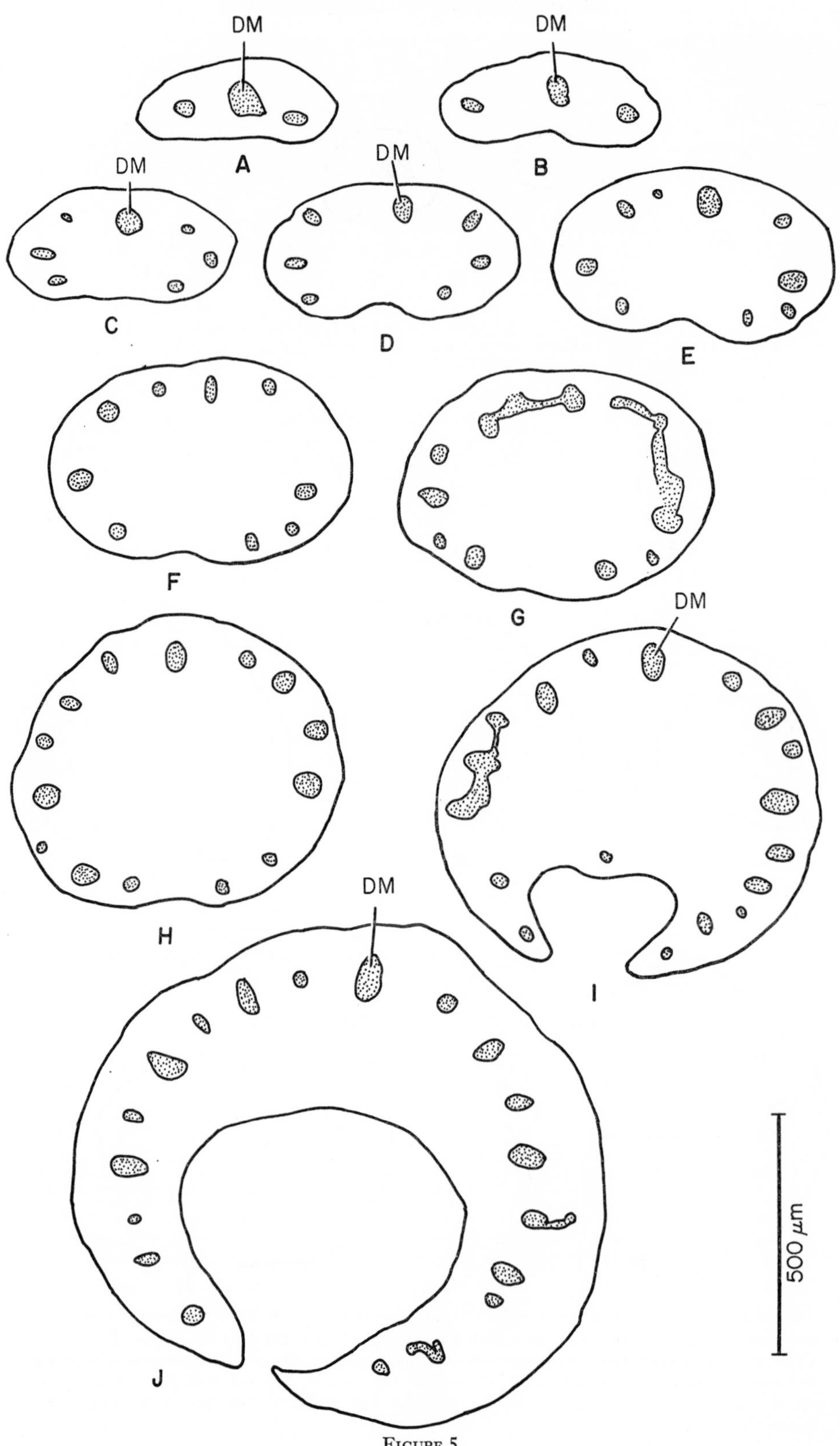

FIGURE 5

While the rosette leaves of *Oxypolis* are characteristically unifacial for most of the length of the upper leaf zone, the uppermost 'internodes' are usually flattened (Plate 1**C**) with a detectable adaxial indentation (similar to that shown for the juvenile leaf in Plate 2**G**). Transverse sections cut through the tip substantiate this bifaciality. A frequently encountered characteristic in the bifacial tip is the occurrence of pairs of appendages at the uppermost nodes (Plate 2**D, G**).

Development of the rosette leaf

While leaf development in *Oxypolis* proceeds in a manner similar to that of *Lilaeopsis*, there are certain features which are distinctive and important to note.

Following its initiation from the apical meristem, growth of the primordium is essentially erect. As in *Lilaeopsis*, accelerated maturation results in greater cell enlargement and vacuolation on the abaxial than adaxial surface of the primordium, but more rapid cell division on the latter side nearly balances cell extension on the dorsal surface so that the leaf does not exhibit a marked hyponastic curvature (Plate 4**B–D**). General maturation proceeds in an acropetal direction and concentrated meristematic activity at the tip of the primordium results in extension and attenuation of the apex as a marked apical pointlet, clearly set-off from the base (Plate 4**C–E**).

The most dramatic histogenetic contrast is the earlier and more conspicuous expression of adaxial meristematic activity. Periclinal division of hypodermal cell lineages forms radial cell files which are responsible for the marked protrusion of the ventral surface when the primordium is only 130–200 μm in length (Plate 4**C**). Continued rapid division in cells and derivatives of this meristem results in an even more accentuated ventral prominence in subsequent stages of development (Plate 4**D, E**).

Transverse sections through leaf primordia at successive developmental stages indicate that the leaf of *Oxypolis*, like that of *Lilaeopsis*, is initiated as a dorsiventral primordium with a marked adaxial furrow (Fig. 4**A–D**). Circumferential initiation and marginal meristematic activity in the leaf base are responsible for the bifacial character of that region (Fig. 4**D, K–N**). Adaxial meristematic activity coupled with central expansion of pith cells promote the rounding of the originally flattened primordium (Plate 5**B–D**; Figs 4**H–J** and 5**F–H**). Adaxial meristematic activity is reduced just above each nodal appendage and this depressed region becomes the adaxial pit subtending each nodal hydathode (Fig. 5**F, H**; Plate 5**C**, arrow). It is also of interest to note that the uppermost region of the leaf remains relatively flat due to an almost complete absence of adaxial growth in that region. This fact can be observed in both longitudinal sections (Plates 4**F** and 5**A**) as well as in transectional series through slightly older primordia (Fig. 5**A–D**) and forecasts the bifacial character of that zone in the mature leaf. Adaxial meristematic activity in the lower regions of the leaf axis continues until the leaf is approximately 900–1000 μm long. It is brought to a close by the initiation of nodal appendages from derivatives of that meristem along the entire adaxial face of the primordium (Plate 4**E, F**).

FIGURE 5. T.S. through a leaf of *Oxypolis greenmanii*, 2040 μm in length, at the following distances below the tip: **A,** 200 μm; **B,** 400 μm; **C,** 600 μm; **D,** 800 μm; **E,** 1000 μm; **F,** 1200 μm; **G,** 1400 μm; **H,** 1600 μm; **I,** 1800 μm; **J,** 2000 μm. Procambial tissue—stippled; DM, dorsal median bundle.

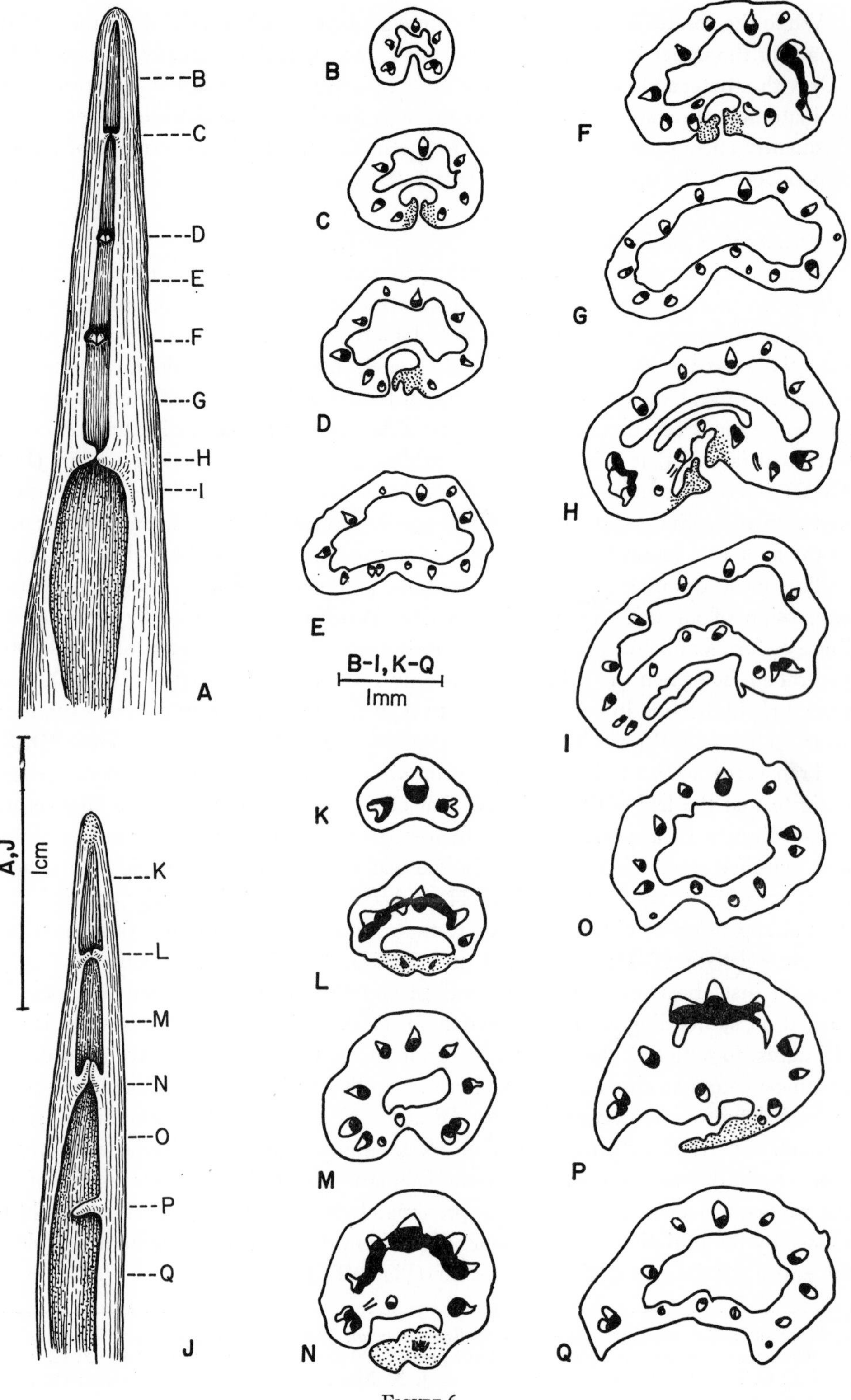

FIGURE 6

The nodal hydathodes in *Oxypolis* also arise in an acropetal sequence from the adaxial surface of the leaf; they are, however, initiated later in development than in leaves of *Lilaeopsis* and arise when the leaf axis is 800–1000 μm long (Plate **4E**). Because of their delayed initiation there is a marked disjunction between the time of septal differentiation and hydathode inception: septa differentiate in an acropetal succession in conjunction with the early phase of apical growth, but they first appear when the leaf is 250–300 μm in length, long before the hydathodes are evident protuberances. Plate **4D** illustrates a primordium 590 μm in length with recently initiated septa (arrows). It is evident that the more primordial septa are less clearly defined toward the apex and that they also become differentiated through differential cell enlargement (Plate **4D, E**). Because of continued adaxial meristematic activity the septa also are less distinct toward the adaxial surface of the leaf (Plate **4D**); however, following hydathode initiation the septa become differentiated across the diameter of the leaf axis (Plate **4E, F**).

Cessation of apical meristematic activity occurs when the leaf is about 1500–1700 μm in length and is signified by differentiation of the terminal hydathode and the most apical septum (Plate **4F**). Following expansion of the apical gland, hydathode enlargement (not initiation!) occurs in a basipetal direction correlated with the later phase of leaf extension. Plate **2D** illustrates the more advanced status of apical hydathodes in a whole leaf 2·3 mm in length. It should be noted that the most distal nodal appendages (arrows, Plate **2D**) are bilobed and that the apical part of the leaf is still somewhat cleft adaxially.

Internodal extension also follows a basipetal course with the uppermost 'internode' manifesting precocious extension when the leaf is about 2 mm long (Plate **4F**), and this is followed in sequence by extension of subjacent 'internodes' (Plate **5A**). As in *Lilaeopsis*, the internodal cavities arise rhexigenously by the breakdown of central pith cells. The most basal 'internodes' are the last to extend (Plate **5A**) and they undergo the greatest growth in length (Plate **1C**).

Provascular differentiation is essentially the same as in *Lilaeopsis* with the dorsal median bundle differentiating first (Plate **4B, C**; Fig. **4C, D**), followed in turn by the more adaxial, lateral bundles (Figs **4G–N** and **5A–J**). However, because of the greater number of vascular strands and the complications involved in the study of their histogenesis, this aspect was not investigated in detail for the leaf of *Oxypolis*.

Comparative morphology of juvenile and stem leaves

Juvenile leaves of an orthotropic lateral shoot differ from the rosette type in the following characteristics: (1) they are shorter, with a much greater proportion of their total length represented in the lower leaf zone; (2) they have a subunifacial to bifacial upper leaf section, and where the leaf axis is bifacial, the nodal appendages are paired. For example, Plate **2E** illustrates a juvenile leaf 10 mm long with a stipular sheath

FIGURE 6. Drawings and accompanying T.S. of juvenile and stem leaves of *Oxypolis greenmanii*. **A,** Distal two-thirds of a juvenile leaf. **B–I,** T.S. of the juvenile leaf illustrated in **A,** taken at the levels indicated for that leaf in **A. J,** Distal two-thirds of a stem leaf. **K–Q,** T.S. of the stem leaf illustrated in **J,** taken at the levels illustrated in that figure. Phloem tissue—unshaded; xylem tissue—black; tissue of lateral and adaxial protuberances (pinnae)—stippled.

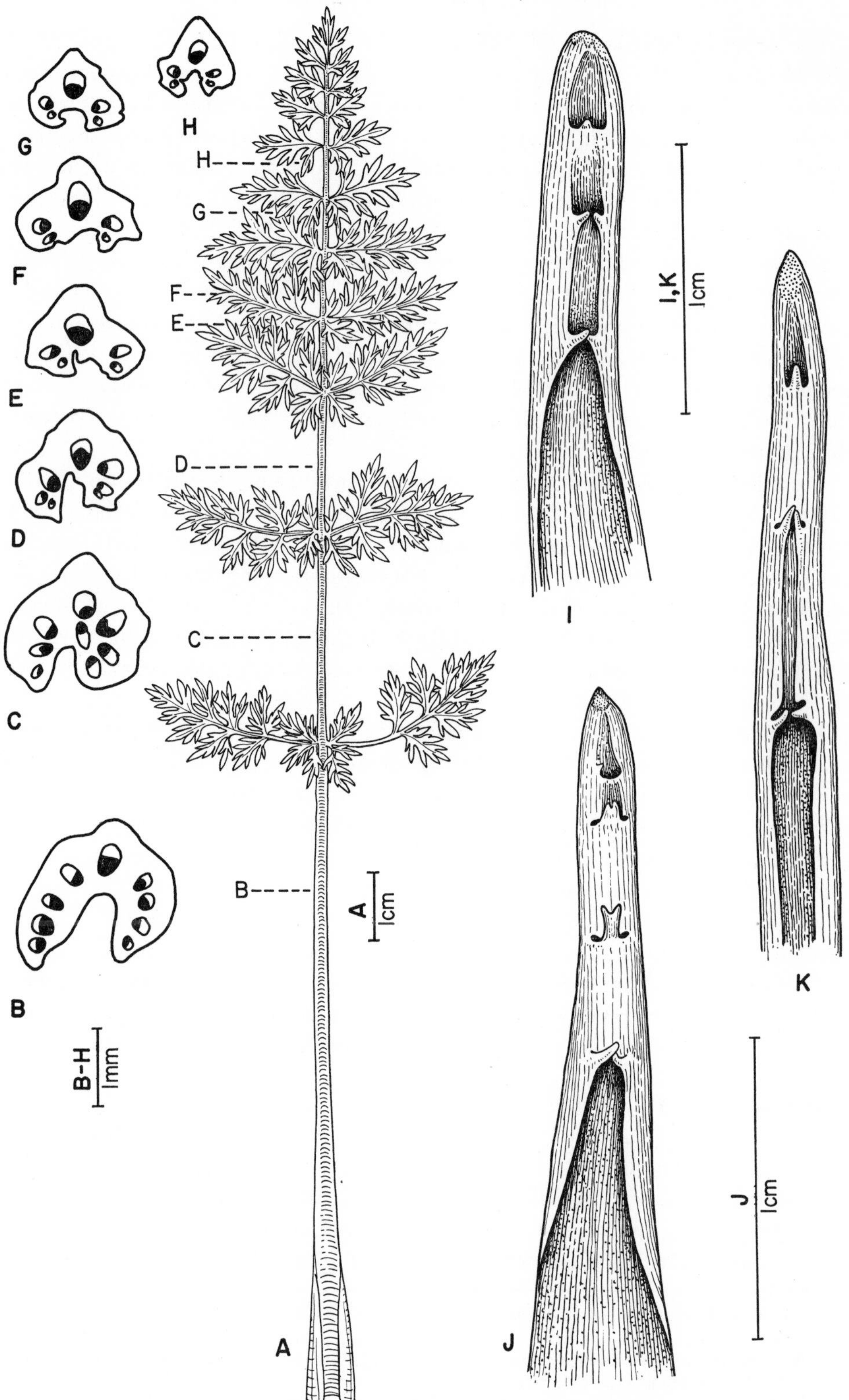

FIGURE 7

9 mm in length and a very reduced upper leaf zone 1 mm long atop the sheathing base. By contrast with the rosette type, this leaf has a detectable adaxial furrow along the length of the 'Oberblatt' and each of the nodal appendages (indicated by arrows, Plate **2E**) is paired (cf. Plate **2E** with Plate **1D**). The third juvenile leaf is usually much longer (30 mm) and the upper leaf zone is also longer (12 mm), but the leaf is still flattened in certain sectors of the 'Oberblatt' and, correspondingly, some of the appendages are double (arrows, Plate **2F**). Another specimen of the third juvenile leaf is illustrated in Fig. 6**A**. This example has a very evident adaxial cleft running the length of the upper leaf zone. Transverse sections of the same leaf taken at the levels indicated in Fig. 6**A** (Fig. 6**B–I**) verify the bifacial nature of the leaf axis—the adaxial indentation demarcates the adaxial surface, and along the length of the leaf the nodal appendages are double, although those at the middle 'nodes' are closely approximated (levels D and F, Fig. 6**A**).

The stem leaves show a similar reduction in length of the upper leaf zone (2–5 cm) and an increase of the leaf base in relation to their total length (4–12 cm). In contrast to the juvenile leaves, however, the base is not as tightly clasping, but is broader and flatter in form. These leaves also show a bifaciality of their upper leaf sectors and an even more extensive elongation of the nodal appendages as lateral projections which resemble reduced, lateral pinnae. Plate **2H, I** shows two, reduced stem leaves; the upper leaf axis consists of only three 'nodes', but the appendages are clearly bipartite structures, lateral in insertion but united to varying degrees to form a median appendage. The leaf shown in Plate **2I** is also illustrated diagrammatically in Fig. 6**J** and is accompanied by transverse sections taken at the levels indicated (Fig. 6**K–Q**). As in the juvenile leaf illustrated in Fig. 6**A**, the bifaciality of the 'internodal' regions is evident from the presence of an adaxial cleft and an essentially open configuration of its vasculature. Unifaciality occurs where the margins of the nodal appendages are united, as for example in Fig. 6**L, N**.

Additional variations in stem leaf structure are illustrated in Fig. 7**I–K**. Depending upon whether the subjacent internodal region is bifacial or unifacial, the nodal appendages are either free and lateral (as in the basal and middle pairs in Fig. 7**I, K**) or they are united, appearing either as bifid or forked appendages (Fig. 7**J**), or even as a single, median appendage at the uppermost 'nodes' (Fig. 7**I, K**).

Development of a juvenile leaf

Plate 5**E–G** illustrates some stages in the development of a prophyll of an axillary shoot; the prophyll is morphologically similar to the second juvenile leaf illustrated in Plate **2E**. The histogenetic pattern in this leaf type is similar in outline to that described for the rosette leaf but shows some marked differences that account for its divergent form. Following initiation, the primordium undergoes apical and intercalary growth in length, the duration of both of which, however, is much shorter than

FIGURE 7. Comparison of mature leaf of *Carum carvi* and representative stem leaves of *Oxypolis greenmanii*. **A,** Mature leaf of *Carum carvi*. **B–H,** T.S. through the leaf axis of *Carum carvi* taken at the levels indicated in **A**. Phloem tissue—unshaded; xylem tissue—black. **I–K,** Distal regions of the stem-leaf axes of *Oxypolis greenmanii*.

in the adult leaf. Adaxial meristematic activity is responsible for some radial extension, but this too ceases earlier in development (Plate **5F, G**). Hydathodes are also initiated from derivatives of the adaxial meristem (NA, Plate **5F, G**) and their inception marks the end of adaxial meristematic activity.

The juvenile leaf differs from the adult most strikingly in its accelerated course of tissue maturation throughout the upper leaf axis and a correlated shift toward extension of the lower leaf zone as the 'Oberblatt' becomes arrested in its growth. Correlated with this precocious maturation is the earlier time of initiation of nodal hydathodes (cf. Plate **5F** with Plate **4E**). Whereas protuberances are first initiated in the rosette leaf when it is approximately 800–1000 μm long, they arise in the juvenile leaf when it is 400–500 μm in length. Similarly, while there was a disjunction between the times of septal and hydathode initiation in the rosette leaf, they occur simultaneously in the juvenile leaf (Plate **5F, G**) and resemble the leaf of *Lilaeopsis* in this regard.

Transverse sections through a young prophyll also show that it is crescentiform and dorsiventral at the time of appendage initiation because it has undergone far less radial thickening by comparison with the rosette leaf at an equivalent stage of development (cf. Plate **5E** with **5B** and cf. Plate **5F, G** with Plate **4D**). Thus, with an accelerated initiation of nodal appendages, adaxial meristematic activity is brought to a close much earlier and hence the juvenile leaf retains to varying degrees its original bifacial character at maturity (Plate **2E**; Fig. **6A**).

DISCUSSION

Interpretation of the nature and position of the nodal appendages of leaves of Lilaeopsis *and* Oxypolis

In terms of their unifacial organization and marked segmentation into 'nodes' and 'internodes', the leaves of *Lilaeopsis occidentalis* and *Oxypolis greenmanii* resemble a defoliated rachis-axis of a pinnately compound leaf. Troll (1934*b*) has provided a very convincing demonstration of this fact by showing not only that the nodal septa correspond to positions of pinna insertion in leaves of other Umbelliferae, but also that the morphology of such 'Rhachisblätter' can be simulated by removing the leaflets from pinnatifid leaves, such as those of *Sium latifolium* and *Cicuta virosa*. Despite the compelling nature of Troll's arguments, the most convincing evidence for rachis homology here would be a demonstration of leaflet initiation and the arrest of their pinnae during subsequent development.

Goebel (1933) attempted such an ontogenetic approach in his limited study of leaf development in *Crantzia lineata* (= *Lilaeopsis novae-zelandiae*). However, since he found that the nodal appendages arose in a median rather than a lateral position, he concluded that they were not equivalent to rudimentary leaflets. Troll (1934*a*), on the other hand, suggested that their median position was related to the unifaciality of the rachis, in the same way that median stipule formation and laminar peltation are correlated with petiolar unifaciality (Troll, 1939). He supported his argument with evidence from the occurrence of adaxial leaflets in other genera of Umbelliferae and emphasized their correlation with rachis unifaciality (Troll, 1934*a*).

Histogenetic data presented in this paper indicate the similarity of the adaxial hydathodes in *Lilaeopsis* and *Oxypolis* to pinnae, both in their manner of initiation from hypodermal cell lineages at the adaxial leaf surface and in their sequence of inception. Furthermore, the present observations of serial leaves of the shoot of *Oxypolis* provide support for Troll's thesis about the relationship of pinna insertion to lateral symmetry of the rachis. It has been shown that where the leaf axis is unifacial, as in adult leaves of *Lilaeopsis* and *Oxypolis*, pinna insertion is median (adaxial). However, those leaves of *Oxypolis* which are bifacial (juvenile and stem leaves) exhibit, correspondingly, pairs of lateral appendages, which, depending upon the extent of unifaciality, may show varying degrees of union into a single, adaxially inserted appendage. Troll (1934*a*) thought of the median position of aborted leaflets in *Lilaeopsis* in terms of a model leaf which possessed *both* lateral and median pinnae, such as the leaf of *Cicuta virosa*. He proposed a derivation of the condition in *Lilaeopsis* through the suppression of the lateral pair of pinnae and arrest of the median leaflets. Observations on leaves of *Oxypolis* suggest, however, that derivation of the median pinna is by fusion of lateral leaflets, rather than by their suppression.

Histogenetic comparisons presented in this paper, particularly between juvenile and adult leaves of *Oxypolis*, provide developmental correlations which support the relationship of pinna position to rachis lateral symmetry. It has been demonstrated that the unifacial leaves of *Lilaeopsis* and *Oxypolis* are initiated as dorsiventrally flattened primordia and acquire their radial symmetry secondarily, through the activity of an adaxial meristem. Troll & Meyer (1955) described a similar pattern of development for unifacial petioles in a number of angiosperm species and have called this type of adaxial meristem a 'rounding meristem' ('Rundungsmeristem') to focus attention on its role in the formation of radial leaf sectors. They have shown that the determination of a given leaf axis as bifacial, subunifacial (adaxially cleft) or unifacial depends upon the proportionate development of rounding versus marginal meristematic activity: in bifacial regions, adaxial meristematic activity is reduced and marginal meristems are active, whereas in unifacial sectors marginal meristematic activity is suppressed in favour of axial thickening. The interaction between marginal and adaxial meristematic activity and relative timing of pinna initiation in juvenile versus rosette leaves appears to play a role in the determination of the lateral versus median positions of leaflets in the respective leaf types in *Oxypolis*. It has been shown that the juvenile appendages in *Oxypolis* differ from the adult in their precocious course of development and tissue maturation. One expression of this developmental acceleration is an earlier initiation of nodal appendages before the leaf axis has become unifacial through adaxial meristematic activity: leaflet inception in these organs occurs at a time when there are still two marginal loci for pinna initiation (Plate 5**E**). By contrast, in the rosette leaf, pinna initiation occurs much later in ontogeny, after adaxial meristematic activity has filled in the adaxial furrow and eliminated these two loci (Plate 5**B**, cf. with **E**): with the assumption of unifaciality only a single meristematic centre remains for pinna initiation (Plate 5**D**) and this accounts for the median position of leaflet initiation in the rosette-type leaf. These correlations would be strengthened if I had equivalent developmental observations for the stem leaves of *Oxypolis*. Unfortunately, I do not have such information at the present time, but considering their morphological similarity to juvenile leaves, I doubt that they would show a marked developmental divergence from the juvenile appendage types. Nevertheless, investigations are in progress on the development of stem leaves in order to test the relationship of development to their adult form as well.

It is surprising that, despite the long history of attempts to interpret the morphology of these 'rachis-leaves', such transitional leaf forms with lateral appendages as observed for *Oxypolis* were not described before this time. A partial answer may lie in the differences in habit between *Oxypolis greenmanii* on the one hand and *Lilaeopsis* and *Ottoa* on the other. In contrast to the very condensed shoots of *Lilaeopsis* and *Ottoa*, which exhibit a more abrupt change from the vegetative to the flowering state (Troll, 1934*b*), the shoot of *Oxypolis greenmanii* is considerably more extended, exhibiting a gradual change in stem and leaf form in transition to the inflorescence region. Similarly, the species of *Oxypolis* studied by Troll (1934*b*) (*Oxypolis filiformis* = *Tiedemannia teretifolia*), by virtue of its smaller stature (Mathias & Constance, 1942), may not have shown the same range of appendage types as the species described here.

Developmental comparisons between 'rachis-leaves' and the leaf of Carum carvi

Additional developmental support for the rachis equivalence of the leaves of *Lilaeopsis* and *Oxypolis* comes from their histogenetic resemblance to conventional pinnatifid leaves in certain other Umbelliferae—for example *Carum carvi*. As described by Troll (1939) the adult leaf of *Carum* is tripinnately compound, consisting of eight to ten pairs of lateral pinnae and a terminal leaflet (Fig. 7**A**). The leaf is differentiated into a sheathing base surmounted by an elongated petiole. Transverse sections along the leaf axis, from the petiole through the basal six 'internodes' of the rachis, indicate that the entire axis of the upper leaf is bifacial, with the typical open configuration of its vasculature and an adaxial furrow (Fig. 7**B**–**H**).

In general, development of the leaf of *Carum* is like that described for *Lilaeopsis* and *Oxypolis*. Following initiation from the shoot apex, it undergoes a protracted period of apical meristematic activity. Lateral pinnae are first initiated when the leaf is 200 μm long and arise in a close, acropetal succession, in a manner identical to that described for the adaxial appendages of *Lilaeopsis* and *Oxypolis* (cf. Plate 6**A, B** with Plates 3**C, E** and 4**E, F**). The major difference is their lateral rather than median position of initiation (Plate 6**C**). In transverse sections it is obvious that the primordium of *Carum* has the same crescentic form as that described for the other two genera (cf. Plate 6**C, D** with Figs 2**A**–**E,** and 4**A**–**D**); however, instead of undergoing thickening in the centre of the leaf axis, meristematic activity is accentuated at its margins in the process of pinna initiation (P, Plate 6**C**). An approximation of this condition occurs, as mentioned previously, in the juvenile leaf of *Oxypolis*, which has a crescentic primordium at the time of pinna initiation. In that case, however, the close proximity of the margins means that in certain cases the rudimentary pinnae are connate basally (cf. PNA, Plate 5**E** with P, Plate 6**C**) even though the leaf is dorsiventral at the time of initiation. In spite of its bifaciality, the leaf of *Carum* does show some adaxial meristematic activity responsible for thickening of the petiolar region in particular (Plate 6**A, B, D**); however, it does not result in loss of bifaciality.

Comparison of longitudinal sections of leaf primordia of *Lilaeopsis* and *Oxypolis* with those of *Carum*, at equivalent stages of development, shows that axis thickening is accentuated over pinna extension in the former genera, whereas in *Carum* the reverse occurs (cf. Plate 6**A, B** with Plates 3**E** and 4**D, E**). Moreover, it is particularly significant that, even though their manner and course of initiation is the same, pinnae in *Lilaeopsis* and *Oxypolis* exhibit an arrest in growth soon after initiation so that they reach an equivalent length irrespective of their time of initiation. This contrasts with the condition in *Carum*, which shows different degrees of leaflet extension in relation to their time of initiation (cf. Plates 3**E**, 4**F** with Plate 6**A, B**).

On the basis of these comparisons I would conclude that developmental evidence supports the interpretation by Troll (1934*a*, *b*, 1939) that the unifacial leaves of *Lilaeopsis* and *Oxypolis* are equivalent to the rachis of a pinnately compound leaf which has been developed at the expense of the pinnae. The latter, in turn, have been suppressed developmentally and differentiated into a series of nodal hydathodes. Their median position seems to be correlated with the unifaciality of the rachis axis and is an expression of the relationship of the timing of leaflet initiation to the timing of adaxial meristematic activity.

It should be emphasized that, while these conclusions appear reasonable for the data available for *Oxypolis* and *Lilaeopsis*, the proposed relationship of leaflet position to rachis symmetry cannot be generalized to other Umbelliferae without further study. For example, Troll (1934*a*) has analysed the morphology of the leaf of *Carum verticillatum*, which also has a unifacial rachis, but by contrast exhibits lateral insertion of pinnae. In order to assess properly the relationship of this leaf type to that of the genera described here, it too will have to be subjected to intensive investigation. Similarly, the leaves of *Lilaeopsis lacustris* from New Zealand pose a particular challenge because they have well-defined rudimentary leaflets that are clearly *lateral* in insertion on a seemingly unifacial rachis (Hill, 1927), and may provide an even more crucial test of the relationship of rachis lateral symmetry and development to leaflet position in another species of *Lilaeopsis*.

ACKNOWLEDGEMENTS

This research was supported in part by NSF grant GB-8104 and NIH Biomedical Science Support grant FR-7006 to the Berkeley campus of the University of California. I wish to express my sincere appreciation to Dr Lincoln Constance for his aid in providing sources of material for investigation, his advice on taxonomic literature of Umbelliferae and his critical reading of the manuscript. Thanks are also due to Miss Christine Hanson for her technical assistance in this project. I also wish to acknowledge the expert collaboration of Mr Alfred Blaker, Principal Scientific Photographer, Berkeley, for his contribution of photographs in Plate 1**A–D** and Plate 2**E–I** and Miss Charlotte Mentges who prepared the drawings in Figs 6**A, J** and 7**A, I–K**. Finally, I would like to thank Drs Adriance Foster, William Jensen and Bruce Sampson for their critical reading of the text and helpful suggestions in preparation of the manuscript for publication.

REFERENCES

BITTER, G., 1897. Vergleichend-morphologische Untersuchungen über die Blattformen der Ranunculaceen und Umbelliferen. *Flora, Jena*, **83**: 223–302.

BRIQUET, J., 1897. Recherches sur les feuilles septées chez les Dicotylédones. *Bull. Herb. Boissier*, **5**: 453–468.

DOMIN, K., 1908. Morphologische und phylogenetische Studien über die Familie der Umbelliferen. *Bull. Int. Acad. Sci. Bohême*, **1908**: 1–46.

DRUDE, O., 1898. Umbelliferae. In Engler, A., & Prantl, K., *Die natürlichen Pflanzenfamilien*, **3** (8): 63–250. Leipzig: Engelmann.

EICHLER, A. W., 1861. *Zur Entwicklungsgeschichte des Blattes mit besonderer Berücksichtigung der Nebenblattbildungen*. Diss. Univ. Marburg.

GAISBERG, E. VON, 1922. Zur Deutung der Monokotylenblätter als Phyllodien. *Flora, Jena*, **115**: 177–190.

GODFREY, R. K. & KRAL, R., 1958. Observations on the Florida flora. *Brittonia*, **10**: 166–177.

GOEBEL, K., 1891. *Pflanzenbiologische Schilderungen*. Zweiter Teil. Marburg: Elwert'sche Verlagsbuchhandlung.

GOEBEL, K., 1900. *Organographie der Pflanzen*. Zweiter Teil. *Specielle Organographie*. 2 Heft: *Pteridophyten und Samenpflanzen. Erster Teil*. Jena: Gustav Fischer.

GOEBEL, K., 1933. *Organographie der Pflanzen*. Dritte Auflage. *Dritter Teil. Samenpflanzen*. Jena: Gustav Fischer.

HILL, A. W., 1927. The genus *Lilaeopsis*: a study in geographical distribution. *J. Linn. Soc.* (*Bot.*), **47**: 525–551.

KAPLAN, D. R., 1970. Comparative foliar histogenesis in *Acorus calamus* and its bearing on the phyllode theory of monocotyledonous leaves. *Am. J. Bot.*, **57**: 331–361.

MATHIAS, M. E. & CONSTANCE, L., 1942. New North American Umbelliferae. *Bull. Torrey bot. Club*, **69**: 151–155.

MATHIAS, M. E. & CONSTANCE, L., 1945. Umbelliferae. *North American Flora*, **28b**: 159–161.

RENNERT, R. J., 1903. The phyllodes of *Oxypolis filiformis*, a swamp xerophyte. *Bull. Torrey bot. Club*, **30**: 403–411.

SASS, J. E., 1958. *Botanical microtechnique*, 3rd ed. Ames: Iowa State College Press.

TROLL, W., 1934*a*. Über den Bau der Rhachis und seinen Einfluss auf die Spreitenbildung von Fiederblättern. *Planta*, **22**: 80–108.

TROLL, W., 1934*b*. Über die binsenähnlichen Blattformen bei Umbelliferen. *Planta*, **23**: 1–18.

TROLL, W., 1939. *Vergleichende Morphologie der höheren Pflanzen. Band 1. Vegetationsorgane, Teil 2.* Berlin: Borntraeger (Reprint, 1967, O. Koeltz, Koenigstein-Taunus).

TROLL, W. & MEYER, H. J., 1955. Entwicklungsgeschichtliche Untersuchungen über das Zustandekommen unifazialer Blattstrukturen. *Planta*, **46**: 286–360.

EXPLANATION OF PLATES

PLATE 1

A. Mature rosette leaf of *Lilaeopsis occidentalis*, ×0·7.

B. Enlarged view of nodal region of leaf of *Lilaeopsis occidentalis*, showing nodal appendage at arrow, ×27.

C. Mature rosette leaf of *Oxypolis greenmanii*, ×0·3.

D. Enlarged view of nodal region of leaf of *Oxypolis greenmanii*, showing a nodal appendage at arrow, ×3·5.

E. T.S. of nodal region of leaf of *Oxypolis greenmanii*, illustrating the hydathode histology of the nodal appendage, ×58.

F. T.S. of nodal region of leaf of *Lilaeopsis occidentalis*, showing the histology of the nodal appendage and the vasculature of the anastomosing, girdling vein, ×165; Hyd, hydathode.

PLATE 2

A–C. Whole mounts of successively-older leaf primordia of *Lilaeopsis occidentalis*: **A,** leaf primordium 550 μm in length, ×70; **B,** leaf primordium 750 μm in length, ×70; **C,** leaf 2·6 mm in length, ×44.

D. Leaf of *Oxypolis greenmanii* 2·3 mm long. Arrows indicate position of bipartite nodal appendages, ×39.

E–G. Juvenile leaf forms in *Oxypolis greenmanii*: **E,** second leaf of an axillary shoot, ×20; **F,** third leaf of an axillary shoot, ×8; **G,** fourth leaf of an axillary shoot, ×8; arrows indicate positions of paired appendages.

H, I. Fixed, upper stem leaves of *Oxypolis greenmanii* with arrows indicating the positions of paired or even lateral nodal appendages in varying degrees of union, ×8; NA, nodal appendage.

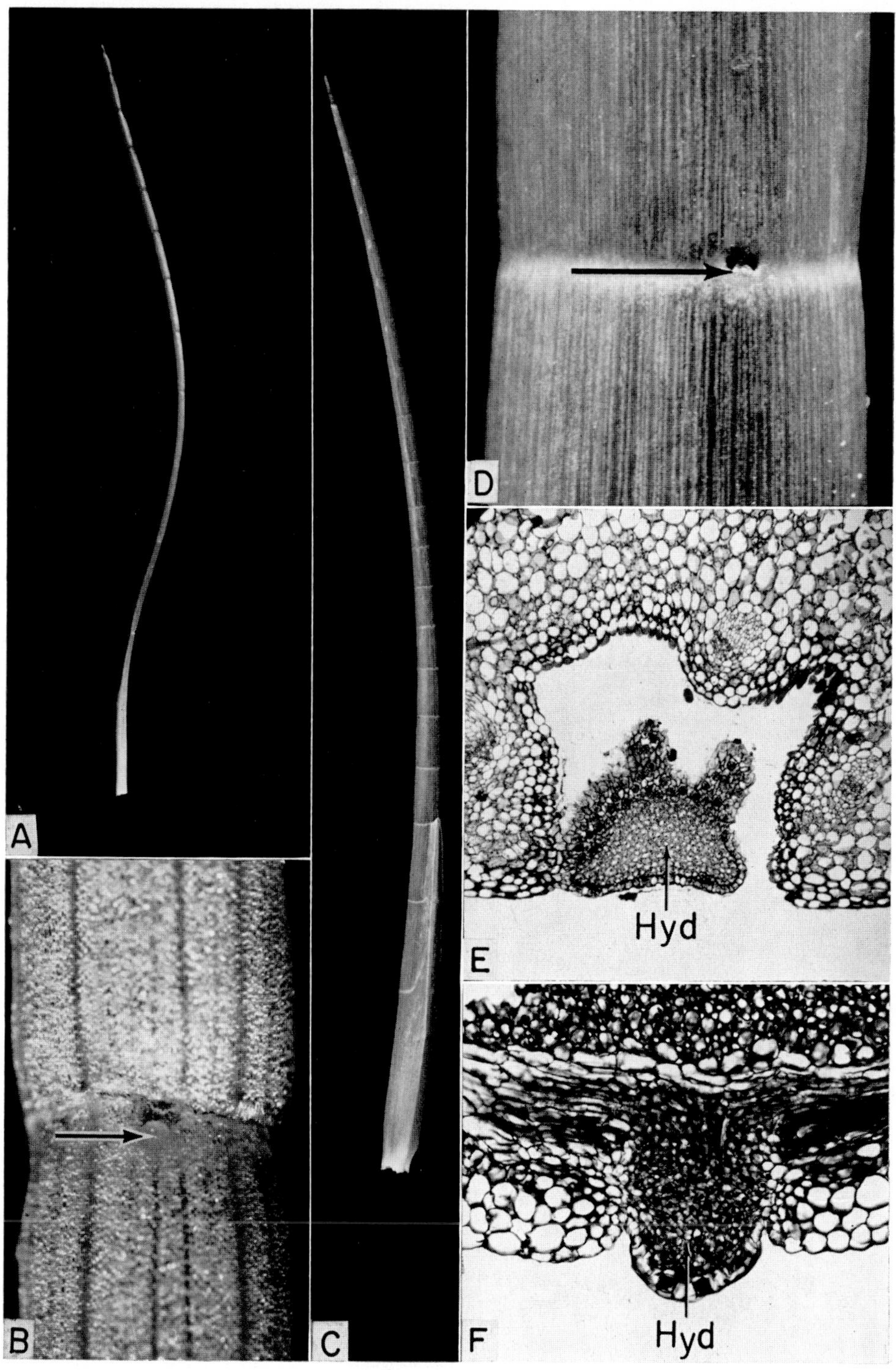

D. R. KAPLAN

(*Facing p.* 124)

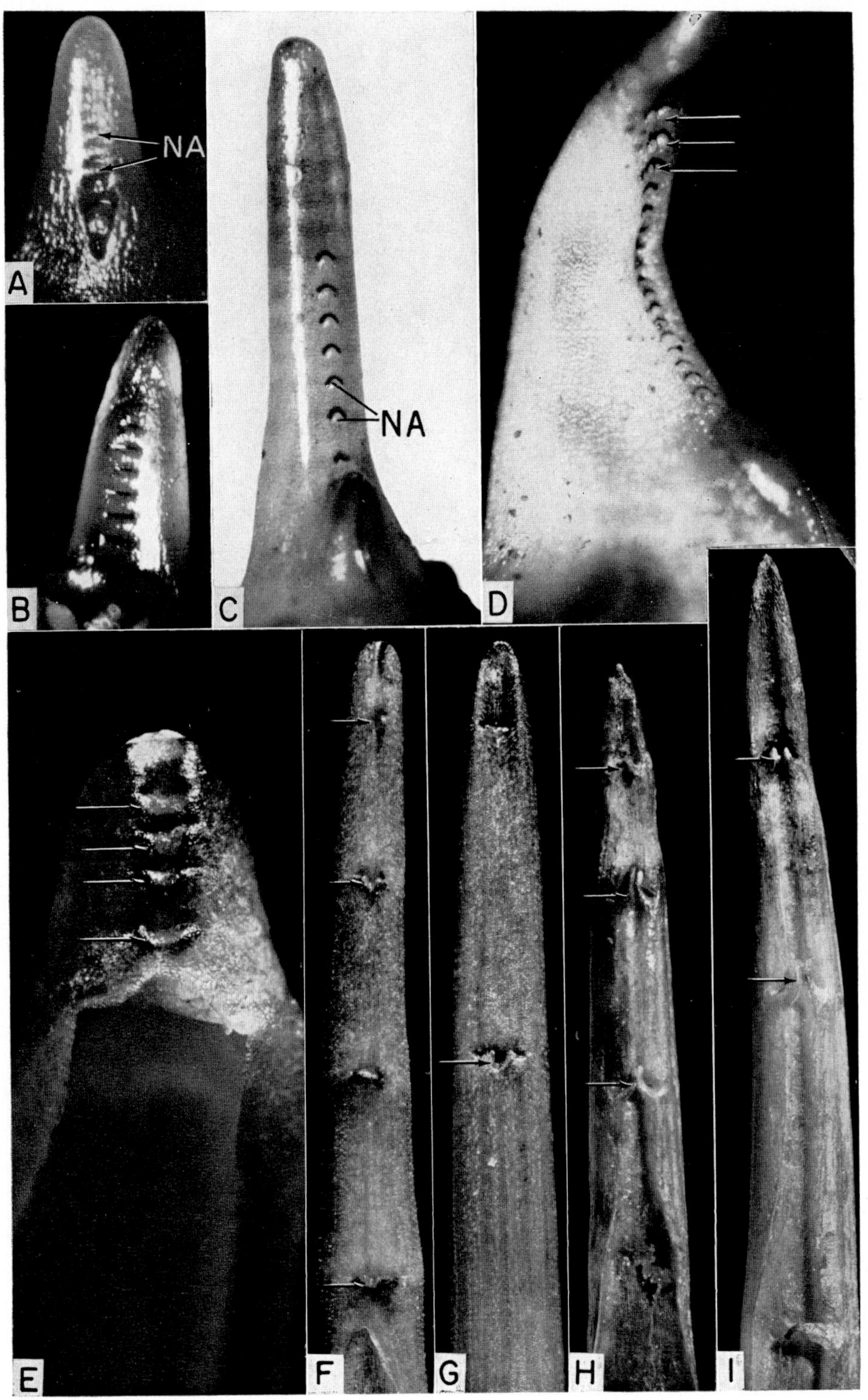

D. R. KAPLAN

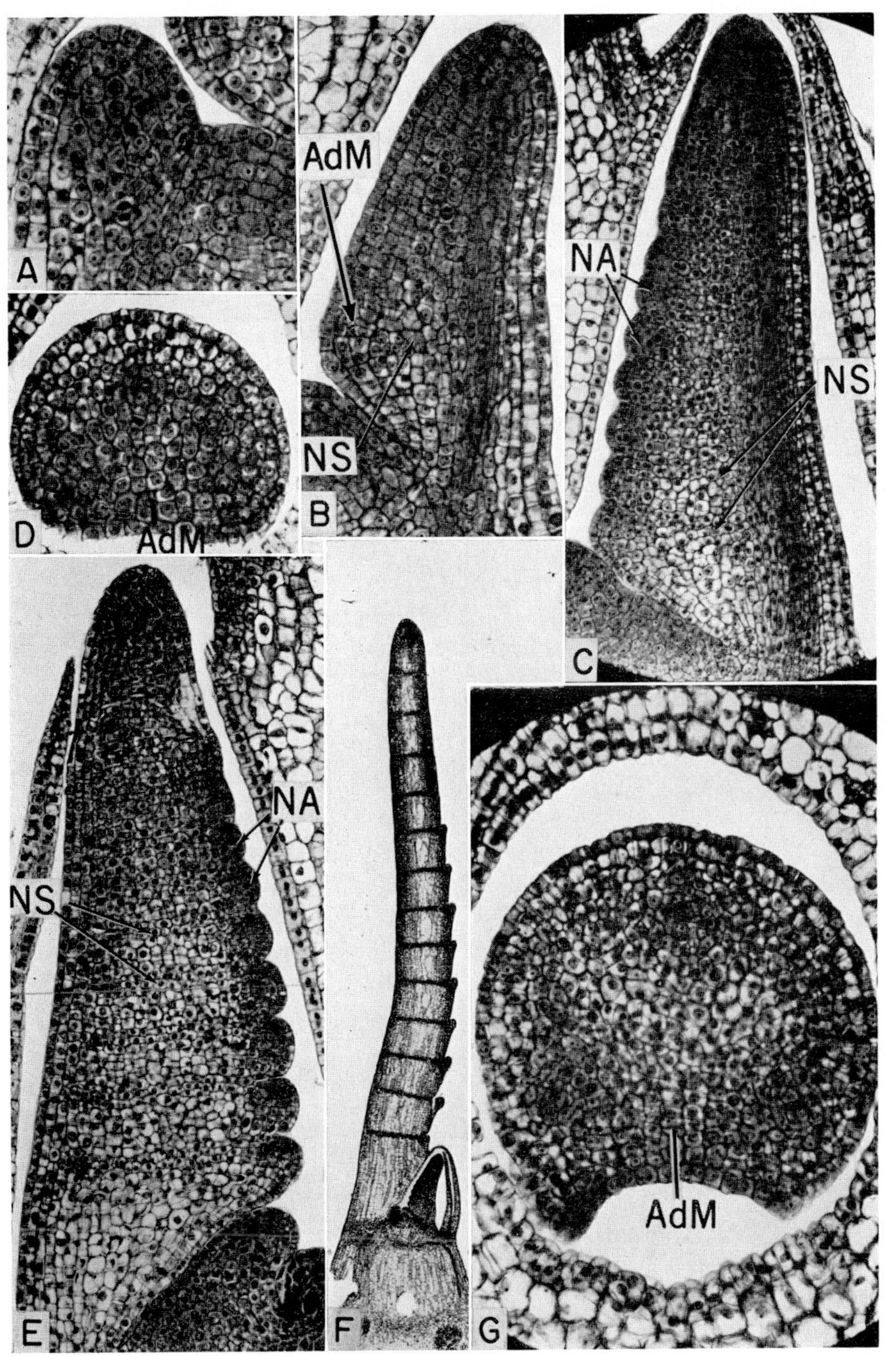
AdM
NS
A
B
NA
NS
C
D
AdM
NA
NS
E
F
G
AdM

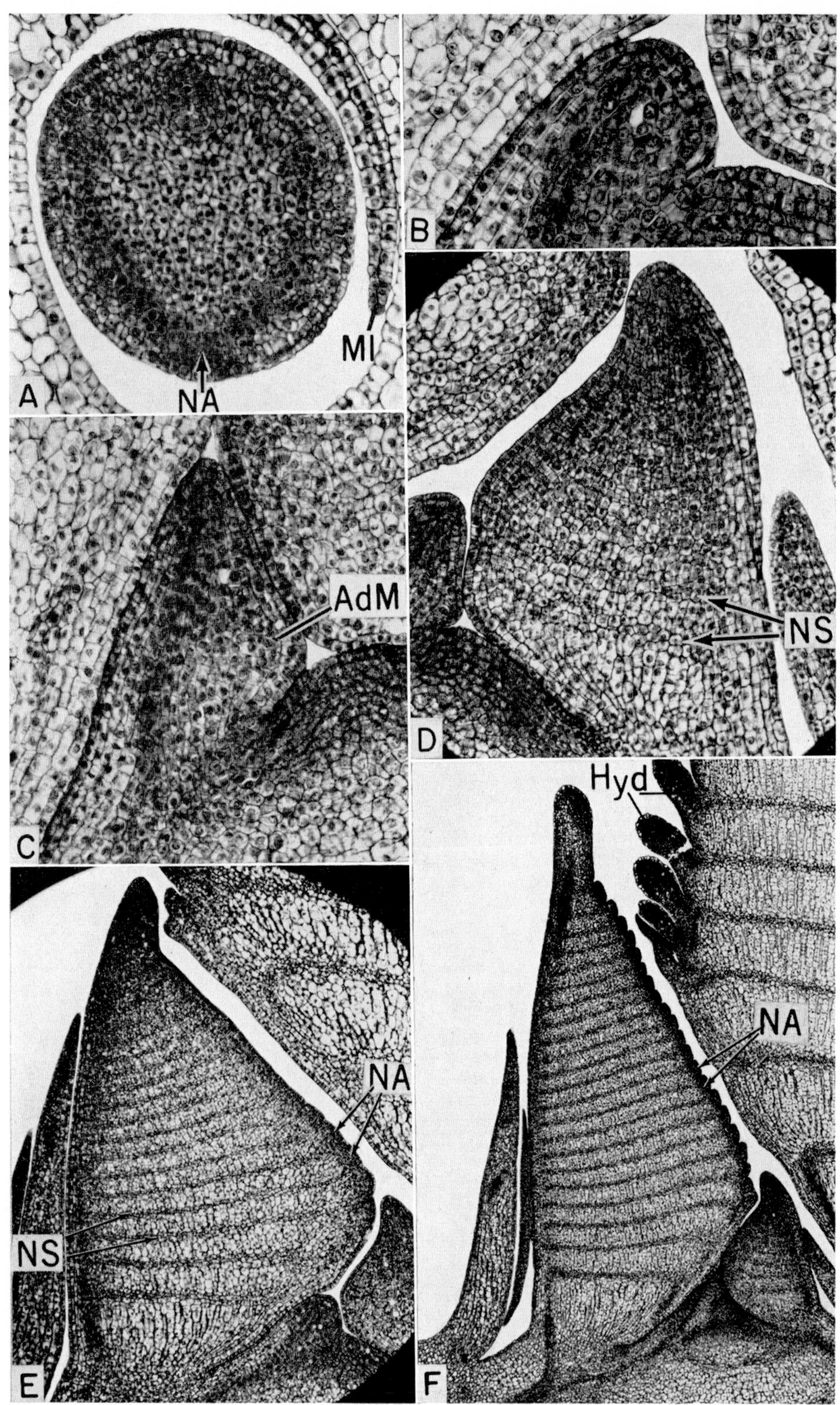

D. R. KAPLAN

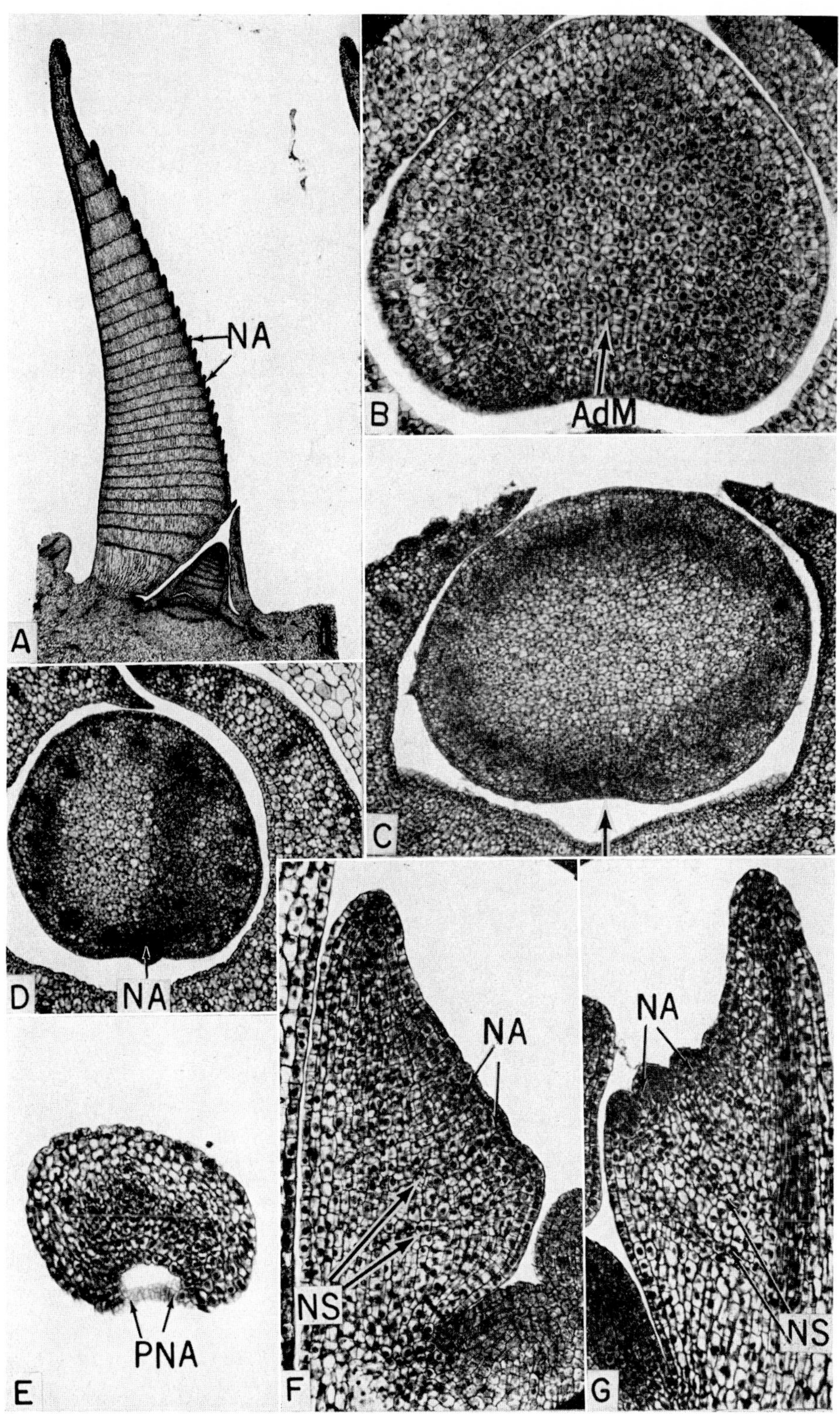
NA
A
AdM
B
C
NA
D
PNA
E
NA
NS
F
NA
NS
G

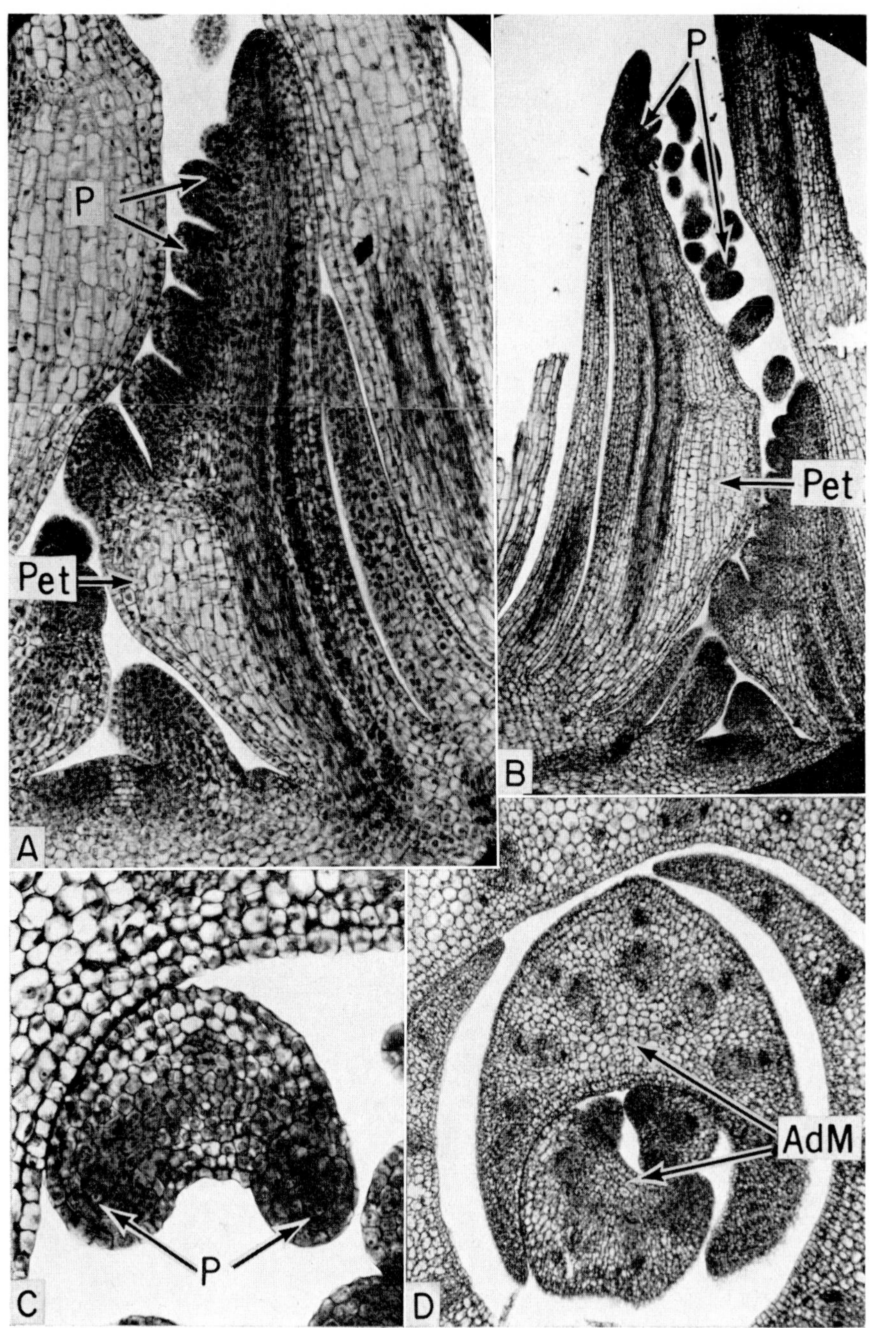

D. R. KAPLAN

PLATE 3

Stages in the development of the rosette leaf of *Lilaeopsis occidentalis*.

A. Recently initiated leaf primordium 50 μm in length, ×300.
B. Median L.S. of leaf primordium 230 μm in length, showing adaxial meristematic activity and course of tissue maturation and procambialization, ×247.
C. Median L.S. of leaf 575 μm in length at the time of initiation of nodal appendages and nodal septa, ×152.
D. T.S. of leaf primordium 190 μm long, cut 140 μm below the tip, showing short cell files of the adaxial meristem, ×260.
E. Near median L.S. of leaf primordium 675 μm in length, following initiation of the complement of adaxial appendages, ×152.
F. Near median L.S. of a leaf 4·15 mm long, showing internode extension and cell stretching as well as breakdown of the central pith during later phases of leaf accretion, ×19.
G. T.S. of a leaf primordium 400 μm long, cut 330 μm below the tip at the juncture of the unifacial upper leaf zone and the sheathing base. Note the radial files of recently divided cells of the adaxial meristem, ×260; AdM, adaxial meristem; NA, nodal appendage(s); NS, nodal septa.

PLATE 4

A. T.S. of a leaf of *Lilaeopsis occidentalis*, 960 μm in length, cut through a primordial nodal septum at a level 520 μm below the tip, ×210.
B–F. Stages in the development and histogenesis of the rosette leaf of *Oxypolis greenmanii*: **B,** median L.S. of a recently initiated leaf primordium 80 μm in length, ×258; **C,** median L.S. of a primordium 190 μm in length, showing precocious adaxial meristematic activity, ×185; **D,** near median L.S. of a leaf primordium 590 μm in length, showing precocious initiation of nodal septa before adaxial appendage initiation, ×111; **E,** median L.S. of leaf primordium 1190 μm in length at the time of nodal appendage initiation, ×60; **F,** near median L.S. of leaf 2·12 mm in length, after cessation of apical meristematic activity and initiation of the complement of nodal appendages, ×34; AdM, adaxial meristematic activity; Hyd, hydathodes; MI, marginal initial cell; NA, nodal appendages; NS, nodal septa.

PLATE 5

A–D. Stages in the development of the rosette leaf of *Oxypolis greenmanii*: **A,** near median L.S. of a leaf 6 mm in length at the beginning of leaf extension, ×13; **B,** T.S. of a leaf 450 μm in length, cut 260 μm below the tip, showing adaxial 'rounding' meristematic activity well before appendage initiation is evident, ×160; **C,** T.S. of nearly fully rounded leaf axis of a leaf 2040 μm in length, cut 1590 μm below the tip (arrow indicates position of adaxial cleft above the nodal appendage), ×83; **D,** T.S. of leaf 2490 μm in length, cut through level of the basal nodal appendage 2270 μm below the tip, ×68.
E–G. Primordia of the first juvenile leaf of an axillary shoot (prophyll) of *Oxypolis greenmanii*: **E,** T.S. of prophyll 480 μm in length, cut 168 μm below the tip through a set of paired nodal appendages. Note the crescentic form of the primordium at this stage of development, ×140; **F,** near median L.S. of a prophyll 430 μm in length, showing stages in acropetal initiation of nodal appendages and concomitant differentiation of septa, ×150; **G,** near median L.S. of prophyll 480 μm in length at a somewhat later stage of appendage initiation, ×150; AdM, adaxial meristem; NA, nodal appendages; NS, nodal septa; PNA, paired nodal appendages.

PLATE 6

Stages in the early development of the leaf of *Carum carvi*.

A. Near median L.S. of a leaf 655 μm in length, showing acropetal sequence of pinna inception, ×152.
B. Near median L.S. of leaf 1170 μm in length, showing some adaxial thickening in the primordial petiole region, ×76.
C. T.S. of leaf primordium 800 μm long, cut 140 μm below the tip, showing a pair of recently initiated lateral pinnae, ×330.
D. T.S. of successively-older leaf primordia, exhibiting some signs of adaxial meristematic thickening in the petiole and the sheathing leaf base, ×98; AdM, adaxial meristem; P, pinnae; Pet, petiole.

Comparative anatomy of *Dioscorea rotundata* and *Dioscorea cayenensis*

EDWARD S. AYENSU, F.L.S.

Department of Botany, Smithsonian Institution, Washington, D.C., U.S.A.

A comparative study of the vegetative organs of two species, *Dioscorea rotundata* Poir. and *D. cayenensis* Lamk., has been made. Although the two taxa are not easily separated on exomorphic grounds, there are clear-cut histological characters in the aerial stems that provide proof that they are two distinct species. There are three types of vascular bundles in each species. In *D. rotundata*, Types I–III consist of eight, four and twelve bundles respectively. In *D. cayenensis* each Type consists of eight bundles. Furthermore, the arrangements of the bundles differ from one another in the two species. It is suggested that the two species may be recently stabilized segregates from a larger complex.

CONTENTS

INTRODUCTION

This article is aimed at providing anatomical and histological information on two of the most important edible dioscoreas of West Africa.

Dioscorea rotundata, commonly referred to as the White Yam, is perhaps the most important species of the genus in cultivation in West Africa, where it is native. The species is also commonly cultivated in the West Indies. *D. cayenensis*, commonly known as Yellow Guinea Yam, is also cultivated in the West Indies. The reader is referred to a recently published account in the Longmans' Tropical Agriculture Series on 'Yams' by Coursey (1967).

We are often reminded by Metcalfe (1963) that 'the anatomy of even some monocotyledons that are of well established economic importance is relatively unknown.' He further urges that 'we ought to learn as much as we can about the anatomical structure of crop plants so as to be able to assist our colleagues who are concerned with the cultivation or physiology of these plants.' Another reason that has prompted the study of these species is that many taxonomists often regard *D. rotundata* as a subspecies of *D. cayenensis*. This view has been held by Burkill (1939), Chevalier (1936) and Miège (1950). However, in his earlier and later publications Burkill (1921, 1960) made them two distinct species. In their study of the West African *Dioscorea*, Hutchinson & Dalziel (1954) used the colour of the flesh of the tubers to distinguish the two species. The use of the colour of tubers has often been considered unsatisfactory because some forms of *D. cayenensis* have almost white flesh, while some forms of *D. rotundata* are appreciably yellow in colour (Coursey, 1967). The structure of the starch grains of the two species was studied by Seidemann (1964), and he concluded that the granules in *D. rotundata* are larger and ovoid-shaped. Those of *D. cayenensis*, on the other hand, are smaller and more or less triangular. Although the two species are close, there are distinct anatomical features in the stem that separate them and therefore support their separation as two distinct species.

MATERIALS AND METHODS

Most of this study is based on fluid preserved materials obtained from Nigeria (Waitt's collection) and collections including the underground organs made by myself in Ghana.

The vegetative organs were fixed in F.A.A. and processed in the usual manner for sectioning on a Reichert sliding microtome, usually at 16 μm. The sections were stained in safranin and counterstained with Delafield's haematoxylin followed by the conventional differentiation, dehydration, clearing in xylene and mounting in Canada balsam.

GENERAL MORPHOLOGY OF THE TUBERS

The tubers generally appear in a variety of forms; it is not unusual to encounter within one species forms that vary between cylindrical, globular or irregularly flattened. Occasionally, many varieties are developed in one plant. The cause for such variety within a single species is not quite understood. However, Archibald (1967) observed during her study of the South African dioscoreas that the final shape of the adult tuber in *Dioscorea sylvatica*, for example, is very dependent upon the physical conditions of the soil. She noted that in loamy soil the tubers were uniformly disk-shaped with few indentations, while similar species on scree slopes exhibited smooth-skinned and variously mis-shapened tubers because of the pressure against rocks. Furthermore, she observed that in shallow soils a tuber may be partially exposed and tessellated above and with irregular lobes in the lower section below the soil level, while in dune sand the tuber becomes enormous with irregularly bifurcating lobes.

The two species under study have been cultivated in West Africa for many years under more or less similar conditions. As a result of the familiarity of these two species to the farmers, most yam cultivators are able to sort out from a heap of tubers those that are *D. rotundata* from those of *D. cayenensis*. Their analyses are based merely on the shape, colour and sculpturing of the bark of the tubers. The tuber of *D. rotundata* is generally cylindrical but often broader at the proximal end and gradually tapers to the distal end. That of *D. cayenensis*, on the other hand, is generally irregular in shape with prominent lobes especially at the distal end. The colour of *D. rotundata* is generally light brown accentuated with streaks of dark brown. That of *D. cayenensis* is often very dark brown with occasional streaks of white. The sculpturing on *D. rotundata* consists of wavy slits, almost vertically arranged and parallel to each other. At the proximal end, the slits go deeper but are shorter and range from 1–5 cm.

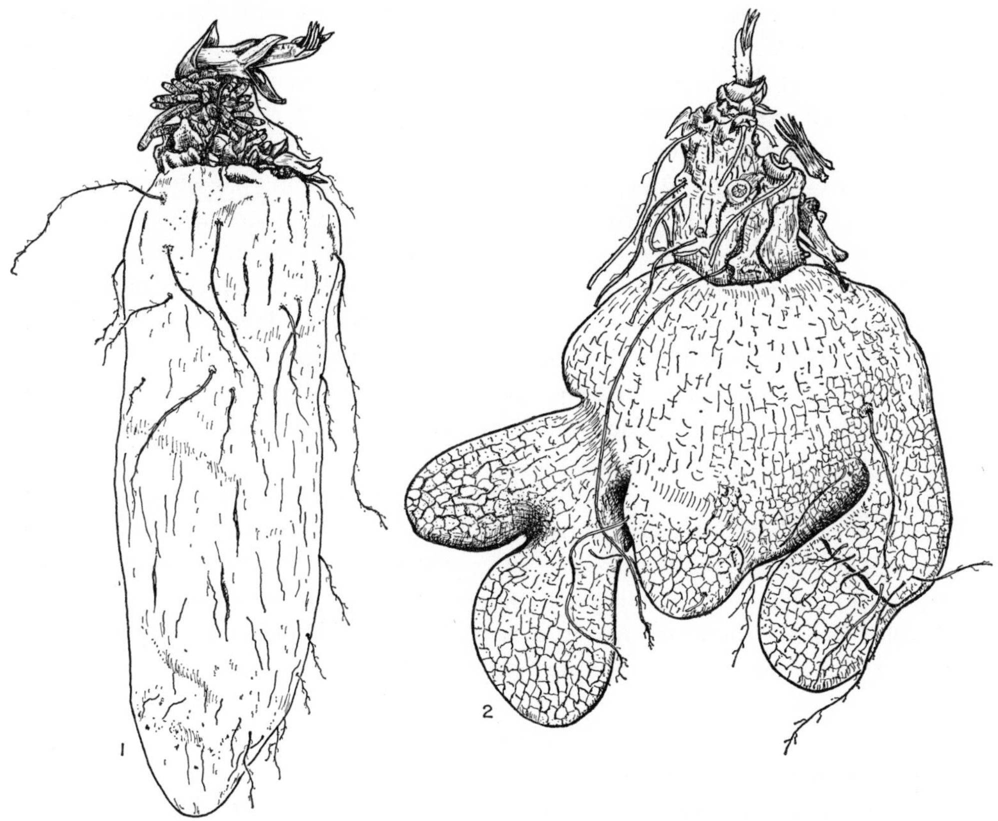

FIGURES 1 and 2. General habit drawings of the tubers of *Dioscorea rotundata* and *D. cayenensis* respectively. ×⅓.

About the mid-portion of the tuber these slits are rather shallow and range from about 1–3 cm. At the distal end, where a certain amount of tapering is encountered on all the lobes on the tuber, the slits are again very shallow and rather short ranging from about 1–2 cm. In *D. cayenensis*, the slits are both vertical and horizontal giving a scale-like appearance to the entire tuber. The scale-like appearance of the bark is much more prominent at the distal end of the tuber than at the middle portion. The accompanying habit drawings of the tubers represent samples of *D. rotundata* (Fig. 1) and *D. cayenensis* (Fig. 2); these are often encountered in the markets of West Africa.

GENERAL ANATOMY OF *DIOSCOREA ROTUNDATA*

Leaf

Trichomes

Hairs absent from mature parts. *Glands* consisting of multicellular bulb-like structures on short unicellular stalks, sporadically distributed in surface view.

Lamina

Dorsiventral. *Cuticle* very thin, slightly thickened around margins and on adaxial and abaxial sides of midveins. *Epidermis* with mainly isodiametric cells; few somewhat rectangular cells irregularly interspersed among the others. Adaxial and abaxial cell walls in surface view slightly sinuous. Abaxial epidermis with distinct cuticular striations in surface view, especially on cells associated with stomata and the intercostal cells. Adaxial twice as large as abaxial cells, but those above and below midveins distinctly small. *Stomata*: 45 × 27 μm, anomocytic, confined to abaxial surface; guard cells without ledges but possessing distinct wall thickenings; pores irregularly orientated in surface view. *Mesophyll*: clearly differentiated into palisade and spongy parenchyma. Palisade cells in two rows confined to adaxial side; cells subjacent to epidermis three to four times as long as those of second row, cells compactly arranged; cells of spongy tissue mainly elongated axially, lying parallel to surface of lamina and occupying half of mesophyll. *Vascular bundles* of midrib much larger than remainder, tracheal elements consisting mainly of tracheids, few vessels and associated parenchyma. Large midvein more or less equidistant from surface layers; small veins occupying lower third of mesophyll. Each vein surrounded by a distinct parenchyma sheath. Secondary veins reticulate, smaller veins including many blind-ending veins, composed of small files of tracheids. Commissural bundles observed, consisting mainly of tracheids and thin-walled parenchyma. Xylem and phloem strands occupying adaxial and abaxial positions respectively; xylem strands alternating with seven or eight phloem units. *Crystals*: idioblasts containing raphide bundles common in mesophyll cells. *Tannin*: quite common in mesophyll, filling idioblastic cells with shapes recalling those of pieces of a composite jigsaw puzzle.

Petiole

Outline pentagonal, with two adaxial wings; wings rather short and not as pronounced as those in *Dioscorea alata*. *Cuticle*: thin and slightly undulating; ridged in places especially on abaxial side. *Epidermis*: in surface view of thin-walled cuboidal cells with stomata similar to those in lamina. Collenchymatous cells in one or two layers subjacent to epidermis, often interrupted by other cortical cells. Cells of cortex of fairly uniform size; those towards central cylinder becoming progressively larger, especially those subjacent to wings. Starch grains present in parenchyma especially in cells near main veins. *Vascular bundles*: generally six, arranged in a circle around the periphery of central cylinder. The largest bundle abaxially positioned. Each bundle partially subtended by three to five layers of fibres. Lignified parenchyma of two to three layers occupying interfascicular regions. Tracheal elements consisting of vessels, tracheids and associated parenchyma; phloem parenchyma. *Ground parenchyma*: uniform thin-walled cells. Few starch grains observed. *Tannin*: tanniniferous substances in idioblastic cells.

Aerial stem

Internode

Hairs absent. *Glands* and *stomata* present, very similar to those of lamina. *Epidermis*: composed of thick-walled, somewhat sclerotic cells with thick cuticle, ridged in places. Walls with distinct pits, conspicuous on radial walls. *Stomata*: quite common, subsidiary cells rather indistinct. Guard cells each with an upper (outer) cutinized ledge more or less symmetrical in transverse section. *Cortex*: narrow, six to ten cells wide, chlorenchymatous, uniform but interrupted by substomatal chambers, and also with conspicuous idioblastic cells containing tanniniferous substances. Cells of innermost cortex distinct (endodermoid), each cell containing a cuboidal or rectangular crystal. *Central cylinder*: fibrous lignified cells with square ends viewed in longitudinal sections subjacent to the endodermoid layer. *Vascular bundles* of three distinct types viewed in transverse sections (Fig. 3**A**): Type I consisting of eight bundles, Type II of

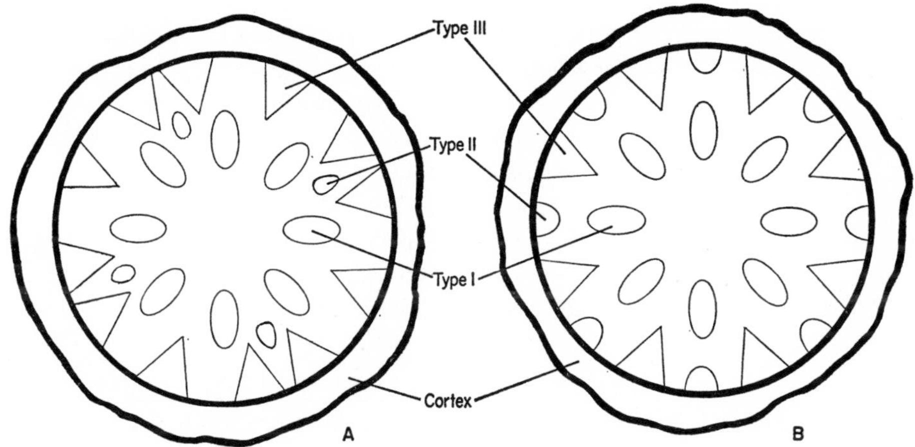

FIGURE 3. Schematic diagrams of the transverse sections of stems illustrating the number and arrangement of the vascular bundles in *D. rotundata* (**A**) and *D. cayenensis* (**B**).

four bundles and Type III of 12 bundles (Plate 2**A**). Type I constituting innermost circle of cauline bundles, with elliptically arranged metaxylem vessels and tracheids. Each bundle with a pair of large metaxylem vessels and below this a pair of phloem units. A pair of phloem units (occasionally only one phloem unit) occurring above the largest pair of vessels, followed by two pairs of metaxylem vessels. The outermost pair of vessels with one phloem unit (Plate 1**C**). Type II composed of one large metaxylem vessel and above it a pair of phloem units followed by three vessels and another phloem unit. One Type II bundle appears between every two bundles of Type I. Type III, constituting the common bundles, exhibiting V-shaped arrangement of metaxylem vessels and tracheids together with one phloem unit at the converging ends of the V and another pair near the flanges. *Ground parenchyma*: thin-walled, filled with starch grains; tanniniferous substances present; cells between vascular bundles rather compact, but those in centre of cylinder with narrow intercellular spaces.

GENERAL ANATOMY OF *DIOSCOREA CAYENENSIS*

Leaf

Trichomes

Hairs absent. *Glands* consisting of multicellular bulb-like structures on short unicellular stalks.

Lamina

Dorsiventral. *Cuticle* thin but thickened around margins and on both sides of midvein. *Epidermis* with mainly isodiametric cells. Adaxial and abaxial cell walls in surface view slightly sinuous, but like *D. rotundata*, abaxial epidermis with distinct cuticular striations in surface view, especially on cells associated with stomata and the intercostal cells. Adaxial three or four times as large as abaxial cells. *Stomata*: 45 × 30 μm, anomocytic, confined to abaxial surface; guard cells without ledges but possessing distinct wall thickenings; pores irregularly orientated in surface view. *Mesophyll* clearly differentiated into palisade and spongy parenchyma. Palisade cells in one row confined to adaxial surface; a second row rather poorly differentiated. Cells of spongy tissue mainly elongated axially, lying parallel to surface of lamina and occupying half of mesophyll. *Vascular bundles* of midrib much larger than others; tracheal elements consisting of few vessels, tracheids and associated parenchyma. Small veins embedded in lower third of mesophyll. Each vein surrounded by a distinct parenchymatous sheath. Secondary veins reticulate, smaller veins ending blindly. Commissural bundles observed, consisting of tracheids and thin-walled parenchyma. Xylem and phloem strands occupying adaxial and abaxial positions respectively; xylem strands alternating with seven phloem units (Plate 1**A**). *Crystals*: idioblasts containing raphide bundles common in mesophyll cells. *Tannin*: quite common in mesophyll.

Petiole

Outline pentagonal. *Cuticle*: thin and slightly undulating. *Epidermis:* in surface view of thin-walled cuboidal cells with stomata similar to those in lamina. Collenchymatous cells in one to three layers subjacent to epidermis. Cortical cells of fairly uniform size; cells becoming progressively larger towards central cylinder. *Vascular bundles*: six, arranged in a circle, the largest bundle abaxially positioned. Each bundle subtended by three to five layers of fibres. Interfascicular region lined with two or three layers of fibres. Tracheal elements consisting of vessels, tracheids and associated parenchyma; phloem composed of large sieve-tubes, companion cells and phloem parenchyma. *Ground parenchyma* of uniform, thin-walled cells. *Crystals*: mainly of raphide bundles. *Tannin*: tanniniferous cells few.

Aerial stem

Internode

Hairs absent. *Glands* and *stomata* present, very similar to those of lamina. *Epidermis*: composed of thick-walled, somewhat conical cells with thick cuticle, ridged in places. *Stomata*: quite common, subsidiary cells rather indistinct. Guard cells with a more or less pronounced upper (outer) cutinized ledge. *Cortex*: wider than in *D. rotundata*, 8–12 cells wide, chlorenchymatous, fairly uniform but idioblastic cells containing tanniniferous material and raphide bundles present. Cells of innermost cortex distinct (endodermoid), each cell containing a cuboidal or rectangular crystal. *Central cylinder*: fibrous lignified cells with square ends viewed in longitudinal sections subjacent to the endodermoid layer. *Vascular bundles* of three distinct types viewed in transverse sections (Fig. 3**B**), each type consisting of eight bundles (Plate 2**B**). Type I constituting the innermost circle of cauline bundles with elliptical arrangement of metaxylem vessels and tracheids. Each bundle with a pair of large metaxylem vessels and below this a pair of phloem units. A pair of phloem units occurring above largest pair of vessels, followed by two pairs of metaxylem vessels. The outermost pair of vessels having one or two phloem units. Type II somewhat U or V-shaped with either one or two relatively large vessel elements and one to three phloem units and appearing immediately above Type I. Type III, which as in *D. rotundata* constitutes the common bundles, having V-shaped arrangement of metaxylem vessels and tracheids together with one phloem unit at the converging ends of the V and another pair on the flanges. *Ground parenchyma*: with slightly thicker-walled cells than those observed in *D. rotundata*; starch grains and raphide bundles present.

Node. The highly complex xylem and phloem glomeruli encountered in the nodal regions of the stem of the Dioscoreaceae have been described in detail elsewhere (Ayensu, 1969; 1970*a*, *b*).

ANATOMY OF THE TUBER

Although the tubers of *Dioscorea rotundata* and *D. cayenensis* are quite distinct from each other in external morphology, their internal structures are very similar.

Epidermis partially replaced by a corky layer, obscuring its structure. Cork primary in origin, many-layered; cells in radial rows and suberized. Cells of outer ground tissue of various size and many-layered. Inner ground tissue composed of thick-walled cells, filled with starch grains. Grains near periphery of inner ground tissue rather small in size, gradually becoming larger towards inner portion of tuber. *Vascular bundles* collateral, sporadically distributed in central ground tissue. Xylem consisting of tracheids and associated parenchyma. Phloem composed of sieve-tubes, companion cells and phloem parenchyma. *Starch grains* mostly ovate or almost elliptical (Plate 1**B**). Hilum located at narrow end of the very distinct ovate grains. Grains uniformly concentrated in inner ground tissue. *Crystals* consisting of raphide bundles. Tanniniferous cells present.

ANATOMY OF THE ROOT

The histological difference between roots of *Dioscorea rotundata* and *D. cayenensis* is almost negligible.

Piliferous layer composed of irregular cells. *Exodermis*: thick-walled, one-layered; cells compactly arranged. *Cortex*: cells of equal size; in *D. cayenensis*, cells compressed and crushed in specimens examined. Cortex constituting about 1/3 of the root diameter. *Endodermis* one-layered; cell walls with characteristic U-shaped thickenings, much more pronounced in *D. cayenensis* than in *D. rotundata*. Passage cells present in both species, but more easily identifiable in *D. rotundata*. *Pericycle* one or two-layered. *Stele* variable depending upon the maturity of the root, but the most mature roots generally 28–30-arch in these two species. Xylem consisting of vessels, tracheids and associated parenchyma. The large vessels of *D. cayenensis* extending to the innermost portion of the root, thus reducing the pith area. Phloem composed of sieve-tubes, companion cells and phloem parenchyma. Xylem strands alternating with phloem units, especially along periphery of central cylinder. Pith parenchyma made up of thick-walled cells. *Crystals* of raphide bundles and *tannin* cells present.

DISCUSSION

Previous taxonomic studies of *Dioscorea rotundata* and *D. cayenensis* have shown that it is almost impossible to determine whether these are two species or whether *D. rotundata* is a subspecies of *D. cayenensis*. As intimated in the introduction, Burkill (1939) and Miège (1950) were unable to come up with very clear-cut exomorphic characters which could be depended upon in assessing the specific delimitations of these species. Lawton (1967) devised a key consisting of 12 characters for the identification of the species in seven out of 24 sections of the genus *Dioscorea*. Although he found the species in Nigeria fairly easy to separate on morphological grounds, *D. rotundata* and *D. cayenensis* could not be separated so easily. Lawton described both species as having, *inter alia*, rounded, markedly glossy green leaves; the stems with thorns either throughout the length or only near the base, and the seeds winged all around.

In the latest taxonomic treatment of the family in West Africa by Miège (1968), *D. cayenensis* and *D. rotundata* were described as having frequently modified leaves, reduced and alternate at the base and opposite above. The lamina is deeply sinuous but with deltoid attachment to the petiole. There are five to seven veins, those on either side of the midrib forming an acute angle. The only difference between the two taxa which Miège indicated is that *D. cayenensis* offers two annual harvests with a short resting period whilst *D. rotundata* offers one annual harvest with a long resting period.

A number of other *Dioscorea* species have been reduced to subspecific level by Miège (1968). These include *D. aculeata* Balbis ex Kunth, *D. berteroana* Kunth, *D. demeusei* De Wild & Th. Dur., *D. liebrechtsiana* De Wild, *D. occidentalis* Kunth, *D. praehensilis* and *D. pruinosa*. Miège further stated that perhaps *D. praehensilis* is the parent of the several varieties of *D. cayenensis*. It is interesting to note that *D. praehensilis* shares with *D. cayenensis* and *D. rotundata* the cuboidal and rectangular crystal type.

The features that have proved to be of great significance in the delineation of the two taxa under study are (a) the numbers of the types of vascular bundles within the stems and (b) their arrangement. These patterns of vascular bundles can be seen in cut stems with a hand lens. As illustrated in Fig. 3**A**, **B**, both taxa have three distinct types of vascular bundle. In *D. rotundata*, Type I, consisting of the innermost circle, numbers eight cauline bundles. Type II, which constitutes the middle circle of bundles, consists of four ovoid-shaped bundles. Type III, with its characteristic

V-shape, numbers 12. In *D. cayenensis*, on the other hand, all three types of bundles consist of eight each. Type II is more or less U-shaped instead of the ovoid shape which characterizes *D. rotundata*.

The arrangement of the bundles in the two taxa differs markedly from one another. While there is basic alternation of the three types of bundles in *D. rotundata*, one can easily observe that this pattern is broken in *D. cayenensis*. Here one can observe a radial arrangement of the bundles that constitute Type I with those of Type II.

The different numbers and arrangement of vascular bundles within the two taxa serve as taxonomically useful characters in considering *D. cayenensis* and *D. rotundata* as two separate species. While discussing *Dioscorea* for the *Anatomy of the Monocotyledons* (Ayensu, 1970*b*), I stressed the fact that the taxonomic study of the genus should take into account the diagnostically useful characters of the stem. There is considerable variety in the anatomy of the stems of *Dioscorea*, which becomes obvious to anyone who has had the experience of handling large collections of this genus. Other vegetative features, unfortunately, are not as useful. The starch grains, for example, are not as different as indicated by Seidemann (1964). These grains from tubers illustrated in Figs 1 and 2 seem inseparable, although there are differences in the external features of the tubers. Furthermore, the colours of the cut tubers of *D. rotundata* and *D. cayenensis* are yellow and white respectively.

The uncertainty surrounding the taxonomic determination of *D. cayenensis* and *D. rotundata* may result from hybridization. The two species may be rather recently stabilized segregates from a larger complex. Similarity in variations in tubers of these and other *Dioscorea* species prompted Miège (1968) to conclude that *D. abyssinica* is 'probably one of the wild parents of *D. cayenensis*' and that *D. lecardii* 'perhaps played a role in the *D. cayenensis* pedigree.'

One of the major problems in the systematics of edible *Dioscorea* is the gross absence of any satisfactory record of the existing varieties of each species. The species occur in a prodigious variation of cultivars. To remedy the unsatisfactory taxonomic situation an attempt should be made to collect and classify living specimens of *D. rotundata* and *D. cayenensis* cultivars. This line of activity will greatly enhance the systematics of this group if it is further coupled with anatomical and other botanical studies.

Our knowledge of plants from the tropical regions of the world is scanty. This is certainly true of an economically important pan-tropical genus such as *Dioscorea*. In order to achieve a deeper understanding of this group, a satisfactory taxonomic system must first be sought. Simple histological techniques can provide adequate data that will lend such support.

ACKNOWLEDGEMENTS

I wish to thank Mr S. K. Avumatsodo for the preparation of the habit drawings. This study was supported in part by a grant from the Smithsonian Research Foundation.

REFERENCES

ARCHIBALD, E. E. A., 1967. The genus *Dioscorea* in the Cape Province west of East London. *Jl S. Afr. Bot.*, **33**: 1–46.

AYENSU, E. S., 1969. Aspects of the complex nodal anatomy of the Dioscoreaceae. *J. Arnold Arbor.* **50**: 124–137.

AYENSU, E. S., 1970*a*. Analysis of the complex vascularity in stems of *Dioscorea composita*. *J. Arnold Arbor.* **51**: 228–240.
AYENSU, E. S., 1970*b*. *Anatomy of the monocotyledons*. VI. *Dioscoreales*. Oxford: Clarendon Press. (In press.)
BURKILL, I. H., 1921. The correct botanic names for the white and yellow Guinea yams. *Gdns' Bull. Straits Settl.*, **2**: 438–441.
BURKILL, I. H., 1939. Notes on the genus *Dioscorea* in the Belgian Congo. *Bull. Jard. bot. Etat. Bruxelles*, **15**: 345–392.
BURKILL, I. H., 1960. The organography and the evolution of the Dioscoreaceae, the family of the yams. *J. Linn. Soc.* (*Bot.*), **56**: 319–412.
CHEVALIER, A., 1936. Contribution à l'études de quelques espèces africaines du genre *Dioscorea*. *Bull. Mus. natn. Hist. nat., Paris, sér.* 2, **8**: 520–551.
COURSEY, D. G., 1967. *Yams*. London: Longmans.
HUTCHINSON, J. & DALZIEL, J. M., 1954. *Flora of West Tropical Africa*. London: Crown Agents.
LAWTON, J. R. S., 1967. A key to the *Dioscorea* species of Nigeria. *Jl W. Afr. Sci. Ass.*, **12**(1): 1–9.
METCALFE, C. R., 1963. Comparative anatomy as a modern botanical discipline. In R. D. Preston (ed.), *Advances in Botanical Research*, **1**: 101–47. London: Academic Press.
MIÈGE, J., 1950. Caractères du *Dioscorea minutiflora* Engl. *Revue int. Bot. appl. Agric. trop.*, **30**: 428–432.
MIÈGE, J., 1968. Dioscoreaceae. In F. N. Hepper (ed.), *Flora of West Tropical Africa*. London: Crown Agents.
SEIDEMANN, J., 1964. Mikroskopische Untersuchung verschiedener *Dioscorea* Stärken. *Stärke*, **16**: 246–253.

EXPLANATION OF PLATES

PLATE 1

A. Transverse section of the leaf midrib of *D. cayenensis* showing the number of the phloem units, ×80.
B. Starch grains of *D. cayenensis* photographed under polarized light, ×45.
C. Transverse section of a cauline vascular bundle of *D. rotundata* exhibiting the number and arrangement of the phloem units and the metaxylem vessels, ×45.
a. ep., Adaxial epidermis; m.v., metaxylem vessel; p.t., palisade tissue; ph.u., phloem unit; sp.t., spongy tissue; st.g., starch grains; s.t., sieve-tube; t., tracheid.

PLATE 2

Transverse sections of aerial stems showing the arrangement of the vascular bundle types that differentiates *D. rotundata* (**A**) from *D. cayenensis* (**B**), ×80.

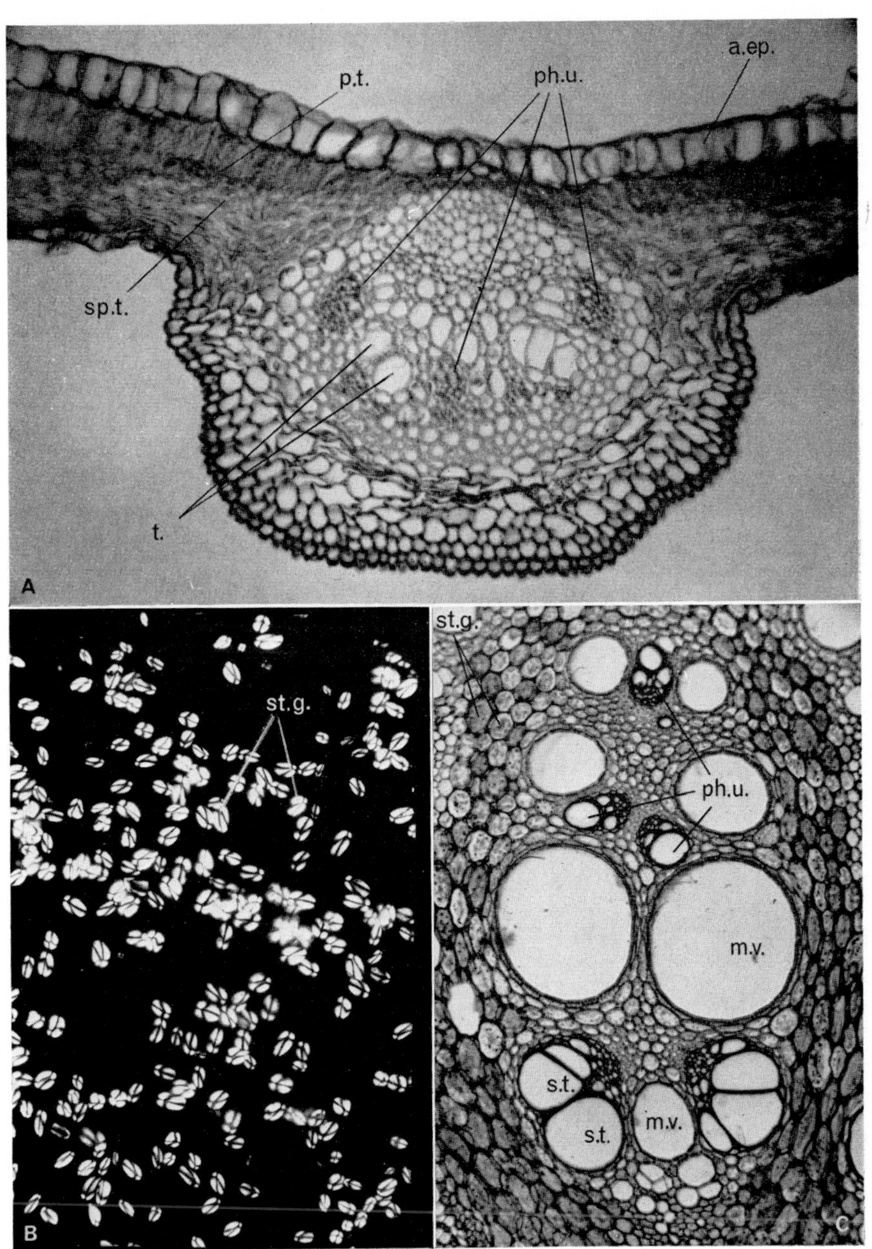
a.ep.
p.t.
ph.u.
sp.t.
t.
A
st.g.
B
st.g.
ph.u.
m.v.
s.t.
s.t.
m.v.
C

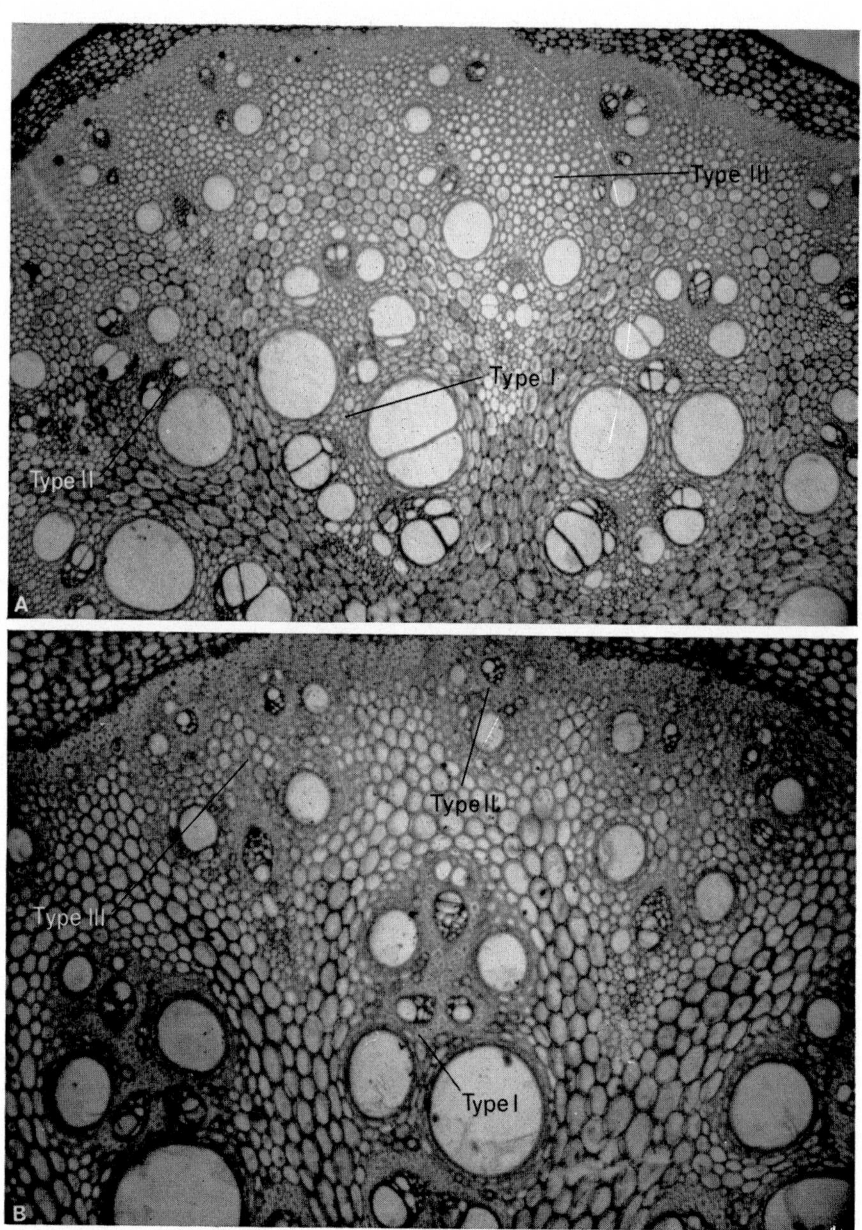

E. S. AYENSU

Vegetative anatomy of *Carex microglochin* Wahlenb. and *Carex camptoglochin* Krech.

I. KUKKONEN

Department of Botany, University of Helsinki, Finland

The structure of the epidermis, and the anatomy of the leaf and culm as seen in transverse and longitudinal sections were studied in five specimens of *Carex microglochin* Wahlenb. and two of *Carex camptoglochin* Krech. (≡ *Carex oligantha* Boott). A number of characters, such as presence or absence of prickles or papillae, size of stomata, and number, size and shape of sclerenchyma girders, was estimated quantitatively. However, qualitative differences were also noted, viz. position of vascular bundles, size and number of air-cavities, and differentiation of chlorenchyma. It is concluded that *C. microglochin* occurs in Tierra del Fuego, and that *C. camptoglochin*, generally included as an infraspecific taxon within *C. microglochin*, is a distinct species. This decision is based on qualitative differences. When more material is available some of the characters here called quantitative may prove to be useful in dividing *C. microglochin* into infraspecific taxa.

CONTENTS

INTRODUCTION

Carex microglochin is a distinct, well-known boreal circumpolar species, with a range divided into several isolated patches (Hultén, 1958). One of these is in northern Finland and the Scandes Mountains; another is in the Alps; in Asia the species occurs in several isolated mountain ranges (Egorova, 1967); in North America it is found from Alaska to Newfoundland and southwards as far as Colorado. In addition, it is one of the much discussed 'bipolar plants' (see e.g. Constance *et al.*, 1963; Löve, 1967) and also occurs in South America, according to Hultén (1958), in an area stretching from Tierra del Fuego and the Falkland Islands to Bolivia and Ecuador. But there has been some doubt as to whether the plant in the Southern Hemisphere is the same species as that in the Northern. Boott (1867) described the South American plant as a new species under the name *C. oligantha*, with the remark that its distinction from *C. microglochin* was uncertain; at the time, the nearest known locality of the latter was in Greenland. Accordingly, *C. oligantha* has since been included in this species as either a variety or subspecies (Kükenthal, 1909). Unfortunately, as Krechetowich (1937: 34) has shown, the name *C. oligantha* Boott is a later homonym for *C. oligantha* Steudel and thus must be rejected; the name *C. camptoglochin* Krech. is adopted here. It was also believed that the South American material was homogeneous, until recently Roivainen (1954) gave it as his opinion that *C. camptoglochin* (*C. oligantha*) is in reality an entity distinct from *C. microglochin*, which, however, also occurs in Tierra del Fuego in localities with a lower precipitation. The aim of this paper is to show the distinctness of *C. camptoglochin* and also to present convincing evidence of the occurrence of the true *C. microglochin* in South America.

MATERIAL AND METHODS

The following specimens have been examined:

Carex microglochin—124. Italy, Fl. Verbano-Lepontica, Valle Fornezza, 1913, *Boggiano* (TUR).—142. Canada, British Columbia, Pink Mtn, Mile 150 Alaska Highway, 1960, *Calder* (26737) *&* *Gillett* (TUR).—194. Chile, Tierra del Fuego, Lago Blanco, 1928, *Roivainen* (H).—195. Finland, Sodankylä, Aska, 1960, *Laaksonen* (H).—196. Finland, Enontekiö, Markkina, Lätäseno, 1966, *Roivainen* (H).

Carex camptoglochin—145. Chile, Fuegia occidentalis, Isla Dawson, 1929, *Roivainen* (H).—197. Chile, Prov. de Magallanes, Isla Clarena, 1929, *Roivainen* (H).

The structure of the adaxial and abaxial epidermis of the leaves and the transverse and longitudinal sections of leaves and culms were studied in permanent slides. The slides were prepared according to the methods in daily use in the Jodrell Laboratory and described by Metcalfe (1960) and used also by Kukkonen (1967). For descriptive terms, see Metcalfe & Gregory (1964).

The stomatal lengths given in Table 1 are each based on 20 measurements.

RESULTS

Carex microglochin

Leaf surface (Fig. 2)

Prickles absent. *Stomata* paracytic, confined to the abaxial surface (Plate 1**E**); subsidiary cells occasionally subdivided. Lengths of the stomata and epidermal cells given in Table 1. *Papillae* present on the adaxial cells of nos. 142, 195 (Plate 1**B**) and 196; almost every cell bearing one prominent papilla in 195, and in the others the papillae less prominent and confined to or near leaf margin. *Silica-cells* numerous among cells above sclerenchyma. Silica-bodies one to five per cell, usually in one row, especially above marginal sclerenchyma, sometimes in two rows or irregularly arranged (no. 196); satellites sometimes very inconspicuous or even absent, especially above marginal sclerenchyma.

Transverse section of lamina (Fig. 6; Plate 1**I**)

Outline subcircular or thickly crescentiform, without keel; margins very thick and rounded, diameter 0·7–1·0 mm. Adaxial surface narrow, especially when outline subcircular. *Epidermis*: adaxial cells much larger than abaxial; outer wall of all epidermal

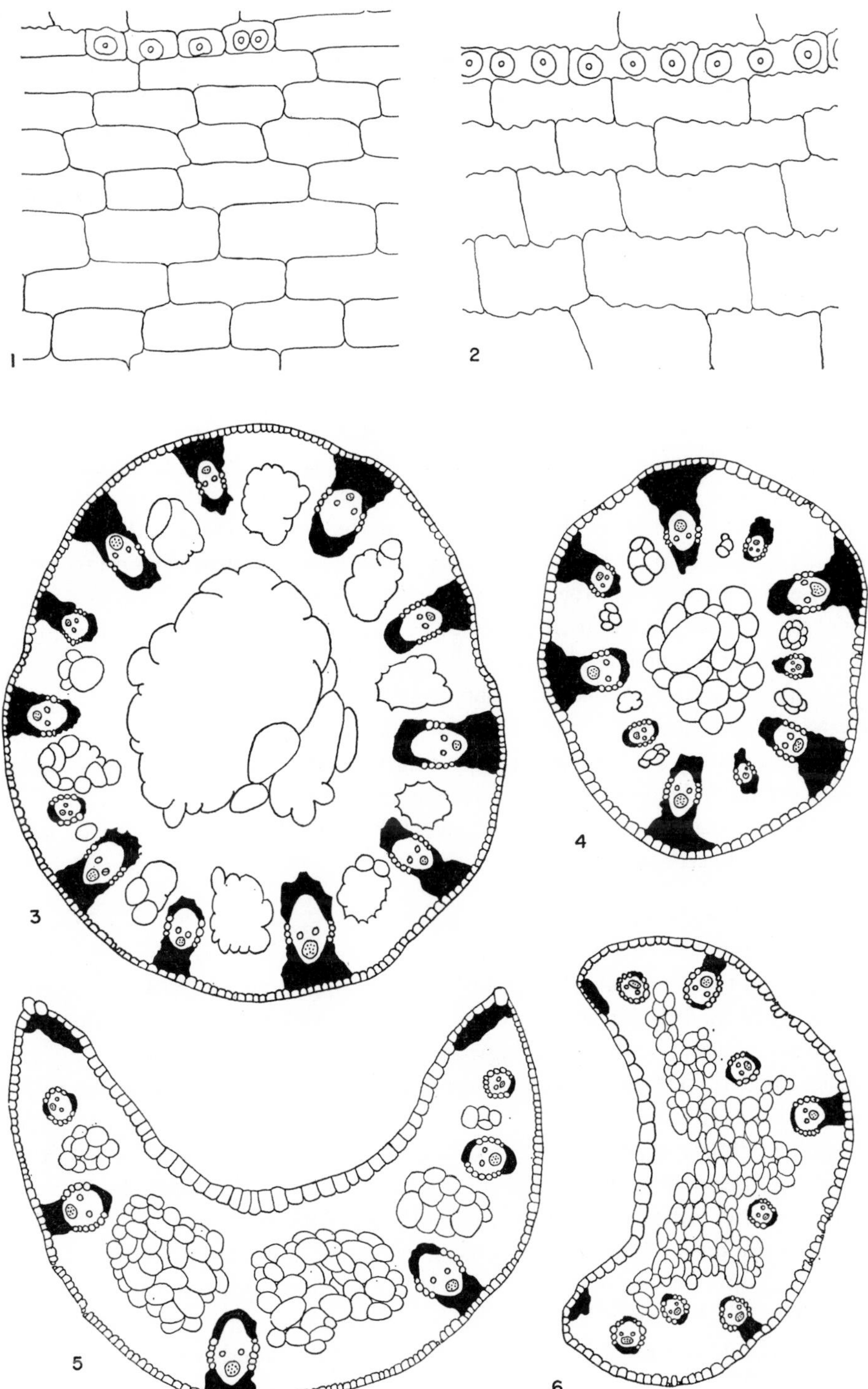

FIGURES 1 to 6. 1. Adaxial leaf surface of *Carex camptoglochin*, showing silica-bodies (×290). 2. Adaxial leaf surface of *C. microglochin*, showing silica-bodies (×290). 3. T.S. culm of *C. camptoglochin* (×70). 4. T.S. culm of *C. microglochin* (×70). 5. T.S. leaf of *C. camptoglochin* (×70). 6. T.S. leaf of *C. microglochin* (×70).
Inner bundle sheath not shown. Solid black, sclerenchyma; dots, phloem.

cells very thick. *Hypodermis* absent. *Sclerenchyma*: usually three of the vascular bundles supported by an abaxial girder, baculiform and always taller than wide; other vascular bundles with an abaxial crescentiform sclerenchyma cap, and all vascular bundles with an adaxial crescentiform cap. Marginal sclerenchyma strand small, more or less oval in outline, formed of about 20–40 cells. *Mesophyll*: chlorenchyma only in peripheral parts, for example, two cell layers adaxially, three or four layers abaxially; however, inner row of vascular bundles connected to abaxial epidermis by chlorenchyma; chlorenchyma cells always at least slightly elongated; differentiated into spongy and palisade parenchyma, best seen in longitudinal sections, especially above the rows of stomata; substomatal cavities very deep, up to *c*. 70 μm (Plate **1G**), size of cells lining cavities varying, 20–32 × 62–112 μm. *Air-cavities*: whole inner part of leaves filled with inflated, translucent cells, but in leaves crescentiform in outline chlorenchyma separating some smaller cavities near the margin, or one single cavity divided into three parts by a strand of slightly lignified, translucent parenchyma cells reaching from the adaxial epidermis (Plate **1C**) to the two vascular bundles of the inner row in the middle of the leaf, thus forming two large air-cavities, one in each wing of the leaf, and a smaller one abaxially between the two vascular bundles and the epidermis; dividing strand frequently only three cells wide. *Bulliform cells* indistinguishable from other adaxial epidermal cells; at least one group of slightly lignified, translucent cells present beneath median parts of the adaxial epidermis, often forming above-mentioned dividing strand. *Vascular bundles*: eight to eleven in number in the whole lamina (in no. 196 only six); located in two rows, alternate bundles distinctly smaller and more remote from abaxial surface. *Bundle sheaths* double, the inner always sclerenchymatous and the outer also interrupted, at least abaxially, by sclerenchyma. *Secretory cells* not observed.

Transverse section of culm

Outline: subcircular or very slightly angular, diameter 0·6–1·0 mm (Fig. 4). *Central ground tissue* formed of large cells diminishing in size towards the periphery, outermost cells slightly lignified or even distinctly thick-walled, connecting sclerenchyma caps of vascular bundles (especially in no. 196). *Air-cavities*: incipient; a group of inflated, translucent cells present between each pair of vascular bundles. *Chlorenchyma* confined to the areas between vascular bundles and around air-cavities; differentiated into spongy and palisade layers, seen especially well in longitudinal sections; substomatal cavities deep (see T.S. of lamina). *Vascular bundles*: ten or eleven in number, alternate ones smaller and slightly more removed from the surface than the larger ones, chlorenchyma layers separating them from epidermis. *Bundle sheaths*: as in the lamina. *Sclerenchyma*: girders supporting larger vascular bundles, but smaller bundles with crescentiform sclerenchyma caps; crescentiform caps present in all bundles on the side towards the central ground tissue. *Secretory cells*: none observed.

Carex camptoglochin

Leaf surface

Prickles: present at or near leaf margin in no. 145 (Plate **1A**); in no. 197 no prominent prickles. Prickles small, very thick-walled, with tip pointing towards leaf apex, hardly extending above leaf surface, or without tip; often in groups. *Stomata*: paracytic, confined to abaxial surface (Plate 1**D**). Length of stomata and epidermal cells, see Table 1. Stomatal papillae absent. *Silica-cells* frequent among cells above sclerenchyma; silica-bodies conical, in one row (Fig. 1), sometimes inconspicuous, one or two per cell in no. 145, one to four in no. 197; satellites frequently inconspicuous.

Transverse section of lamina

Outline: thickly crescentiform or subcircular, 0·8–1·0 mm in diameter; no keel, margins thick, pointed (Fig. 5; Plate 1**H**). *Epidermal cells* isodiametric, outer and radial walls thickened (Plate 1**F**); adaxial cells larger than abaxial. *Hypodermis* absent. *Sclerenchyma*: median vascular bundle and median bundles in each half of lamina supported by an abaxial girder, always wider than tall; these bundles adaxially and all others both adaxially and abaxially with crescentiform caps; adaxial marginal strand wide, formed of *c.* 45–60 cells. *Mesophyll*: chlorenchyma not restricted to peripheral parts, but surrounding air-cavities; not differentiated into spongy and palisade layers; substomatal cavities small, for example, *c.* 10 μm deep; cells lining cavities not different from other chlorenchyma cells (Plate 1**F**). *Air-cavities*: four or five groups of inflated, translucent cells in mesophyll, surrounded by chlorenchyma. *Bulliform cells* indistinguishable from other adaxial epidermal cells; no slightly lignified, translucent cells beneath adaxial epidermis. *Vascular bundles*: five to seven in one row near abaxial surface, median bundle and bundles in middle of each half of lamina larger than remainder. *Bundle sheaths*: inner sheath sclerenchymatous, outer parenchymatous, but interrupted by sclerenchyma adaxially and abaxially; more around the larger vascular bundles. *Secretory cells* not observed.

Transverse section of culm

Outline subcircular or very slightly angular, 0·8–1·2 mm in diameter (Fig. 3). *Central ground tissue* formed of very large cells, gradually diminishing in size towards the periphery; cells in the outermost part of ground tissue with slightly thickened walls, e.g. around the triangular intercellular spaces. *Air-cavities*: incipient, group of inflated, translucent cells between vascular bundle pairs. *Chlorenchyma* present around air-cavities; cells isodiametric, not differentiated into spongy and palisade cells. *Vascular bundles*: eight to twelve in number, in one ring. *Bundle sheaths*: as in the lamina. *Sclerenchyma*: girders supporting all vascular bundles, except occasionally one or a few of the smaller ones; girders wide, usually wider than tall; the side of the vascular bundles facing the pith supported by thick, crescentiform caps. *Secretory cells* not observed.

DISCUSSION AND CONCLUSIONS

The material used for this study consisted of only a few rather small herbarium specimens, and it was not possible to be absolutely consistent in making the anatomical slides, which partly explains the considerable variation observed within the material. Fortunately, this, in a way, emphasizes the qualitative differences. In the following, some of the characters are treated as quantitative, because of the small amount of material, although a larger body of material may show them to be true qualitative differences.

Quantitative characters

Prickles were observed on leaves of *Carex camptoglochin* only; however, in one of the specimens studied, these were not definitely observed, although the structure of the epidermal cells was otherwise similar. Papillae, again, were noted only in the Finnish and Canadian *C. microglochin* specimens, whereas none were found in specimens from the European Alps and Tierra del Fuego. These differences may turn out to be qualitative when more extensive material is examined.

Measurements of stomatal length revealed an interesting sequence in stomatal size. *C. camptoglochin* has the smallest size and narrowest variation range, the means being only 36·5 and 39·2 μm. In *C. microglochin* the stomata are larger, the means ranging from 44·7 to 47·3 μm; however, in the Finnish specimens they were found to be 48·6 and 58·2 μm. A similar sequence could be found in the sizes of epidermal cells.

Table 1. Length of stomata and abaxial epidermal cells. The variation range is given, in the case of stomata, with the mean in the middle

Specimen	Stomata (μm)	Epidermal cells (μm)
Carex microglochin		
no. 124	42–46·0–50	37–88
no. 142	44–47·3–54	50–94
no. 194	38–44·7–52	42–74
no. 195	36–48·6–58	40–108
no. 196	(42) 54–58·2–68	38–110
C. camptoglochin		
no. 145	32–36·5–42	24–86
no. 197	35–39·2–44	30–98

Another set of quantitative characters is found in transverse sections of the leaf and culm. The size and shape of the supporting sclerenchyma girders seem to show differences between *C. microglochin* and *C. camptoglochin*; in the former they are always taller than wide, mostly baculiform, but in the latter usually wider than tall. There is also a marked difference in the size and shape of the marginal sclerenchyma strands.

A considerable variation was found in the outline of the transverse section of the lamina, but no conclusions can be drawn from the present material. The same applies to the diameter of the culms. *C. camptoglochin* has thicker culms, in the material examined; this may reflect a real difference between the species, because it is considered to be the more robust and taller of the two species.

Qualitative characters

Transverse sections of lamina and culm, however, reveal differences of a more fundamental nature, which are here called qualitative. These characters seem to be assignable to either *Carex microglochin* or *C. camptoglochin* alone and no gradation between them was observed.

(1) *Vascular bundles*. In *C. camptoglochin*, in transverse sections of both culm and leaf, the vascular bundles occur in a single row or circle. Their distance from the surface is the same, and they are usually supported by a girder, especially in the culm, except for a few very small bundles. In *C. microglochin* alternate vascular bundles are more remote from the surface and, in addition, separated from it by chlorenchyma. The other bundles are large and supported by a girder. It therefore seems that two sizes of vascular bundles are present, which tend to be situated in two rows in the leaves.

(2) *Air-cavities*. In *C. microglochin* usually half or more than half of the mesophyll in the lamina is transformed into one or a few large air-cavities, filled with inflated, translucent cells. In the lamina of *C. camptoglochin* there are several distinct, smaller air-cavities separated from each other by chlorenchyma. Usually in the leaves of *C. microglochin* the central large air-cavity is divided by a strand of special, slightly lignified parenchymatous cells, obviously absent from the leaves of *C. camptoglochin*; this is another qualitative difference between these species.

(3) *Differentiation of chlorenchyma*. The most fundamental difference between *C. microglochin* and *C. camptoglochin* is the differentiation of chlorenchyma in leaves and culms of *C. microglochin* (but not of *C. camptoglochin*) into spongy and palisade parenchyma. The palisade cells seem to occur only below the stomata, the longest cells forming the walls of the substomatal cavities; the surrounding palisade chlorenchyma gradually merges into spongy parenchyma. The palisade parenchyma is accordingly restricted in the leaves to the abaxial side. In the chlorenchyma of *C. camptoglochin* no tendency for differentiation was observed; the compact, homogeneous chlorenchyma very closely resembles that of *Uncinia* (Kukkonen, 1967).

Taxonomic notes

I wish finally to draw attention to some taxonomic corollaries based on the above findings. They seem to throw new light on the morphological and ecological differences mentioned by Roivainen (1954: 198). Roivainen stated that the perigynia of *Carex camptoglochin* (*C. oligantha*) are finally completely reflexed and are few in number (female flowers 2–4, male 2–3), whereas *C. microglochin* has more flowers (female 4–12, male 5–6). In addition, there are differences in the size of the perigynia and in the colour of the whole plant. Furthermore, *C. camptoglochin* is, according to Roivainen, a characteristic species of the western humid areas of Tierra del Fuego and is oligothrophic in its edaphic requirements. *C. microglochin*, on the other hand, occurs in the drier central and eastern parts of Tierra del Fuego and has specialized edaphic requirements, a fact which is well-known in the other parts of its range. The two species seem to differ in general habit, and this is borne out by the material studied here. Even Boott (1867: 174) referred to *C. microglochin* as the 'smaller species'.

No one has so far specifically mentioned that even the mode of growth is different in these two species. *C. microglochin* has typically straight culms; the stolons send out roots, and at the base of the culm the internodes are short. In *C. camptoglochin* the basal internodes are long and the culms are ascending, a character also typical of *C. pauciflora*, another species of this section of *Carex*. These features are well illustrated by Boott in his plate no. 589, where *C. microglochin* and *C. camptoglochin* are drawn side by side. An exactly similar illustration would be produced from the specimens 194 and 197 of this paper. All the evidence points to the existence of two distinct taxa at specific level.

Returning to the question raised in the introduction of whether or not *C. microglochin* occurs in Tierra del Fuego, this has been shown to be so (sample no. 194); however, its range in South America is at present unknown. The distinction between the species is based on qualitative characters. Samples of *C. microglochin* originating from different parts of its range showed differences which are classified here as quantitative. Characters such as the presence of papillae and the large stomatal size found in the Finnish specimens indicate that some of the isolated populations of *C. microglochin* may in future best be treated as separate infraspecific taxa. It is unfortunate that chromosome numbers ($2n = 56, 58$) are known for only certain northern populations of *C. microglochin* (Löve & Löve, 1961), and none have been recorded for *C. camptoglochin.*

ACKNOWLEDGEMENTS

This paper is dedicated to my teacher, Dr C. R. Metcalfe, on his retirement from his post as Keeper of the Jodrell Laboratory of the Royal Botanic Gardens, Kew. Its intention is to recall my stay at the Jodrell Laboratory and to proffer once more my humble thanks to Dr Metcalfe for his kindness to me at that time.

Professor J. Jalas has read the manuscript, Mr M. Korhonen has taken the photographs and Mrs Jean Margaret Perttunen has revised the language. My sincere thanks are due to all of them.

This study has been supported by grants from the National Research Council for Sciences.

REFERENCES

Boott, F., 1867. *Illustrations of the genus* Carex, Part IV: 127–233, Tables 412–600. London: L. Reeve & Co.

Constance, L., Heckard, L. R., Chambers, K. L., Ornduff, R. & Raven, P., 1963. Symposium on amphitropical relationships in the herbaceous flora of the Pacific Coast of North and South America. *Q. Rev. Biol.*, **38**: 109–177.

Egorova, T. V., 1967. Cyperaceae-Juncaceae. In *Plantae Asiae Centralis*, **3**: 1–120. Leningrad: Inst. Bot nom. V. L. Komarov, Acad. Sci. U.S.S.R.

Hultén, E., 1958. The amphi-Atlantic plants and their phytogeographical connections. *K. svenska VetenskAkad. Handl.* (*ser.* 4), **7** (1): 1–340.

Krechetowich, V. I., 1937. Cyperacearum novitates. *Not. syst. Herb. Bot. Acad. Sci. U.S.S.R.*, **7**: 27–37.

Kükenthal, G., 1909. Cyperaceae-Caricoideae. In A. Engler (ed.), *Das Pflanzenreich*, IV, **20**: 1–824.

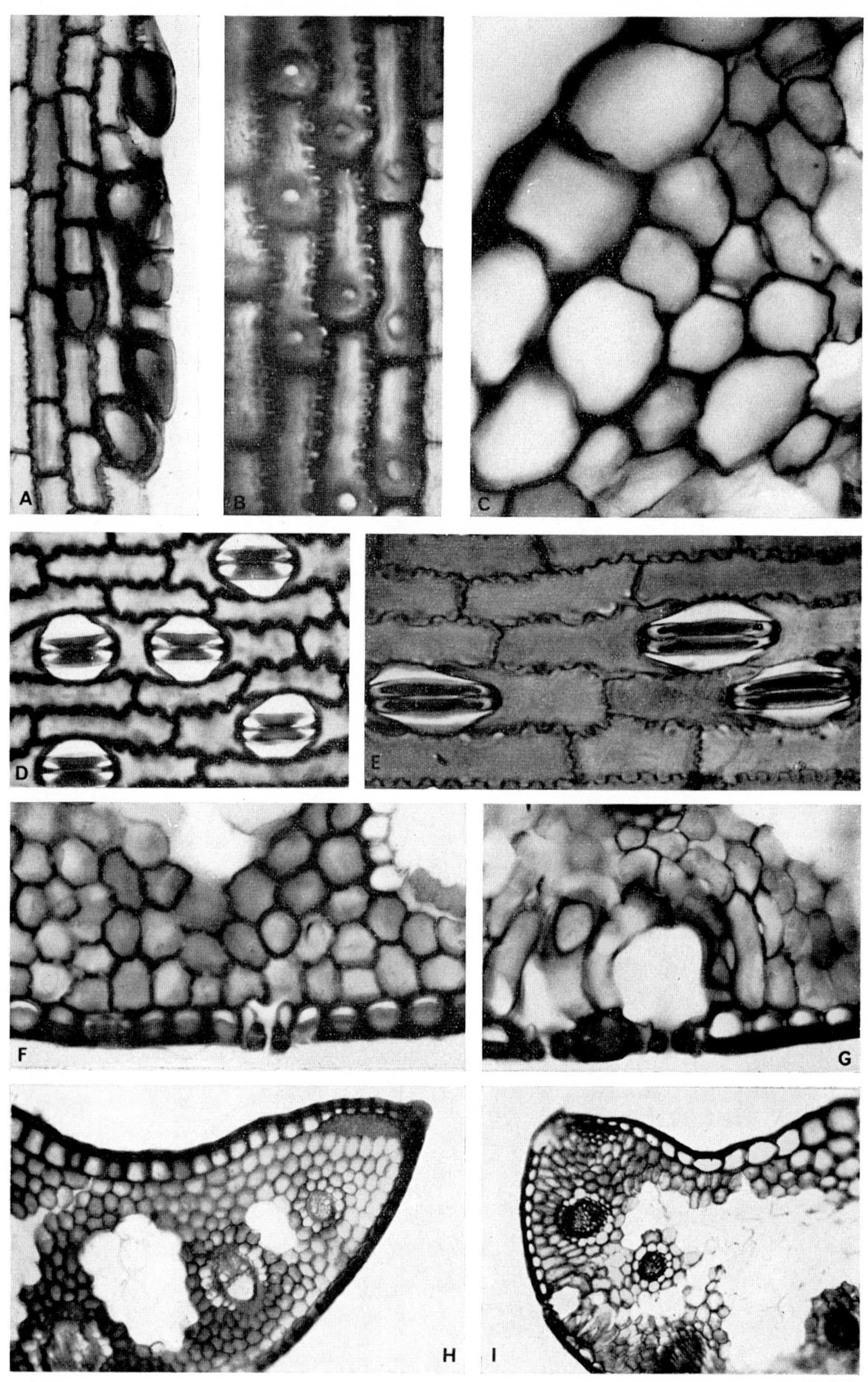
A
B
C
D
E
F
G
H
I

Kukkonen, I., 1967. Vegetative anatomy of *Uncinia* (Cyperaceae). *Ann. Bot., N.S.*, **31**: 523–544 (incl. an 'Addendum' by C. R. Metcalfe).

Löve, A., 1967. The evolutionary significance of disjunctions. *Taxon*, **16**: 324–333.

Löve, A. & Löve, D., 1961. Chromosome numbers of Central and Northwest European plant species. *Op. bot. soc. bot. Lund*, **5**: 1–581.

Metcalfe, C. R., 1960. *Anatomy of the monocotyledons*. **I**. *Gramineae*. Oxford: Clarendon Press.

Metcalfe, C. R. & Gregory, M., 1964. Comparative anatomy of monocotyledons. Some new descriptive terms for Cyperaceae with a discussion of variations in leaf form noted in the family. *Notes Jodrell Lab.*, **1**: 1–11.

Roivainen, H., 1954. Studien über die Moore Feuerlands. *Ann. bot. Soc. zool.-bot. fenn. Vanamo*, **28**(2): 1–205.

EXPLANATION OF PLATE

Plate 1

A. Prickles on leaf margin of *Carex camptoglochin* (×350).
B. Papillae on adaxial surface of lamina; *C. microglochin* no. 195 (×400).
C. T.S. translucent cells under middle part of adaxial surface; *C. microglochin* no. 195 (×500).
D. Abaxial leaf surface of *C. camptoglochin* (×400).
E. Abaxial leaf surface of *C. microglochin* no. 196 (×400).
F. T.S. chlorenchyma and stoma in abaxial leaf surface of *C. camptoglochin* (×400).
G. T.S. chlorenchyma and stomata in abaxial leaf surface of *C. microglochin* (×350).
H. T.S. leaf margin of *C. camptoglochin*; note prickle at margin (×125).
I. T.S. leaf margin of *C. microglochin* (×100).

Anatomy of *Petrosavia stellaris* Becc., a saprophytic monocotyledon

MARGARET Y. STANT, F.L.S.

Jodrell Laboratory, Royal Botanic Gardens, Kew, England

The form of the plant is briefly described in relation to its mode of nutrition. The morphology and histology of the shoot apex together with the inception and development of leaf primordia throw some light on the growth habit and method of perennation. Microscopic observations of leaf, stem and root are recounted in detail. Macerations of stem tissue provide information on variations and specialization of cell types within the conducting and supporting tissues. All the anatomical data are assessed in relation to postulated affinities with other groups, and the saprophytic condition.

CONTENTS

INTRODUCTION

Petrosavia is the sole genus of the family Petrosaviaceae. It is a rare saprophyte confined to the dense tropical rain forests of Malaya, Borneo, southern China and Japan.

This peculiar plant was originally described as parasitic on roots, but it is known now to be a saprophyte, one of several saprophytic families belonging to the monocotyledons. *Petrosavia* is a small perennial, 10–15 cm in height, consisting of a narrow erect stem bearing alternate colourless scale-like leaves and a terminal inflorescence. Towards the base of the stem, the internodes shorten, become thicker and merge with an underground rhizome covered with sheathing leaf scales. From this basal stem arise a few long thin roots, mainly concentrated at the junctions with aerial stems.

The whole plant is devoid of chlorophyll; its structure is simple and adapted to the specialized mode of nutrition. Leaves are extremely reduced and endotrophic mycorrhizae proliferate in the root cortex. The primitive state of differentiation of the vascular tissue can be related to the saprophytic habit.

The only previous investigation of this genus is that of Groom (1895), which contains a perfunctory account of its morphology and structure. A more detailed study of the anatomy of this curious and interesting plant is therefore long overdue. On receipt of suitable material, I have been able to rectify this omission and an account of the work follows.

TAXONOMY

Petrosavia was placed in the Melanthaceae, a group of the Liliaceae, by Beccari (1871), who first described the genus. This determination was subsequently confirmed by Bentham & Hooker and Engler & Prantl. Groom (1895) regarded *Protolirion*, which he separated from *Petrosavia*, as a link between the Liliaceae and Triuridaceae. However, *Protolirion* is considered now to be synonymous with *Petrosavia*. Hutchinson (1959) considers the true affinity of *Petrosavia* to be with *Scheuchzeria* and the Triuridaceae, which are also saprophytes. He awards family status to *Petrosavia* as the Petrosaviaceae with two species, *P. stellaris* Becc. and *P. sakuraii* (Makino) J. J. Sm. ex Van Steenis. *P. borneensis*, previously separated as another species, is included now in *P. stellaris*.

Erdtman (1952), following Engler's system, describes the pollen of *P. sakuraii*, a species from Japan, under the Liliaceae, including the Petrosaviaceae in the Melanthoideae adjacent to Tolfieldieae.

MATERIALS AND METHODS

A morphological examination was made of all specimens of the genus *Petrosavia* in the Herbarium at Kew.

The material examined anatomically consisted of whole plants of *Petrosavia stellaris* Becc. from Mt. Kinabalu, North Borneo, fixed in formalin-acetic-alcohol. Some specimens labelled *P. borneensis* were also included in the investigation together with a selection of revived herbarium material. Material of *P. sakuraii* was not available for anatomical examination.

Leaf, stem and root were sectioned transversely and longitudinally on a Reichert sledge microtome. Sections were stained in a mixture of safranin and Delafield's haematoxylin, dehydrated and mounted in Canada balsam. Stem tissue was macerated in a mixture of equal parts of 10% chromic acid and 10% nitric acid at a temperature of 50°C. After washing, macerations were teased out and mounted in dilute glycerine.

FORM AND STRUCTURE OF THE SHOOT APEX

Flowering stems arise successively in the axils of the leaf scales at the base of a preceding stem or from the basal stem itself. Longitudinal sections through the basal

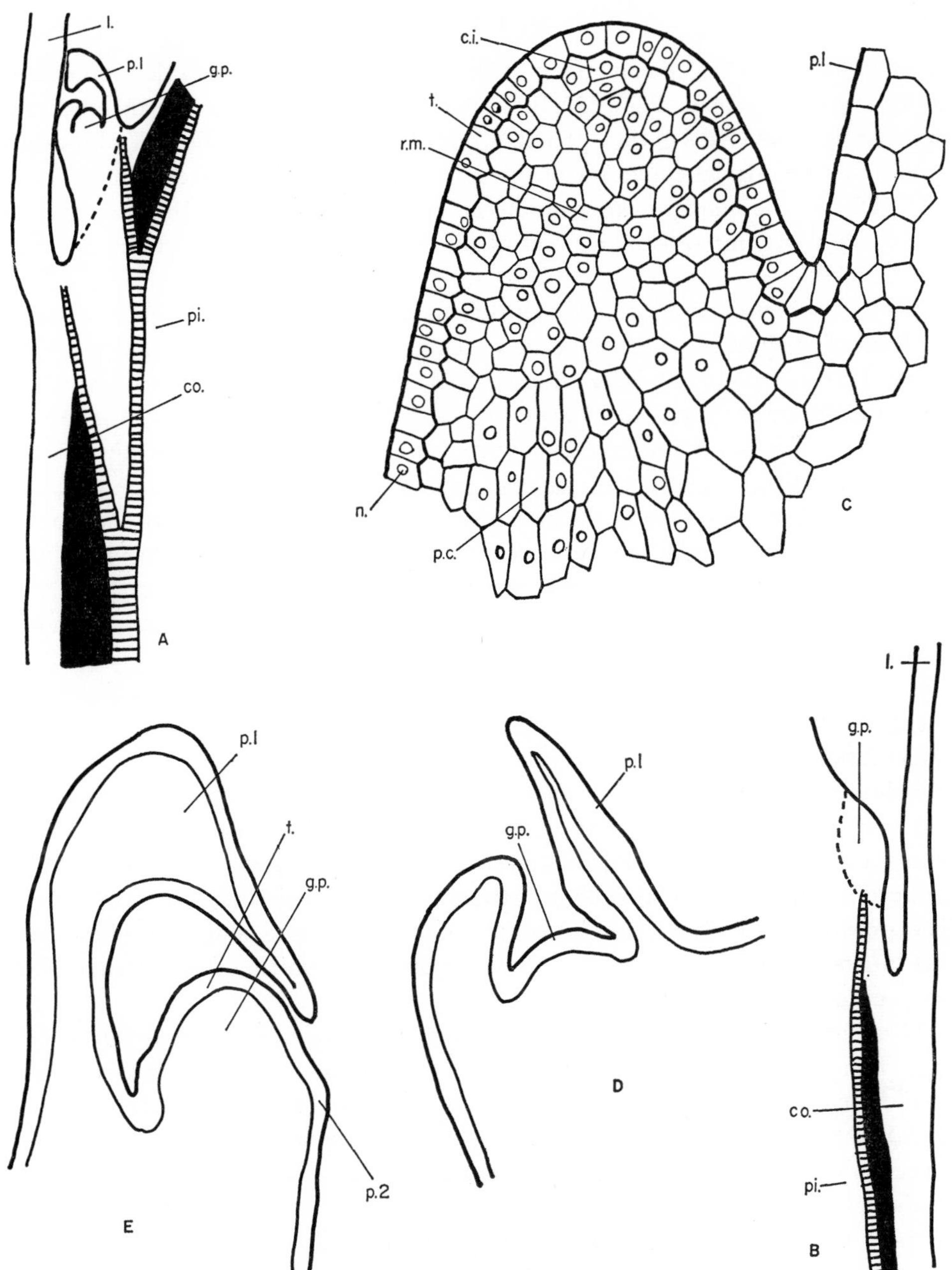

FIGURE 1. *Petrosavia stellaris*. L.S. shoot apex.

A. Main stem with shoot apex in axil of leaf. Vascular bundle (horizontal lines) of stem with branches to leaf and axillary shoot apex. Sclerenchyma cylinder (solid black) with leaf trace gap. Meristematic tissue demarcated by dotted line (×35).

B. Early stage in initiation of new stem apex. Main stem with lateral hump of meristematic tissue in axil of leaf. Shading as in **A** (×35).

C. Growing point meristems. Tunica and corpus with initial cells, rib meristem and procambial strand (×350).

D. Early growing point with leaf primordium (×175).

E. Growing point with 2 leaf primordia (×175).

c.i., Corpus initial; co., cortex of stem; g.p., growing point; l., leaf; n., nucleus; p.1, first leaf primordium; p.2, second leaf primordium; p.c., procambial strand; pi., pith; r.m., rib meristem; t., tunica.

leaves frequently revealed something of the origin, morphology and development of the vegetative shoot apex (Plate **2A**). It is initiated as a cell proliferation in the epidermal and subepidermal layers of the leaf axil, forming a slight bump on the side of the stem opposite to the adaxial epidermis of the scale leaf (Fig. **1B**). Development proceeds (Fig. **1D**), and the stem growing point becomes deeply domed (Fig. **1E**), the organization of its meristems resembling that of other monocotyledons described by Stant (1952, 1954).

The leaf primordia are formed in distichous succession, the first being initiated on the flank of the growing point opposite to the leaf in the axil of which the new shoot arises. The growing point is enveloped by a uniseriate tunica layer in which all cell divisions are anticlinal, i.e. at right-angles to the surface (Fig. **1C**). At the sides of the growing point tunica cells divide more frequently, the cells are narrower with the anticlinal walls closer together, while across the top of the apex, cell divisions are less frequent and the anticlinal walls spaced further apart. As a result of this differential rate in cell division within the tunica the outer tangential walls at the tip are greater in surface area than those at the sides of the apex. Cell walls of the tunica are thin and densely stained and all the protoplasts have prominent nuclei.

A leaf primordium is initiated on one side of the growing point at a distance of about 70–80 μm from the tip. The site is always directly opposite to the preceding primordium and is probably determined by its influence which may be exerted as a combination of spatial and nutritional factors. The leaf initial orginates as a series of anticlinal cell divisions in the tunica, forming a slight fold. This is followed by less regular cell divisions in the subjacent layers of the flank meristem, enlarging the bulge. Cell proliferation spreads laterally, constructing a collar encircling the apex (Fig. **1E**). The growth of the leaf rudiment is maintained at a higher rate on the side of inception and in later plastochrones this activity inaugurates the unilateral appearance of the procambial strand.

There are indications of a second tunica layer over the more vigorous flanks of the apex (Fig. **1C**). If there is an inner tunica, its cells are more irregular in arrangement than those of the outer tunica proper and the layer becomes completely indeterminate at the tip of the growing point. Cells of this second layer occasionally divide tangentially, functioning in the development of leaf initials as part of the flank meristem.

It is possible to distinguish a small group of corpus initial cells immediately within the tunica of the tip (Fig. **1C**). Behind these is a series of cells in which transverse divisions predominate, constituting the rib meristem. At the level of the second leaf primordium, cells to the side of the rib meristem begin to elongate, marking the distal end of the procambium. This merges basipetally with the provascular strand of the new shoot, maintaining continuity with the inner vascular tissue of the basal stem through a large leaf gap in the sclerenchyma sheath (Fig. **1A**).

At a later stage the new axillary shoot elongates, bearing the scale leaves aloft. This is followed by morphogenetic changes in the apical meristems as the vegetative apex is converted to the reproductive phase. An inflorescence terminates the shoot and further growth is sustained by subsequent axillary budding.

ANATOMICAL DESCRIPTIONS OF THE PLANT ORGANS

Leaf

Leaf small, length 3–5 mm, breadth 1–2 mm, scale-like, sheathing at the broader base, sessile.

Surface view

Hairs absent. *Epidermis* (Fig. 2**A**); cells rectangular, elongated and orientated parallel to long axis of leaf; cell walls slightly beaded. *Stomata* confined to abaxial epidermis and distal half of leaf; limited to central midrib region superimposed over vascular strand. Subsidiary cells absent, stomata anomocytic. Guard cells short and wide, pore elliptical.

Transverse section (Plate 1**C**)

V-shaped, triangular in outline, adaxial face deeply grooved. Cells in six layers except at midrib; margins consisting of adaxial and abaxial epidermis only (Fig. 2**D**). *Epidermis* (Fig. 2**B**); cells rectangular in shape, outer tangential walls convex, more thickened than inner walls and slightly cuticularized. *Stomata* confined to midrib region of abaxial epidermis. Guard cells slightly protruding with marked outer and inner ledges of wall thickening (Fig. 2**B**). *Mesophyll* consisting of one to four layers of rounded parenchyma cells with small intercellular air-spaces. Cell walls slightly thickened and lignified. *Vascular bundle* of midrib (Fig. 2**C**) consisting of one central strand of unspecialized vascular tissue. Xylem consisting of thin-walled spiral tracheids. Phloem consisting of sieve-tubes; cells in vertical series separated by narrow oblique sieve-plates. Xylem and phloem intermingled, not in discrete collateral bundles. In some leaves vascular tissue distinguished from surrounding mesophyll by smaller cell diameter only. In herbarium material labelled *P. borneensis*, vascular strand separated from mesophyll by intermittent ring of solitary lignified cells, probably sclerenchyma. *Crystals* not observed in material available. Groom (1895) recorded raphides in mucilaginous mesophyll cells between vascular bundle and abaxial epidermis.

Axis

Stem in T.S.

Scape bearing flowers, diameter 0·5–1·0 mm; circular in outline (Fig. 3**B**), becoming triangular towards the terminal raceme (Fig. 3**A**).

Epidermis: outer tangential cell walls lenticular, slightly thickened, lignified and cuticularized (Plate 1**B**). *Stomata* absent. *Cortex* consisting of three to six layers of parenchyma; subepidermal cells angular in shape in T.S., cells of middle layers more rounded, becoming radially flattened and tangentially stretched towards stelar zone (Fig. 3**C**). Cells rectangular in L.S., with a length/breadth ratio of four to six. Walls slightly thickened and lignified with numerous simple pits. *Endodermoid layers* forming a continuous ring bounding inner cortex. Thickening of inner and tangential walls extensive and stratified, C- or U-shaped. Endodermoid zone frequently two cells deep (Fig. 3**E**) and occasionally with cells of a third layer in material of *P. borneensis*. Passage cells sometimes present; with thinner but evenly thickened walls (Fig. 3**F**).

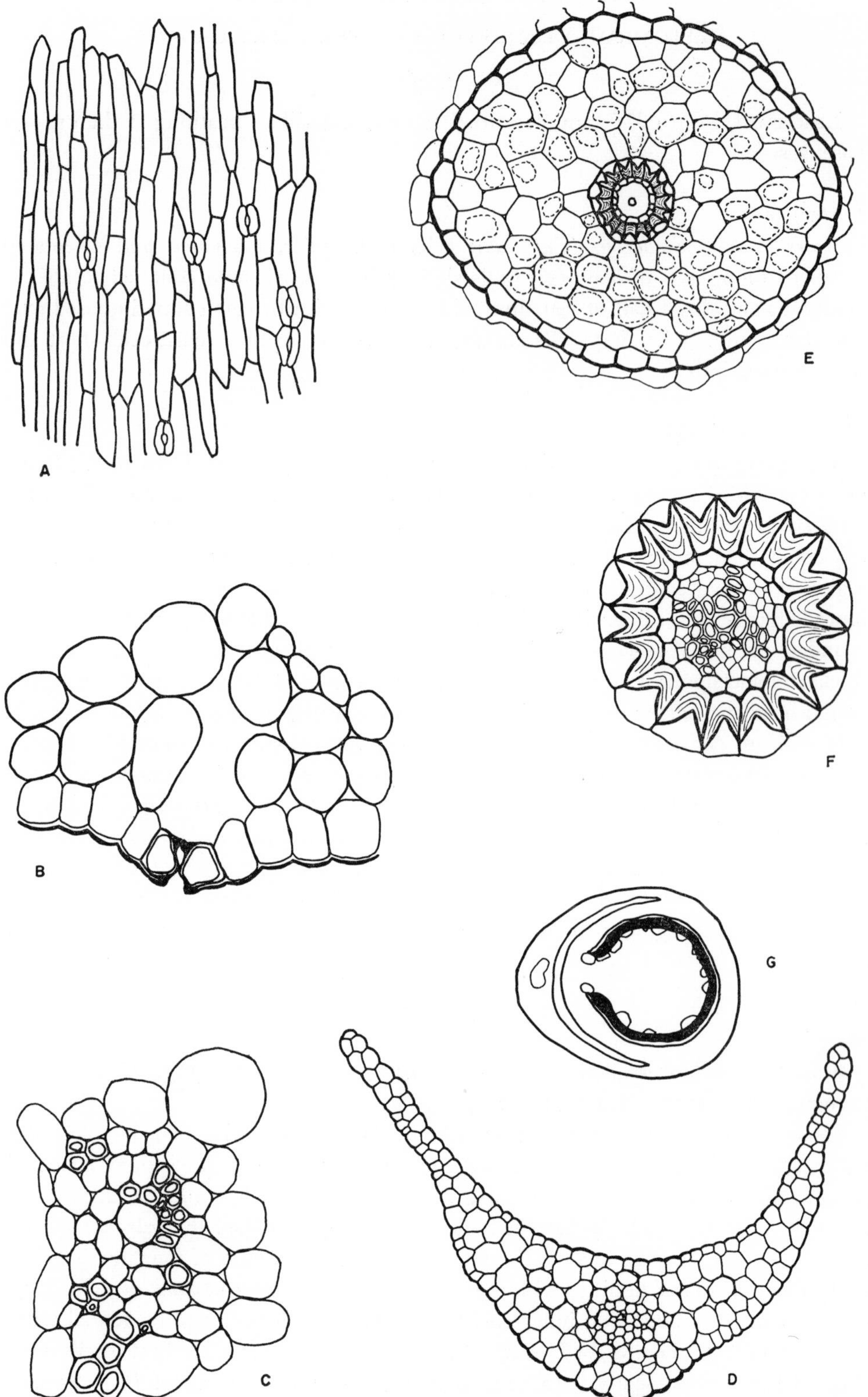

FIGURE 2

Cell walls less thickened in some specimens and in younger or more terminal parts of the stem. *Sclerenchyma* forming a cylindrical sheath within endodermoid layer (Fig. 3**A, E**), consisting of four to six cell layers of thick-walled, heavily lignified elements with small lumina and frequent pitting. The nature of these cells and others included in the vascular tissue will be discussed in more detail in a later section. *Vascular bundles* generally small, numbering 7–14 (Fig. 3**B**); arranged in a circle attached to inner face of sclerenchyma sheath. Size and number varying considerably and difficult to determine since individual strands not always clearly defined, showing a tendency to merge laterally (Plate 1**B**). Larger vascular bundles may be composite, consisting of several smaller ones. Intermediate vascular bundles small and rudimentary, defying specification. Organization of vascular bundles irregular, making the interpretation of their tissues difficult. Phloem consisting of small groups of sieve-tubes with transverse or slightly oblique sieve-plates, and parenchyma cells; tissue frequently surrounded by lignified cells, sclerenchyma or xylem (Fig. 3**C**). Protoxylem consisting of long narrow tracheids with thin cellulose walls and loose spiral thickening; cells frequently disintegrating to form irregular cavities. Metaxylem indistinctly delineated, generally between phloem and protoxylem, sometimes surrounding phloem; consisting of tracheary elements with terminal scalariform thickening or pitting, and bordered pits along parts of the side walls. The composition of this tissue is described in more detail below. *Pith* consisting of parenchyma cells, broader and larger than those of cortex with a length/breadth ratio of 2–4. Cells polygonal in T.S., rectangular in L.S. with thickened lignified walls and frequent conspicuous simple pits. Tissue more or less solid, occasionally breaking down to form air-spaces. *Crystals* absent. No other cell inclusions present.

Basal stem or rhizome in T.S.

Diameter 1·0–1·5 mm. Circular in outline with crowded overlapping leaf scales (Fig. 2**G**). Structure resembling that of more aerial flowering stem. Tissues similarly disposed with an increase in number of cell layers. *Cortex* sometimes wider on one side, possibly under side; tissues much expanded, composed of large squarish sclereid-like cells with very thick lignified walls (Fig. 3**H, I**). Material insufficient to determine frequency and cause of this abnormality. *Sclerenchyma* sheath wider (Fig. 3**D**), cells with thicker walls and more heavily lignified. Phloem more extensive than in upper stem; groups of sieve-tubes and parenchyma cells expanding laterally, sometimes uniting with adjacent groups to form an intermittent ring within sclerenchyma sheath. Small isolated patches of phloem embedded within sclerenchyma. Xylem more

---

FIGURE 2. *Petrosavia stellaris* anatomy.

A–C, Leaf: **A**, surface view of abaxial epidermis with stomata (×110); **B**, T.S. stoma from abaxial epidermis and midrib mesophyll cells (×330); **C**, T.S. vascular bundle from midrib, with scattered fibres (×330).

D. T.S. lamina (×110).

E, F. T.S. root: **E**, cortical cells with mycorrhizal hyphae (broken lines); exodermis; piliferous layer; stele with central vessel-like elements (×110); **F**, stele. Endodermal cell walls with U-shaped stratified thickening; pericycle; four phloem and four xylem groups (×330).

G. T.S. lower stem with sheathing base of scale leaf; leaf trace gaps in sclerenchyma cylinder (×24).

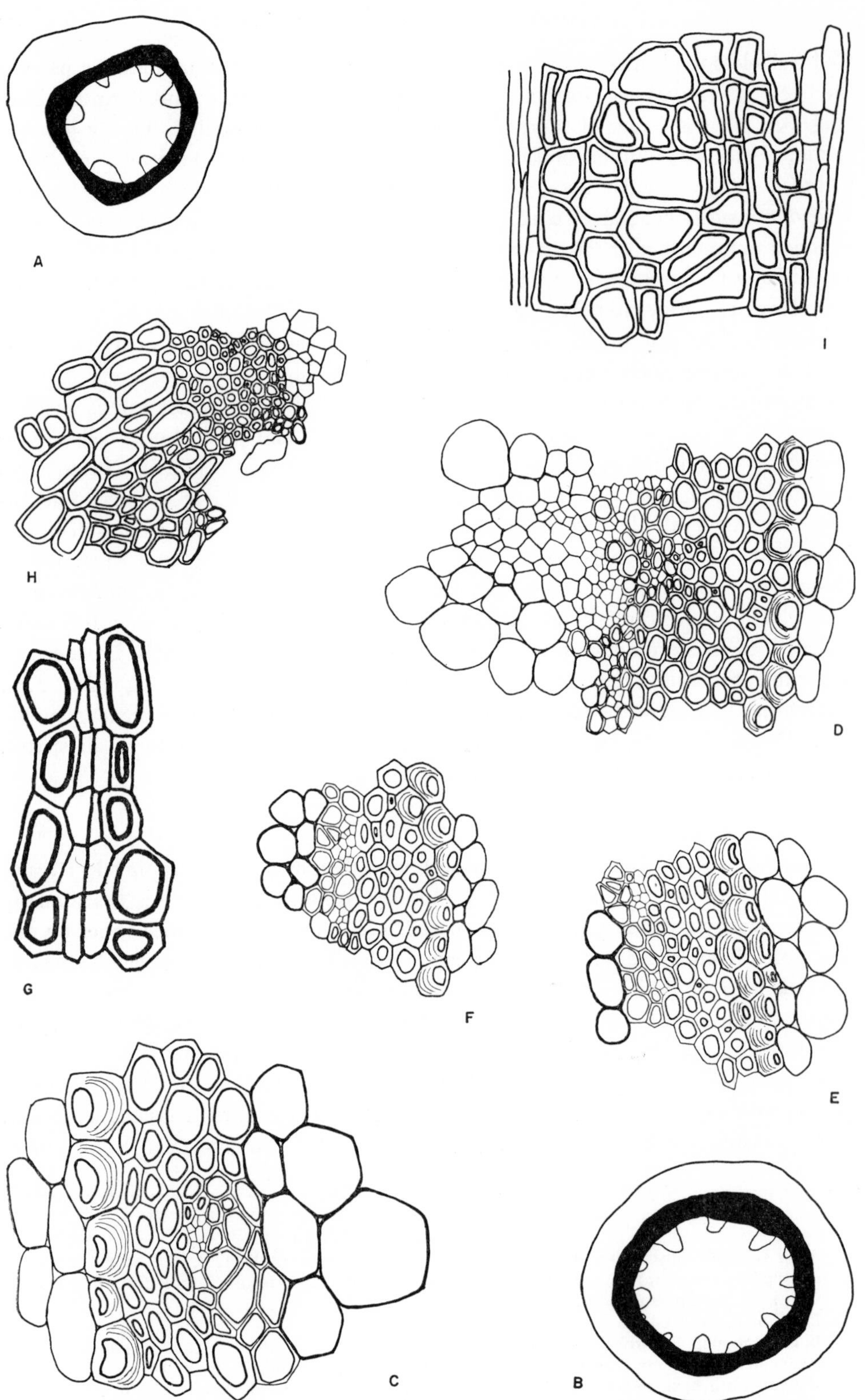

FIGURE 3

developed than in upper stem, bounding inner face of phloem and extending towards centre at some points; consisting of tracheary elements with bordered pits, scalariform and scalariform-reticulate types of thickening more common. Spiral protoxylem elements mostly disintegrated, leaving small irregularly shaped spaces in T.S.

Root in T.S.

Root system sparse. Roots long and fine, diameter *c.* 0·5 mm. *Piliferous layer* partly persistent, composed of thin-walled cells tangentially extended in T.S. (Fig. 2**E**). *Exodermis* consisting of a single layer of cells radially alternating with those of the piliferous layer; anticlinal and outer tangential cell walls thickened, lignified and suberized. *Cortex* consisting of four to six layers of large thin-walled parenchyma cells containing mycorrhizal hyphae (Plate 1**A**). *Endodermis* (Fig. 2**F**); cells elongated longitudinally and orientated in T.S. with radial diameter greater than tangential width; anticlinal and inner cell walls greatly thickened and lignified with marked stratification, thickening occupying more than 50% of the total cell volume, leaving a small triangular lumen. *Pericycle* (Fig. 2**F**) consisting of a single layer of thin-walled living cells radially alternating with endodermal cells. *Stele* small in diameter; vascular tissue not highly differentiated (Fig. 2**F**). Phloem consisting of four groups, each composed of about three narrow cells, one appearing to be an immature sieve-tube with elongated nucleus. Xylem occupying centre of stele; consisting of immature tracheary elements with small bordered pits and lignified parenchyma cells with large simple pits (Plate 3). In the centre, vessel-like elements with elongated scalariform end walls or perforations and small inconspicuous bordered pits or occasional scalariform thickening.

CELL MORPHOLOGY

Stem macerations revealed cells of many types; a complex and continuous variation in cell shape, size and structure. In the region of the cylindrical sheath of lignified cells, including the xylem and adjacent parenchyma of cortex and pith, there is a wide range of cells which do not always fall easily into the normal distinct categories. Many of these cells are lignified, some conspicuously with thick stratified walls, others more lightly with thinner walls. There is a gradual transition between these types and an infinite variety in the nature of their wall sculpturing.

All the elements of this zone are elongated to some degree although there are considerable differences in length and width and in the relative proportions of these two dimensions.

FIGURE 3. *Petrosavia stellaris* anatomy.

A–H. T.S. stem: **A,** upper stem. Sclerenchyma (solid black) with vascular bundles along inner face (×48); **B,** lower stem, as **A** (×48); **C,** inner cortex; endodermoid cells with stratified thickening and small lumina; sclerenchyma; small vascular bundle with phloem and xylem; lignified parenchyma of outer pith (×330); **D,** basal stem; endodermoid layer; sclerenchyma; vascular tissue; pith (×210); **E,** double endodermoid layers (×210); **F,** endodermoid layer with thin-walled passage cells (×210).

G. Radial seriation: thin-walled cambial-like cells between sclerenchyma and xylem (×700).

H. Basal stem with thickening of under side composed of sclerosed cortical cells; sclerenchyma with outline of vascular tissue.

I. L.S. sclerosed cortical cells from basal stem (×110).

This study of the cells mainly concentrates on their walls but the state of their protoplasts is also significant. Some of the cells are obviously dead, but the majority show evidence of living contents, usually a nucleus and frequently cytoplasm as well. There follows a description of the cells and an attempt to interpret them according to location.

Sclerenchyma fibres

The cells are long, narrow and of varying length, the longest being three times the length of the shortest. The end walls differ in shape, one or both sloping or pointed, occasionally bifurcated. Sometimes one or both end walls are transverse or straight and oblique. Cytoplasmic contents were not observed usually but a nucleus is frequently present. The walls are thick and lignified with minute pitting. The pit apertures are small and round, each pit occupying a sloping slit-like groove. The slits are parallel to one another at an acute angle to the side wall and follow a spiral sequence so that pits on opposite walls appear crossed in the same focal plane. The angle of pitting reflects the angle of thickening of the cell wall and is steeper in more elongated elements. The pits are simple or slightly bordered.

Dimensions: mean length 496·7 μm (range 278·8–754·5 μm), mean breadth 20·2 μm (range 9·1–30·3 μm).

Parenchyma

All cells are approximately rectangular; some with rounded, oblique or pointed end walls. The ratio of length/breadth is extremely variable, from 3–30. The longest cells are of the same order of length as the shorter fibre cells, and similar in shape. They are located in the immediate vicinity of the sclerenchyma and vascular tissue. Cells towards the centre of the pith are shorter and wider, and those derived from the cortex are always longer and narrower. Cytoplasmic contents and nucleus are invariably visible. All cells have round simple pits, much larger than fibre pits and not slit-like. Cell walls may be thick and lignified but less thick than in cells of the preceding category. Wall thickening is frequently laid down in a double spiral approximating to an angle of 45° with the side wall. This occurs markedly in elongated elements and in wider ones with a completely transverse end wall. The opposing bands, which may be of wall material or cytoplasmic origin, cross and recross each other, forming a precise pattern with interstices which are square or diamond-shaped. In these regular gaps large rounded or oval pits are situated. The cells usually have prominent nuclei and it is interesting to speculate on their function. They may provide some mechanical support but their appearance is more suggestive of affinity with conducting tracheary elements.

Tracheary elements (Plate **2B–D**)

The cells are long and narrow with overlapping tapering end walls, occasionally bifurcated. Sometimes one or both ends of the cell are slightly greater in diameter than intermediate parts. The terminal parts are specialized. In some cells only one end is differentiated, the other being straight, transverse or oblique. The wall thickening of some elements is transversely banded in the form of a tight double spiral which opens out distally revealing gaps between successive coils. Median parts of the cell may have more continuous wall thickenings with rows of small narrow bordered pits. The angle of spiral and pitting is variable along the length of the cell and there are frequent changes in the angle of inclination. Where wall thickening is more continuous, the thickening becomes more transverse and, terminally, the spaces between the bars of thickening form narrow elongated bordered pits usually orientated at right-angles to the side wall. Such cells can be described as having scalariformly pitted end walls. According to the amount of secondary wall laid down and the proportionate size of the gaps, the end wall is defined as scalariformly pitted or thickened. Both types may be present in the same cell and in close juxtaposition. Pitting or thickening may become scalariform at other intervening positions along the cell or, occasionally, may occur along the whole length. Where scalariform thickening or pitting is confined to one or both terminal zones, it can vary enormously in proportion to the total length of the cell. Scalariform end wall thickening may be branched to form an intricate criss-cross pattern, and the interstitial pits may or may not have well-defined borders. Usually the scalariform zones occupy only part of the circumference of the cell, the opposing face exhibiting small slit-like bordered pits which may be in vertical series or alternating. The scalariform end walls taper off proximally, forming an elongated elliptical shape.

Dimensions: mean length 483·9 μm (range 245·5–812·1 μm), mean breadth 20·4 μm (range 12·1–27·3 μm).

The problem of the interpretation of these cells must be considered. Are they long narrow vessel elements with overlapping end walls providing long scalariform perforation plates? Or, are they tracheids in which scalariform thickening or pitting is confined to a well defined terminal zone? When Fahn's method (1954) of applying pressure to the cells was tried, it was not found possible to displace the scalariform bars. According to this test it can be assumed that an inner cell membrane is present and therefore the walls are not completely perforated. Hence by definition this type of metaxylem element could be called a tracheid.

The length is short in comparison with the tracheids of other families of monocotyledons (Fahn, 1954). Tracheids of members of the Alismataceae are five to ten times longer (Stant, 1964). The cell diameter is also less than usual, about half that in Alismataceae although only slightly less than that of *Luronium*, one of the simplest members of this family. The size suggests that these elements can be defined most aptly as vessel-tracheids since their length is more of the order of a vessel element than a tracheid. Usage of this expression avoids determining criteria between vessels and tracheids and it is a convenient collective term for all the variations and intermediates which appear in *Petrosavia*.

DISCUSSION

The above classes under which cell types have been described, are slightly artificial as there is no clear separation between them and many cells fall into intermediate categories. In spite of this blurring, there are cells at the extremes of the range of variation which are clearly fibres or tracheids. There are others which might be classified as fibre-tracheids and the term vessel-tracheid has also been used. The anomalies seem to be concentrated in the parenchymatous range. At the maximum radial distance from the sclerenchyma, centripetally and centrifugally, cells are simple and obviously parenchymatous. Near the vascular strands and sclerenchyma, parenchyma cells become more complex in structure, diverging towards a tracheidal or fibre-like configuration.

The distinction between sclerenchyma and parenchyma is a fine one and not always easy to determine. Sclerenchyma has been defined as: 'a complex of thick-walled cells often lignified whose principal function is mechanical' (Esau, 1953), 'tissue composed of cells with thickened secondary cell walls, lignified or not, whose principal function is support and sometimes protection' (Fahn, 1967). Both writers proceed to specify the tissue under two constituents, fibres and sclereids, whilst admitting that parenchyma cells may become elongated, developing thick lignified walls and assuming the characteristics of fibres. It seems to me that liberal interpretation of the definition of sclerenchyma could include parenchyma of this type. It is often accepted that fibre cells have pointed ends and parenchymatous ones transverse end walls. It has been demonstrated above that in *Petrosavia* there are fibre-like cells with oblique or approximately transverse end walls and conversely parenchyma cells with sloping ends, together with a host of intermediate forms.

CONCLUSIONS

The significance of the microscopic observations outlined above can be correlated with both the nutrition and the taxonomic affinities of the genus.

Habit

Some of the salient features of *Petrosavia*'s structure can be related directly to its saprophytic habit. The reduced morphology of the leaves is reflected in their simple histology. The tiny scales are provided with a limited number of stomata, facilitating the gaseous exchanges of respiration and transpiration. Elaborate organic compounds normally metabolized in leaves are here fabricated by mycorrhizal activity in the roots, the plant having dispensed with the biologically sophisticated chlorenchyma of the mesophyll and the distribution paths for its products.

Presumably the plant absorbs water and mineral nutrients in the usual manner, and vascular tissue, although indifferently organized, is fairly conspicuous. Soluble organic materials and minerals need to be transported upwards from the fungal-infested roots, the former in reverse of the regular distribution pattern in other plants. It would be interesting to learn how far the phloem participates in this inverted flow and the exact nature of the role played by the xylem elements. It is perhaps significant that the separation of these two conducting tissues is not altogether clear and their arrangement within the vascular strand is unusual and variable. The composition of other tissues shows some degree of reduction or primitiveness which may be related to the plant's way of life.

Linkage with dicotyledons

The most remarkable aspect of the anatomy of *Petrosavia* is revealed in the general topography of the stem section. The structural appearance and arrangement of the vascular bundles in one ring surrounded by a sclerenchymatous sheath is characteristic of the dicotyledonous stem and differs from the usual formation in monocotyledons, with the exception of members of the Triuridaceae. Furthermore, there is a discernible trend towards lateral continuity of the vascular tissue, as larger strands extend, and smaller ones appear in the interfascicular regions.

Within the vascular strands, xylem and phloem are not always collateral and sometimes intermingle. There is no cambium, but occasionally thin-walled, radially flattened cells of cambial-like appearance are observed within the conducting tissue (Fig. 3**G**). The disposition of these cells demonstrates a radial polarity of division, although the divisions are too few to generate the seriation of secondary growth found in dicotyledons. Traces of intrafascicular cambial activity have been recorded by Arber (1918) in many families of monocotyledons and have been interpreted as evidence indicating the derivation of the monocotyledons from the dicotyledons.

Without becoming embroiled in arguments about this hypothesis at the present time, it can be stated that the presence of cambium, ephemeral, vestigial or otherwise, is a feature associated with the dicotyledonous stem. The stem of *Petrosavia*, therefore, can be described briefly as pseudodicotyledonous in structure. This theme will be exploited further in a subsequent publication on the anatomy of the Triuridaceae. It may be possible then to postulate on the order of precedence of the two major groups of flowering plants. As Metcalfe & Chalk (1950) rightly remark with reference to the Ranunculaceae, 'the division of the angiosperms into dicotyledons and monocotyledons is not so absolute as has at times been supposed or implied'.

Phylogenetic trends

Xylem

The reduced nature of the xylem has been referred to. The composition of this tissue in *Petrosavia* is phylogenetically primitive, consisting mainly of elements of the tracheid type or those intermediate between tracheids and vessels. The tracheary elements of the root are closer to a true vessel than those found in the stem. In contrast, the spiral tracheids of the reduced leaves are extremely simple. This situation is reminiscent of the xylem constitution observed in Alismataceae and Butomaceae (Stant, 1964, 1967), where vessels are confined to the root. However, the tracheids or vessel-tracheids of *Petrosavia* are generally shorter than those of many other monocotyledons (Fahn, 1954), and a decrease in xylem element length is accepted as a criterion of evolutionary advancement. The wall architecture of the tracheary elements of the stem and root appears to be relatively complicated. Elements of a similarly specialized type have been noted also in *Scheuchzeria palustris* and may be fairly widespread in the monocotyledons.

Stomata

Stomatal structure is another character which may indicate phylogenetic trends. Although the leaf of *Petrosavia* is so reduced, it has conspicuous stomata of the anomocytic type, in which there are no subsidiary cells accompanying the guard cells. According to Stebbins & Khush (1961), this type of stoma is more advanced than and derived from one with two or more subsidiary cells and may have originated under the influence of xerophytic conditions. One locality of *Petrosavia* is described as dry hill woods at an altitude of 1000–3000 feet. Most of the primitive aquatic groups of monocotyledons have at least two subsidiary cells, and *Scheuchzeria* is equipped with four per stoma. Morphologically, *Petrosavia* and *Scheuchzeria* bear a superficial resemblance to one another, but there is little anatomical similarity, apart from the configuration of their metaxylem cells. *Scheuchzeria* is semihydrophytic and shows anatomical affinities with the aquatic families of Helobiae.

With respect to its stomatal complex, *Petrosavia* belongs to the same category as most families of the Liliales, supporting the systematic position assigned to the genus by Engler. It may also be significant that the saprophytic group Burmanniales has anomocytic stomata, as do the orchids.

The affinity of *Petrosavia* with the lower groups of aquatic monocotyledons and Liliales has been assessed briefly. Consideration will now be given to the concept of a close association with the Triuridaceae.

Relationship with Triuridaceae

My recent investigations of this group have revealed a striking identity between the anatomical structure of stem, root and leaf of the Triuridaceae and Petrosaviaceae. The resemblance is so complete that I would have no hesitation in placing *Petrosavia* in the family Triuridaceae on the basis of anatomical evidence. As both families are saprophytic, the influence of parallel evolution must be taken into account in any taxonomic verdict. Anatomical investigation of two other families of saprophytes, Burmanniaceae and Thismiaceae, may throw further light on the problem.

The issue of distinguishing between anatomical characters that are related to habit and environment and those that point to taxonomic affinity is constant and complicated. Interpretation is particularly difficult when dealing with specialized groups such as saprophytes or hydrophytes.

ACKNOWLEDGEMENTS

I have great pleasure in dedicating this article to Dr C. R. Metcalfe, with whom I have had the honour of working for many years.

I am indebted to Miss D. M. Catling for technical assistance and to Mr T. Harwood who prepared the photomicrographs.

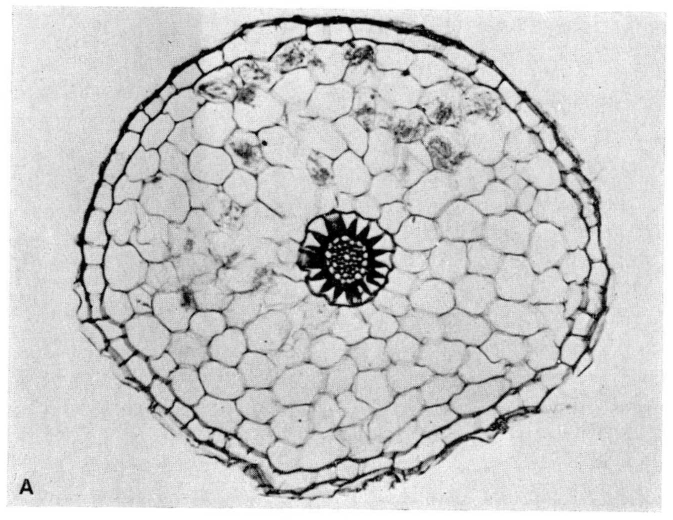
A

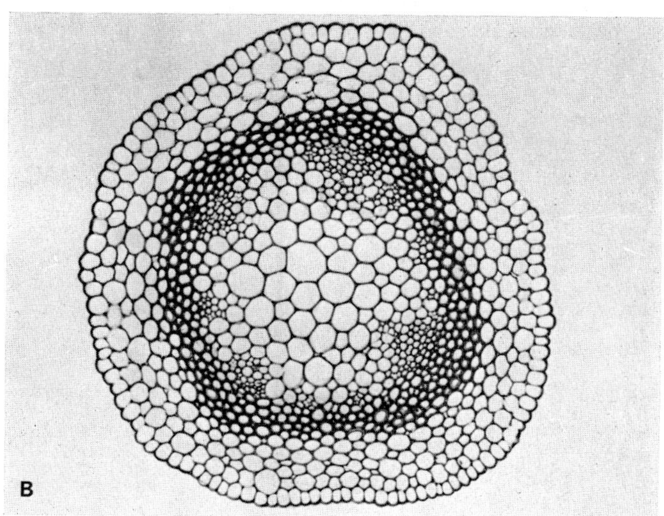
B

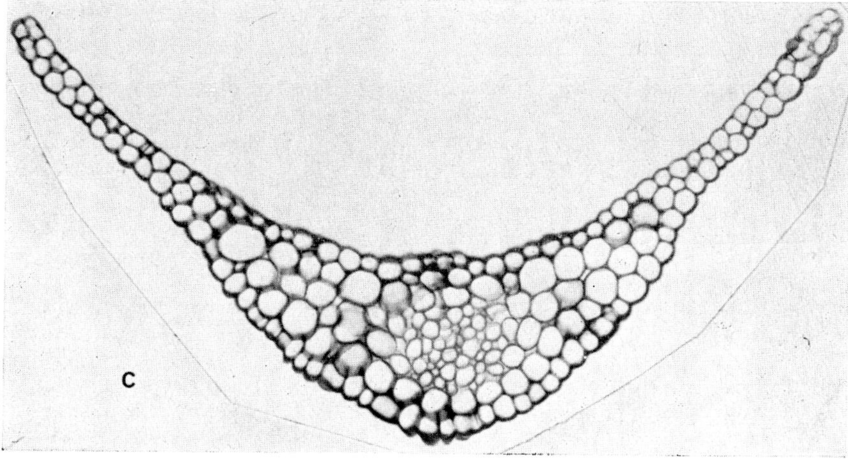
C

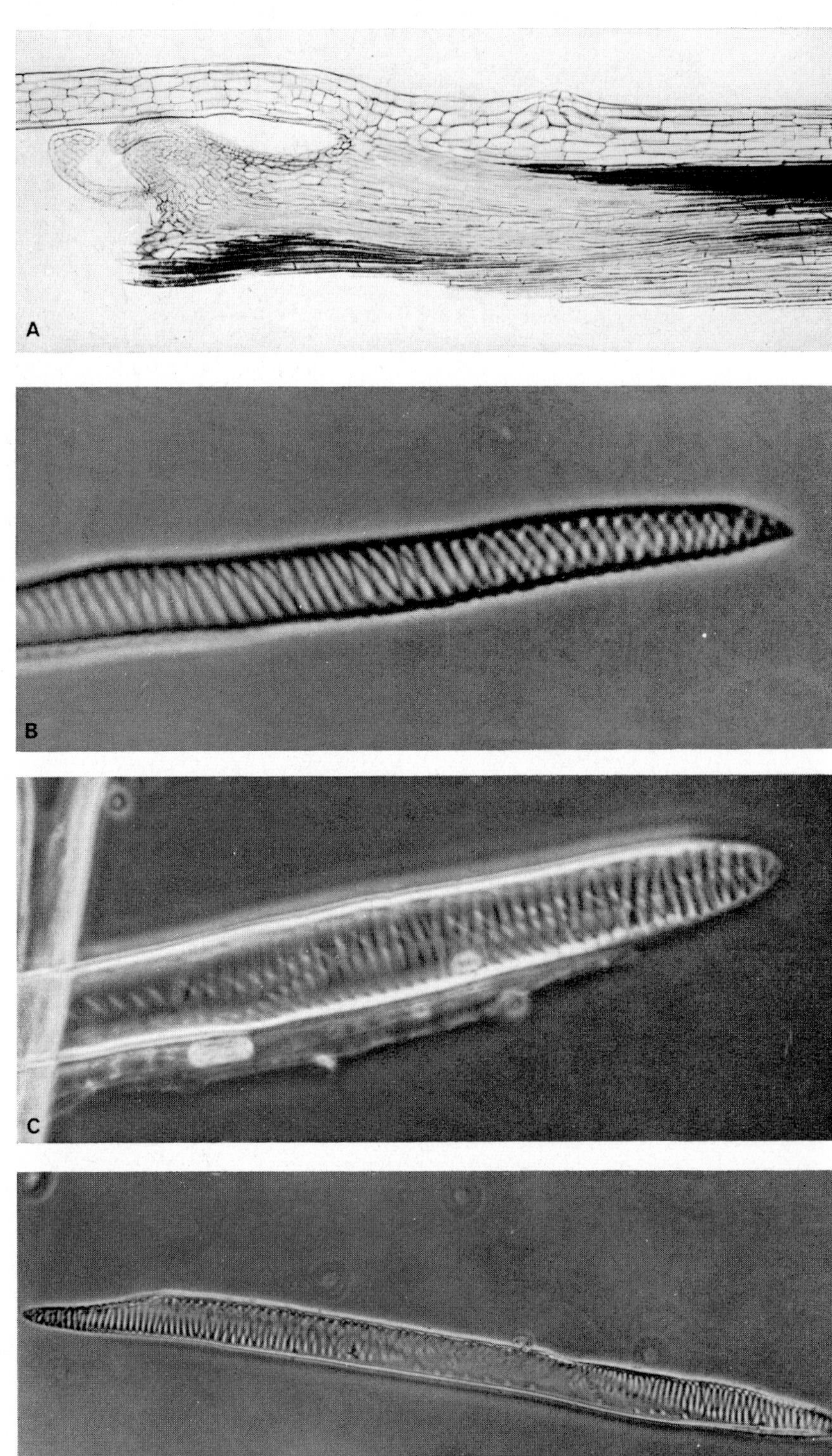

M. Y. STANT

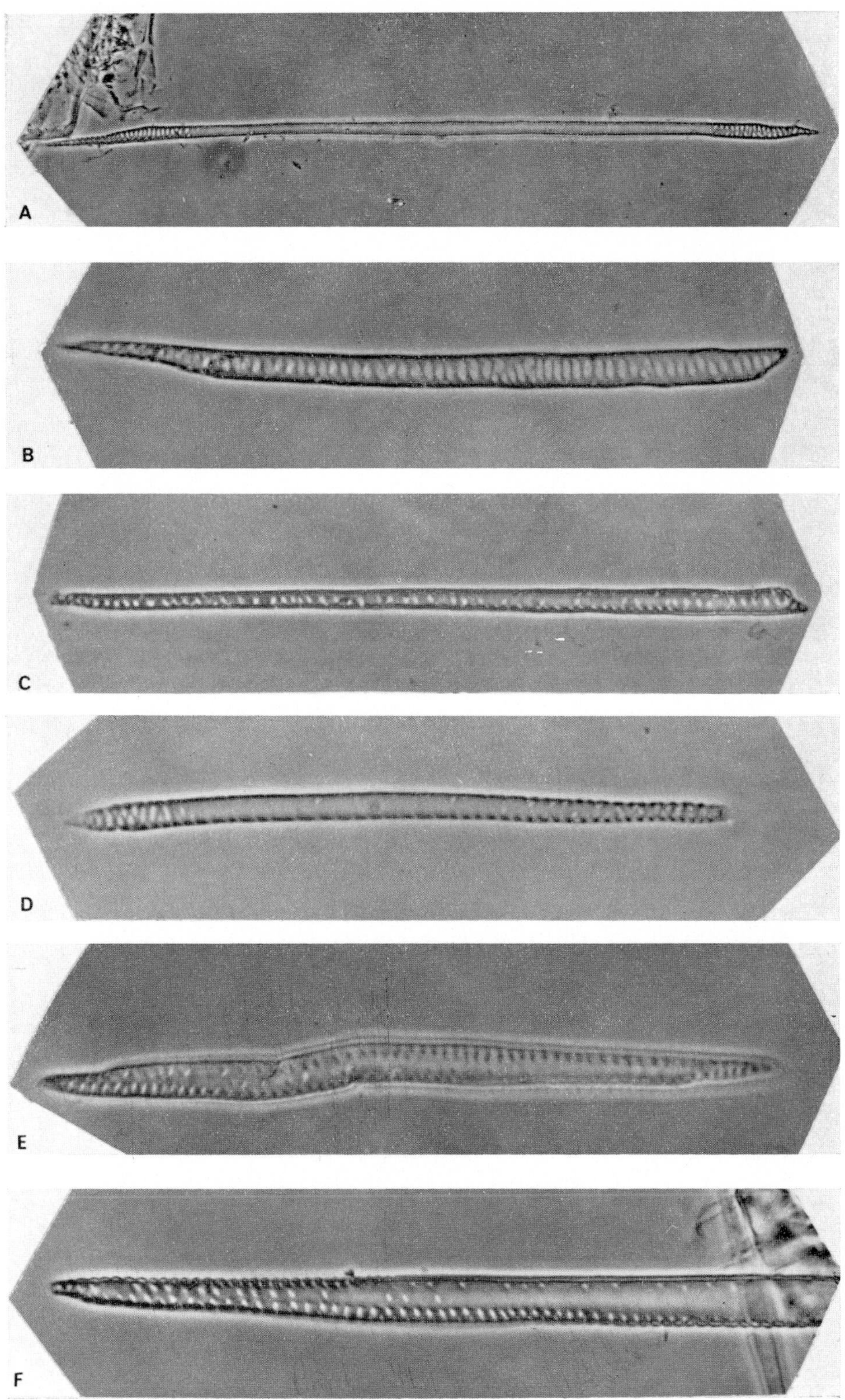

M. Y. STANT

REFERENCES

Arber, A., 1918. Further notes on intrafascicular cambium in monocotyledons. *Ann. Bot.*, **32**: 87–89.

Beccari, O., 1871. *Petrosavia*, nuovo genere di piante parassite della famiglia delle Melanthaceae. *Nuovo G. bot. ital.*, **3**: 7–11.

Erdtman, G., 1952. *Pollen morphology and plant taxonomy*. Stockholm: Almquist & Wiksell.

Esau, K., 1953. *Plant anatomy*. New York: Wiley.

Fahn, A., 1954. Metaxylem elements in some families of the Monocotyledoneae. *New Phytol.*, **53**: 530–540.

Fahn, A., 1967. *Plant Anatomy*. London: Pergamon Press.

Groom, P., 1895. On a new saprophytic monocotyledon. *Ann. Bot.*, **9**: 45–58.

Hutchinson, J., 1959. *The families of flowering plants*. II. *Monocotyledons*, 2nd ed. Oxford: Clarendon Press.

Metcalfe, C. R. & Chalk, L., 1950. *Anatomy of the dicotyledons*, Vol. I Oxford: Clarendon Press.

Stant, M. Y., 1952. The shoot apex of some monocotyledons. I. Structure and development. *Ann. Bot., N.S.*, **16**: 115–128.

Stant, M. Y., 1954. The shoot apex of some monocotyledons. II. Growth organization. *Ann. Bot., N.S.*, **18**: 441–447.

Stant, M.Y., 1964. Anatomy of the Alismataceae. *J. Linn. Soc.* (*Bot.*), **59**: 1–42.

Stant, M. Y., 1967. Anatomy of the Butomaceae. *J. Linn. Soc.* (*Bot.*), **60**: 31–60.

Stebbins, G. L. & Khush, G. S., 1961. Variations in the organization of the stomatal complex in the leaf epidermis of monocotyledons and its bearing on their phylogeny. *Am. J. Bot.*, **48**: 51–59.

EXPLANATION OF PLATES

Plate 1

Petrosavia stellaris

A. T.S. root with fungal mycelium in cortical cells (×100).
B. T.S. young stem (×100).
C. T.S. leaf (×140).

Plate 2

Petrosavia stellaris, stem.

A. L.S. stem with leaf and axillary shoot apex (×40).
B. End wall of xylem tracheary element with bands of wall thickening; phase contrast (×560).
C. End wall of xylem tracheary element with scalariform thickening and bordered pits; phase contrast (×540).
D. Vessel-tracheid from metaxylem with small bordered pits and scalariform ends; phase contrast (×190).

Plate 3

Petrosavia stellaris, root.

Xylem elements of varying types, phase contrast; **A** (×210), **B** (×350), **C** (×340), **D** (×420), **E** (×350), **F** (×420) end wall only.

Short fibres with clearly defined intrusive growth, with special reference to *Fraxinus*

LAURENCE CHALK, F.L.S.

33 Belsyre Court, Observatory Street, Oxford, England

The very short fibres with abruptly constricted ends of intrusive growth which have been described for *Fraxinus excelsior* L. occur also in the xylem of many, possibly all, the other species. In *Fraxinus nigra* Marsh. the mean length of these cells in a pore zone was 450 μm compared with 1280 μm for the fibres in the preceding latewood. The vessel members also were distinctly shorter, mean 250 μm, in the pore zone than in the preceding latewood, mean 328 μm. Apparently similar short fibres have been reported or observed in some genera of the Sonneratiaceae and Melastomataceae.

CONTENTS

INTRODUCTION

In a paper on the ultramicroscopic structure and development of the cell wall, Bosshard (1952) made use of the abruptly narrowed and often forked ends of the fibres in the pore zone of *Fraxinus excelsior* L. as parts of the wall clearly formed by intrusive growth, the broader central parts representing the cambial initials from which the cells were derived. Though he described and illustrated these cells, the present author thought that it would be useful to describe their occurrence in other species and to make further comparisons of the changes in length of the fibres and other cells in the inner and outer parts of the growth ring.

MATERIAL AND METHODS

Sections of the following specimens of *Fraxinus* were examined: *Fraxinus americana* L., C.F.I. 134, Y 37685 A and B (root); *F. chinensis* Roxb., R.B.G. Kew; *F. excelsior* L., C.F.I. 12; *F. longicuspis* Sieb. et Zucc. var. *latifolia* Nakai, C.F.I. 2085; *F. nigra* Marsh., C.F.I. 150, P.R. 13A–27, Y 11740; *F. oregana* Nutt., C.F.I. 11243; *F. ornus* L., R.B.G. Kew; *F. oxycarpa* Willd., R.B.G. Kew; and also of *Duabanga sonneratioides* Buch.–Ham., C.F.I. 10322 and *Sonneratia griffithii* Kurz, I.F.I. 2709.

Macerations were prepared of *F. nigra* Marsh., C.F.I. 3247, *Duabanga sonneratioides* Buch.–Ham., C.F.I. 10322 and *Sonneratia alba* Sm., C.F.I. 1852, both the latter from Malaya. The method of preparing the sample of *F. nigra* is of relevance. A narrow strip of wood was dissected out so as to include a growth ring boundary. This was then split along the boundary so as to give two thin strips, one of the pore zone and the other of the latewood of the *preceding* ring, formed from the same initials.

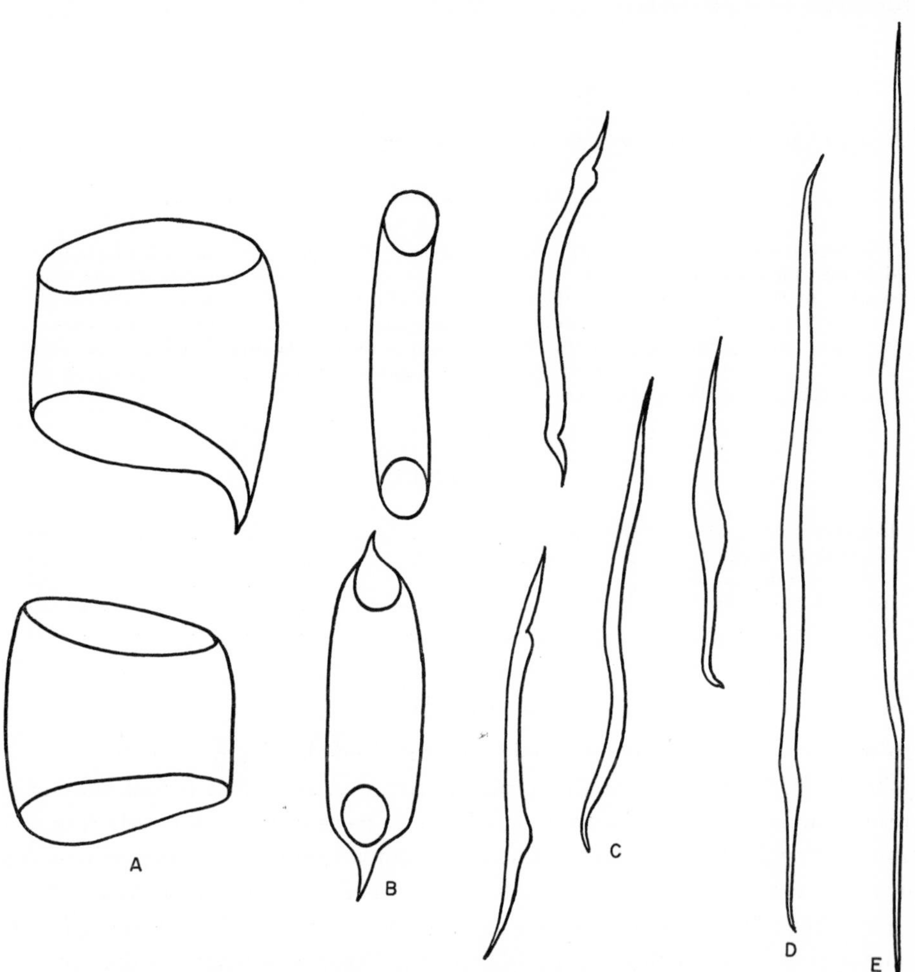

FIGURE 1. *Camera lucida* drawings from *Fraxinus nigra* Marsh. **A,** Pore zone vessel members; **B,** vessel members from preceding latewood; **C,** short fibres from pore zone; **D,** fibres from outer edge of pore zone; **E,** fibres from preceding latewood. × 100.

RESULTS AND DISCUSSION

In tangential sections that include a pore zone, the short fibres may be distinguished by their shape, which resembles that of fusiform parenchyma cells (Plate 1**A**). Indeed, Record & Hess (1943) have described them as such in American species. Such cells were observed by the author in the pore zones of all the species of *Fraxinus* examined.

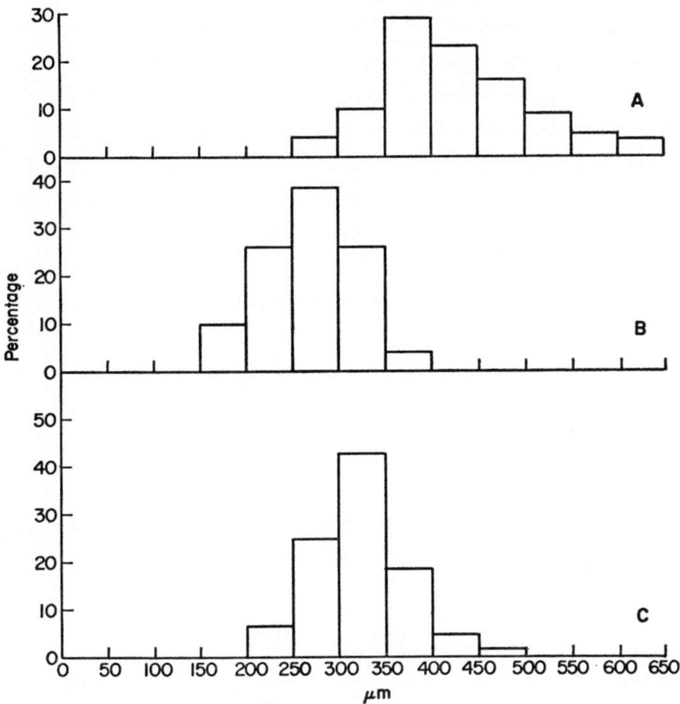

FIGURE 2. The distribution of lengths in *Fraxinus nigra* Marsh. **A,** Short fibres from pore zone; **B**, pore zone vessel members; **C,** latewood vessel members from preceding ring.

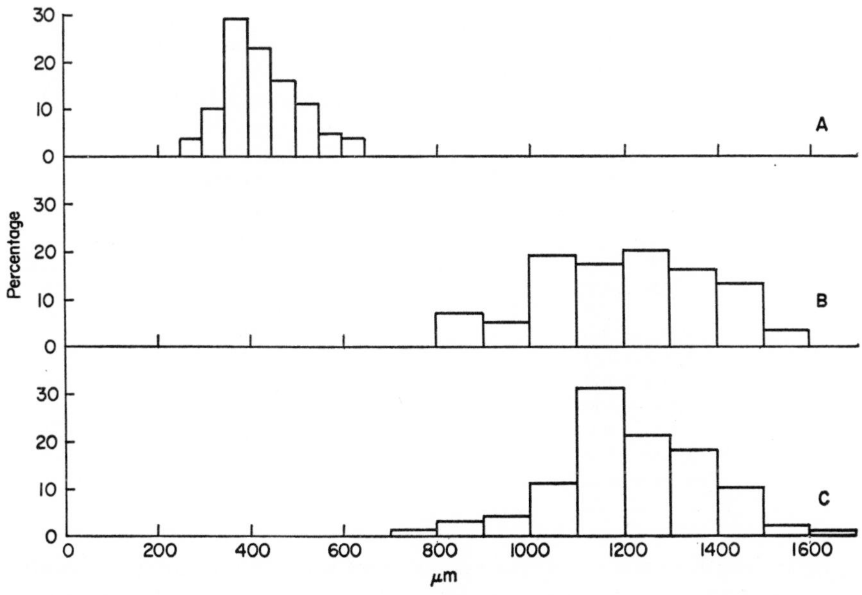

FIGURE 3. The distribution of lengths in *Fraxinus nigra* Marsh. **A,** Short fibres from pore zone; **B,** earlywood fibres from outer edge of pore zone; **C,** latewood fibres from preceding ring.

They were most obvious in wide rings of the stems of the American species *F. americana* and *F. nigra*, and were most difficult to find in the narrowest rings of the one root specimen examined (*F. americana*). Plate **1B** shows the characteristic narrow ends of intrusive growth and some cells with rounded ends and no intrusive growth.

In Fig. 1 *camera lucida* drawings are shown of these short fibres together with drawings of vessel members from the pore zone in which the fibres occurred and from the preceding latewood. Outlines of ordinary fibres from the latewood and early wood are also included. Apart from the short length and the ends of the fibres, the most interesting feature is the decrease in member length between the latewood and the succeeding pore zone. This difference is better illustrated in the histograms in Figs 2 and 3.

Table 1. Lengths of short, pore zone fibres and of vessel members

Length (μm)	Vessel members Pore zone	Vessel members Latewood (preceding ring)	Short Fibres
150–200	5		
200–250	13	4	
250–300	17	15	4
300–350	13	26	10
350–400	2	11	29
400–450		3	23
450–500		1	16
500–550			9
550–600			5
600–650			4
No. measured	50	60	100

Table 2. Lengths of ordinary fibres grouped in classes for histograms

Length (μm)	Fibres Early wood	Fibres Latewood (preceding ring)
700– 800		1
800– 900	7	3
900–1,000	5	4
1,000–1,100	19	11
1,100–1,200	17	29
1,200–1,300	20	21
1,300–1,400	16	18
1,400–1,500	13	10
1,500–1,600	3	2
1,600–1,700		1
No. measured	100	100

The mean length of the short, pore zone fibres was 450 μm, compared with mean fibre lengths of 1240 μm for the immediately following early wood and 1280 μm for the preceding latewood. Bosshard also found longer vessel members in the latewood, but this was in the succeeding latewood. He attributed it to the well established trend for length to increase outwards from the pith at any one level. This trend, however, could not account for the drop in length from the latewood to the succeeding pore zone, which is also difficult to reconcile with the belief that the cambium does not decrease in length except over long periods of time. The difference (about 16%) seems to be too great to be the result of chance and is much greater than that observed by Bosshard (about 4%).

Short fibres of this shape are apparently not restricted to *Fraxinus* or even to ring-porous woods. Moll & Janssonius (1914) refer to short fibres with thin walls and abruptly tapered ends in *Sonneratia acida* L.f., with lengths of 55 to 700 μm compared with 800 to 1100 μm for the ordinary fibres. Pearson & Brown (1932), however, failed to observe such cells in Indian species of *Sonneratia*. The author has observed occasional cells of this type in sections of *S. griffithii* Kurz from Burma and a large number of short fibres without suddenly constricted ends in *S. alba* Sm. from Malaya. Fibres with suddenly constricted ends were observed in *Duabanga sonneratioides* Buch.-Ham., also of the Sonneratiaceae, but these were not restricted to the very short fibres. This, however, may be accounted for on the basis of the wide range of vessel member length, from 430 to 1070 μm compared with a maximum fibre length of about 1600 μm. Moll & Janssonius also record what appear to be similar cells in *Astronia*, *Medinella* and *Melastoma*, all of the Melastomataceae, but not associated with the vessels.

ACKNOWLEDGEMENTS

The author wishes to thank the Professor of Forest Science for permission to use the facilities of the Commonwealth Forestry Institute, Oxford during the preparation of this paper, particularly for access to the collections of slides and woods and for assistance in preparing the illustrations, and Miss M. Gregory of the Jodrell Laboratory, Royal Botanic Gardens, Kew for valuable criticism of the manuscript.

REFERENCES

Bosshard, H. H., 1952. Elektronenmikroskopische Untersuchungen im Holz von *Fraxinus excelsior* L. *Ber. schweiz. bot. Ges.*, **62**: 482–508.

Moll, J. W. & Janssonius, H. H., 1914. *Mikrographie des Holzes der auf Javavorkommenden Baumarten*, Vol. III. Leiden: E. J. Brill.

Pearson, R. S. & Brown, H. P., 1932. *Commercial timbers of India*, Vol. II: 549–1150. Calcutta: Govt. of India Central Publication Branch.

Record, S. J. & Hess, R. W., 1943. *Timbers of the New World*, p. 640. New Haven: Yale University Press.

EXPLANATION OF PLATE

Plate 1

A. Tangential section of wood of *Fraxinus americana*, ×100.
B. Maceration of wood of *F. nigra*, showing fusiform fibres with characteristic narrow ends, ×100.

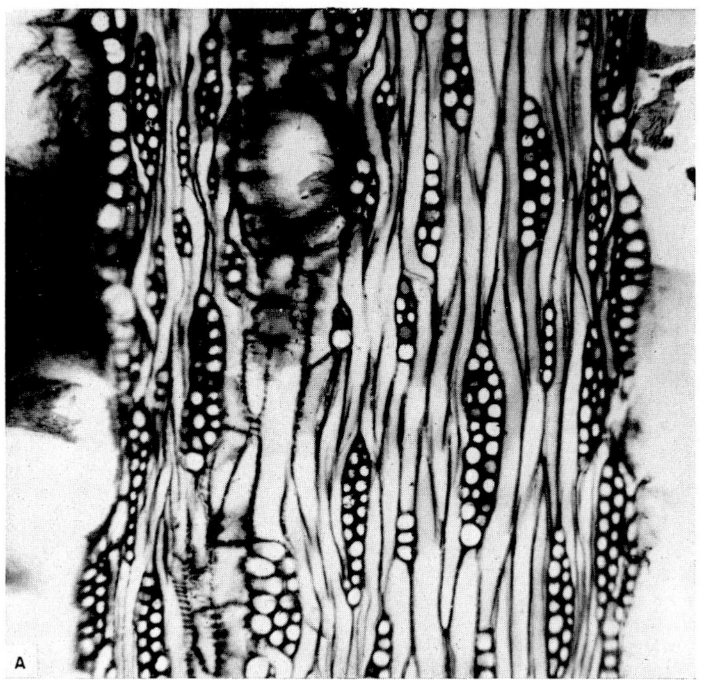

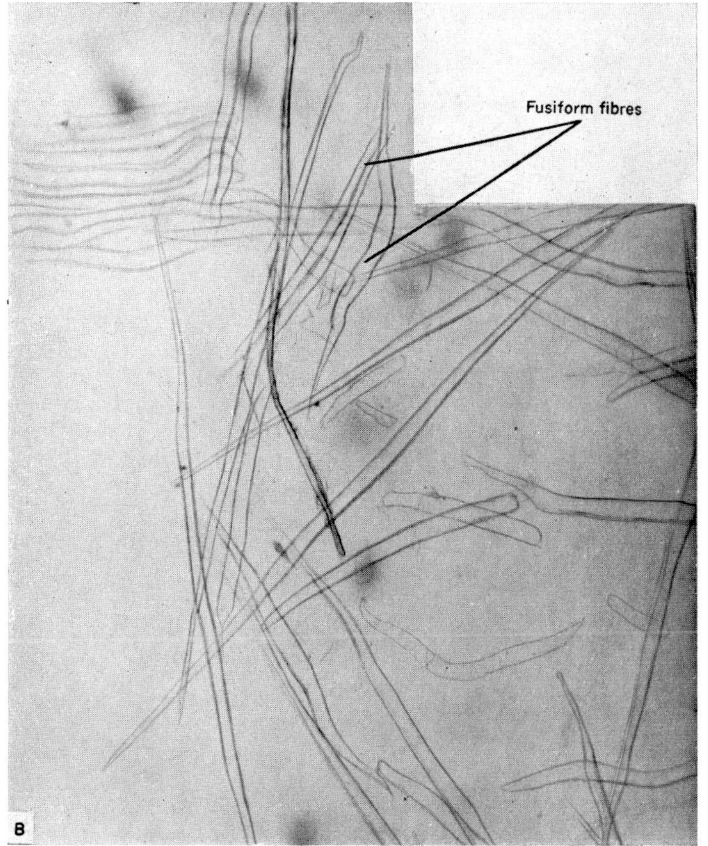

L. CHALK

Seed coat structure and anatomy of Indian pulses

K. A. CHOWDHURY
AND
G. M. BUTH

Department of Botany, Aligarh Muslim University, Aligarh, India

---

Macroscopic and microscopic anatomy of the seed coat of Indian pulses is reported here. These include 14 species belonging to 10 genera of the Papilionaceae. Although they have some common features, there are differences amongst them in anatomy. It has been possible to prepare a key for their classification, based on these differences.

CONTENTS

INTRODUCTION

The object of the present study is to find out whether Indian pulses, all of which belong to the Papilionaceae, could be classified and identified with the aid of their seed coat anatomy. Such a study has not so far been made, probably because the colour of the pulses is so characteristic that, in spite of some similarity in shape and size amongst them, neither the seller nor the buyer experiences any difficulty in distinguishing them. The archaeological plant remains, which are usually recovered in a charred state, do not show their original colour. Identification of these charred pulses is therefore a problem for botanists. At present there is an attempt by archaeologists and botanists to identify the plant remains from pre-historic sites, with a view to tracing the origin of cultivated plants. In this context, this investigation has some practical value. Plant breeders have also been interested in this type of research.

The past work on the anatomy of seed and seed coat can be placed under two main headings. Firstly, there is the study of development of seeds, mostly by embryologists. Here the emphasis is on the sequence of development of different parts of the seed. Pammel (1899), Netolitzky (1926) and Zimmerman (1936) have given considerable information regarding development of the ovule and its integuments in leguminous seeds. Kondo (1913) has also reported on the seed anatomy of some pulses. In a recent review on angiosperm seeds of India, Bahadur Singh (1964) has dealt with some papilionaceous seeds without giving much anatomical detail of the testa.

Secondly, some seeds of economic value have been investigated with different objects in view. For instance, Coe & Martin (1920) and Martin & Watt (1944) determined the structure and chemical nature of the seed coat in relation to impermeable seeds of some leguminous forage plants. A somewhat similar study was made by Ott & Ball (1943) on the seed coat of *Phaseolus vulgaris*. Again, the comparative internal morphology of seeds has been dealt with by Martin (1946). In the same year, Reeve published results of his investigation on the chemical composition and structural characteristics of the walls of sclereids in the seeds of *Pisum sativum* (Reeve, 1946*a*, *b*). The latest comprehensive work on the classification of leguminous seeds has, however, been carried out by Corner (1951). He showed that the structure of seed coat of legumes can be utilized for separating these seeds from seeds of other families. He put emphasis on the structure of 'outer epidermal palisade of the outer integument and its hypodermal layer of hour-glass cells'. He pointed out the importance of characteristic structure of hilum and its associated tissues in papilionaceous seeds. In this connexion Hyde (1954) has also made an intensive study of the hilum in relation to its functions in some fodder plants. Sterling's work (1954) has also thrown further light on the development of seed coat of lima bean. Lately, the importance of identification of isolated seeds for agricultural and commercial purposes has been highlighted by the publication of *Seed identification manual* by Martin & Barkley (1961) and *Identification of crop and weed seeds* by Musil (1963).

Our investigation has shown that it is possible to classify and identify isolated seeds of Indian pulses with the aid of the structure and the anatomy of their seed coat. Here one may need to make a statistical analysis of the height of the palisade cells present in the seed coat.

MATERIAL

Mature seeds of 14 species belonging to 10 genera were obtained from five sources, viz. (1) Indian Agricultural Research Institute, New Delhi; (2) Agricultural College, Kampur; and (3) Agricultural Research Institute, Coimbatore. In addition to these, our study included seeds from (4) the Museum of the Botany Department and (5) some fresh seeds collected from local fields. The list of different species (Maheshwari & Singh, 1965) with their sources, is given below:

(1) *Cajanus cajan* (L.) Millsp. (1, 2, 3, 4)
(2) *Cicer arietinum* L. (1, 3, 4, 5)
(3) *Cyamopsis tetragonolobus* (L.) Taub. (2, 5)
(4) *Dolichos biflorus* L. (2, 5)

(5) *Dolichos lablab* L. (3, 4, 5)
(6) *Lathyrus sativus* L. (1, 5)
(7) *Lens culinaris* Medic. (1, 5)
(8) *Phaseolus aconitifolius* Jacq. (2, 5a, 5b)
(9) *Phaseolus aureus* Roxb. (2, 3, 5)
(10) *Phaseolus mungo* L. (2, 3, 5)
(11) *Pisum sativum* L. (1, 4, 5)
(12) *Vicia faba* L. (4, 5)
(13) *Vigna catjang* Walp. (2, 4, 5)
(14) *Vigna sinensis* (L.) Savi ex Hassk. (3, 4, 5)

The last two species have recently been placed according to the rules of nomenclature under a single species, *Vigna unguiculata* (L.) Walp. (Backer & Bakhuizen, 1963; 1968; Santapau, 1967). We have, however, received these seeds from different agricultural research institutes in India under the names given here, and find some justification, from the anatomical point of view, for keeping the names as sent by these institutes.

METHODS

Five seeds of each species from different sources were selected at random. These were softened in equal parts of glycerine and hot water, and the liquid was changed four or five times a day. After treatment for four to five days, the seeds were pressed between the fingers to find out whether the seed coat would slip off. When this stage of softening was reached, the seeds were ready for embedding. A piece 3 × 3 mm was taken from about 1 mm away from the hilum and a piece of the same size about 2 mm away from the hilum. After the air had been removed using a vacuum pump, these pieces were dehydrated in tertiary butyl alcohol and embedded in paraffin (Histowax, granular, M.P. 56–58°C. Matheson, Coleman & Bell) by the usual method. Serial sections 10–15 μm thick were cut by Spencer's rotary microtome.

Staining was a difficult problem. No single or combined stain gave satisfactory results. Ultimately, crystal violet and erythrosin, or safranin and light green proved to be fairly satisfactory. Seed coats of different genera, and even of different species of the same genus, did not respond equally well to any particular combination of stains. For instance, *Dolichos* and *Cicer* were well stained with crystal violet and erythrosin, while *Cajanus* reacted with safranin and light green. In the case of *Phaseolus*, two species, namely *P. aureus* and *P. mungo*, were well stained with safranin and light green, but for *P. aconitifolius* we had to use crystal violet and erythrosin. We then attempted to measure the height of palisade and hour-glass cells from these serial sections. Although there were more than a thousand sections from each species, yet only a few cells showed clear demarcation between the palisade cell and hour-glass cell to enable accurate measurement. It was then decided to macerate the seed coat and measure the cell dimension therefrom. Three different parts of seed coat were macerated: near the hilum, 3 mm away from the hilum (each piece being 3 mm square) and the entire piece of seed coat leaving out the hilum. Maceration was done with dilute nitric acid and a few crystals of potassium chlorate. The material was then repeatedly washed and centrifuged, and mounted on slides for measurement. The mean height and its standard deviation of palisade cells was determined. Height of hour-glass cells could not, however, be dealt with in the same manner, because in some species the hour-glass cells completely or partially dissolved during maceration.

RESULTS AND DISCUSSION

Size, shape and surface of seeds

Size and shape of pulses vary a great deal. The maximum axial length varies from 4·3–14·3 mm. Seeds belonging to the same genus do not always have the same dimensions. For instance, seeds of *Dolichos lablab* are 11·3 × 7·9 mm while those of *D. biflorus* are 6·5 × 4·5 mm. In another genus, namely *Phaseolus*, the maximum axial length varies between 4·6 and 5·1 mm. In view of these facts it has not been possible to classify these seeds under generic groups based on their size (Table 1). In this context the shape of the seeds is more helpful. There is considerable similarity in the shape of seeds belonging to the same genus, e.g. *Dolichos biflorus* and *D. lablab*; *Phaseolus aconitifolius*, *P. aureus* and *P. mungo*; and *Vigna catjang* and *V. sinensis*. When we compare the colour, we are once again faced with difficulties. The *Dolichos lablab* seed coat is brown or black, while in *D. biflorus* it is light brown and shiny. In the genus *Phaseolus*, the species *mungo* is dark blackish or greenish in colour, *aureus*, light green in colour, and *aconitifolius*, yellow in colour with a greenish tinge. In *Vigna sinensis* it is brownish in colour and in *V. catjang* is whitish to darkish brown or both. These seeds can be further categorized into those that are multicoloured, e.g. *Phaseolus mungo*, *Lens culinaris* and *Lathyrus sativus*, and those that are uniform in colour, whether white, black or brown.

There is also some difference with regard to the surface of the seed coat. In some, the surface is covered with blisters while in others it is smooth. *Cicer arietinum* and *Cyamopsis tetragonolobus* (Table 1) have a blistered surface, more prominent on the former than on the latter. Another important point is that in *Cicer arietinum* the surface of the seed is undulating throughout. This will be further discussed elsewhere. While studying the pulses, one must obtain fully mature seeds. Some immature seeds develop wrinkles during drying. These wrinkles may be confused with the undulating surface of seeds of *Cicer arietinum*.

Hilum

The hilum of the pulses shows a great deal of variation in shape and size. It is largest in *Vicia faba* and *Vigna catjang* and smallest in *Cyamopsis tetragonolobus* (Table 1). The shape varies from round to oblong, oval or elliptical. In the genus *Phaseolus*, the shape is almost similar in all the species, but in *Vigna* the two species dealt with here show a difference (Table 1).

Often, the hilum is obscured by what has been described by Musil (1963) as a corky material. Amongst the 14 species studied by us, four, namely *Lathyrus sativus*, *Lens culinaris*, *Pisum sativum* and *Vicia faba* do not show any corky growth (Table 1).

The remaining ten species all show whitish tissue either partially or completely obscuring the hilum, which is at a lower level than the seed surface. Amongst these, two species show some peculiarity. In *Cicer arietinum* the hilum occurs below the level of seed surface surrounded by a pouch, while in *Cyamopsis tetragonolobus* it can be seen through a circular hole (Table 1).

Table 1. Seeds and seed coat anatomy of Indian pulses

No.	Species	Seeds			Hilum		Palisade cells		Cuticle
		Shape	Max. axial length (mean mm)	Surface	LxB (mm)	Position	Shape	Height	
1	*Cajanus cajan* (L.) Millsp.		6·2 x 4·8	Smooth	3 x 2	Partially covered with whitish hard tissue raised above the level of seed surface		89 ± 13	Smooth, thin
2	*Cicer arietinum* L.		7·2 x 5·5	Prominently blistered	2·3x1·6	In a sunken pouch below the level of seed surface		137 ± 27	Slightly rough
3	*Cyamopsis tetragonolobus* (L.) Taub.		4·3 x 4	Lightly blistered	1 x 0·5	Below the level of seed surface partially covered with whitish tissue		94 ± 5	Smooth
4	*Dolichos biflorus* L.		6·5 x 4·5	Smooth	1·8x0·5	Completely covered with whitish hard tissue		64 ± 10	Smooth
5	*Dolichos lablab* L.	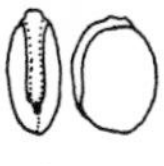	11·3x7·9	Smooth	2·5x2	Above the level of seed surface along prominent raphe running many mm		134 ± 8	Smooth
6	*Lathyrus sativus* L.		4·8 x 4·5	Smooth	2 x 1	In level with seed surface whitish tissue absent		96 ± 8	Dentate
7	*Lens culinaris* Medic.		4·8 x 4·6	Smooth	1·5x0·5	Almost in level with seed surface whitish hard tissue absent		47 ± 5	Dentate
8	*Phaseolus aconitifolius* Jacq.		4·7 x 2·7	Smooth	2 x 1	Completely covered with whitish hard tissue		56 ± 9	Smooth
9	*Phaseolus aureus* Roxb.		4·6 x 3·8	Smooth	2·6x2	Completely covered with whitish hard tissue		56 ± 7	With papillae or papillae-like outgrowth
10	*Phaseolus mungo* L.		5·1 x 4·2	Smooth	3 x 2	Partially covered with whitish hard tissue raised above the level of seed surface		59 ± 6	Rough
11	*Pisum sativum* L.		8·6x6·6	Smooth	3x3	In level with seed surface with whitish tissue absent		89 ± 8	Rough
12	*Vicia faba* L.	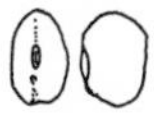	8·4x7	Smooth	4·6x2	Almost in level with seed surface whitish tissue absent		171 ± 12	Slightly rough
13	*Vigna catjang* Walp.	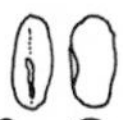	8·8x5·4	Smooth	4·6x2	Completely covered with whitish hard tissue raised above the level of seed surface		76 ± 10	Rough
14	*Vigna sinensis* (L.) Savi ex Hassak.	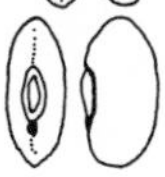	14·3x7·5	Smooth	3x2	Completely covered with whitish hard tissue		64 ± 11	Rough

Cuticle

This is the outermost layer of the seed coat and best seen in the longitudinal sections. In the species studied, only one, namely *Cajanus cajan* (Table 1), has a very thin cuticle, the rest have a fairly thick cuticle. In the latter, there may be some slight variation in the thickness of the cuticle but not prominent enough for classification into subgroups. Again, the surface of the cuticle may be smooth or rough. Amongst those with a smooth or nearly smooth cuticle can be placed *Cajanus cajan*, *Cyamopsis tetragonolobus*, *Dolichos biflorus*, *D. lablab* and *Phaseolus aconitifolius*. Two of the species with a rough cuticle show a dentate structure. These are *Lens culinaris* and *Lathyrus sativus* (Table 1). In *Phaseolus aureus*, the cuticle has papillae or papilla-like outgrowths. Some longitudinal sections of this species show clear papillae with bulbous bases attached to the cuticle (Fig. 1**A**, **B**). To detect this structure one must have uniformly cut thin sections.

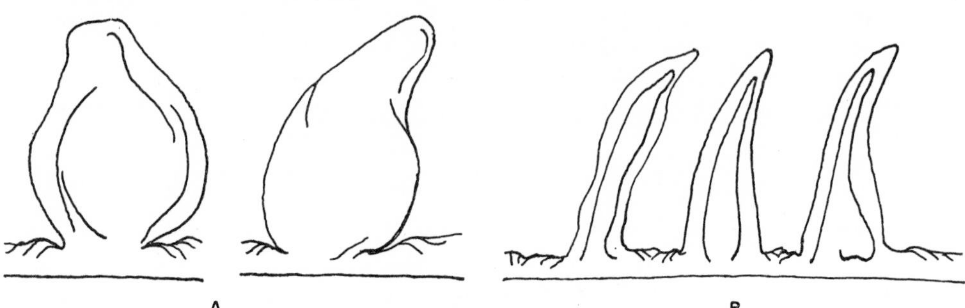

FIGURE 1. *Phaseolus aureus*: **A,** papillae; **B,** papilla-like outgrowths. (×500)

Palisade cells

The palisade cells occur next to the cuticle on the outer part of the seed coat. These are derived from the outer epidermis of the outer integument and are found in two layers—palisade and counter-palisade (Fig. 2**A**), adjacent to the hilum. At some distance from the hilum, the counter-palisade tapers out and the layer of well-packed palisade cells of medium height becomes superficial. Where the counter-palisade layer disappears, the hour-glass cells make their first appearance (Fig. 2**A**). The height of the palisade and hour-glass cells is most uniform at a distance of about 1 mm from the point of disappearance of counter-palisade. As the palisade cells extend towards the centre of the seed, there is a reduction in their height.

All the palisade cells do not show uniformly thick walls. Only in a few is the cell wall uniformly thick throughout; in the rest, thickness varies at different places in the cell. Study of the palisade cell walls in longitudinal sections of the testa and in macerated material shows that it is possible to group the palisade cells under three types. In type I, the wall thickness is usually uniform throughout, e.g. *Cyamopsis tetragonolobus* and *Dolichos lablab* (Fig. 3**A**). In type II, the end of the palisade cells away from the cuticle is bulbous. This is due to less deposition of cell wall material in that particular place (Reeve, 1946; Fig. 3**B**). Included in this type are *Cajanus cajan*, *Dolichos biflorus*, *Lens culinaris*, *Phaseolus aconitifolius*, *P. aureus*, *P. mungo*, *Vigna catjang* and *V. sinensis*. In type III, the inner wall shows a distinctly corrugated structure at the lower end of the cell. Under this type can be placed *Cicer arietinum*, *Lathyrus sativus*, *Pisum sativum* and *Vicia faba* (Figs 2**B**; 3**C**). It may be emphasized here that the walls of palisade cells need careful examination of longitudinal sections of seed coat and macerated material in order to determine the condition present in each species.

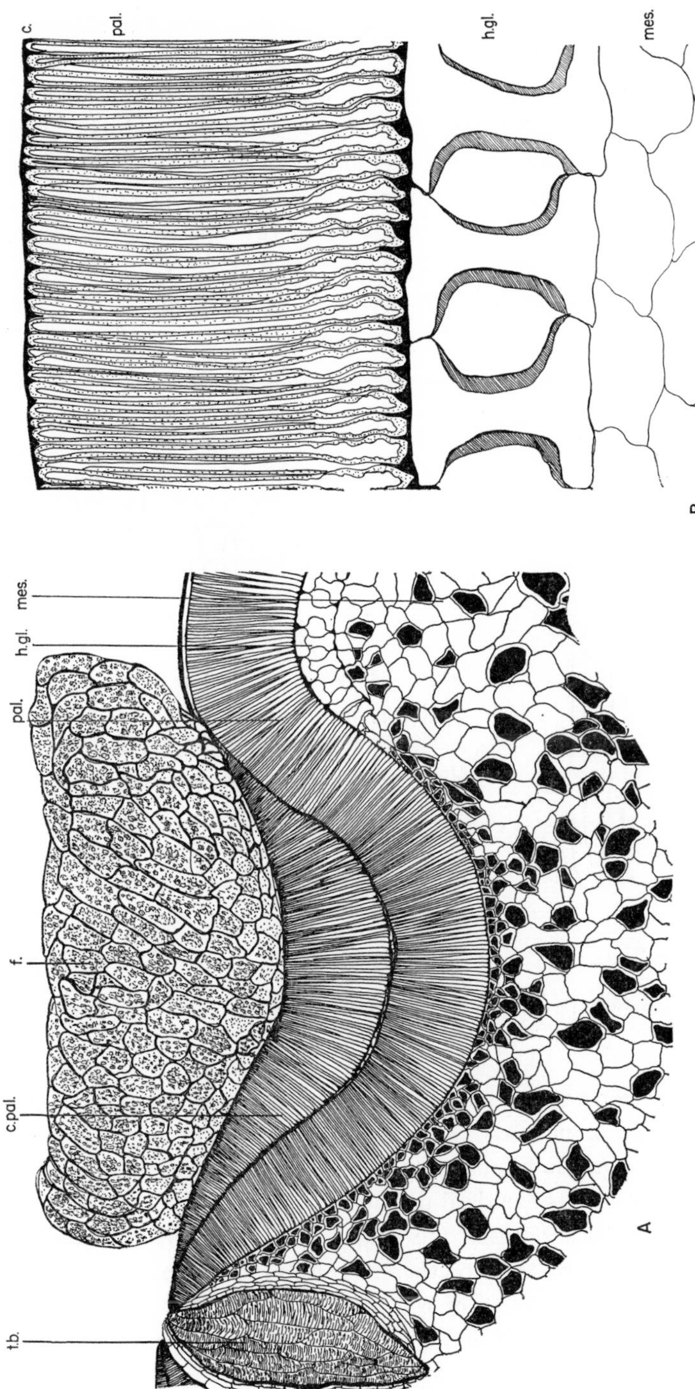

FIGURE 2. **A.** *Phaseolus aureus*: median L.S. of testa through hilum. **B.** *Vicia faba*: L.S. of testa away from hilum. c.pal., Counter-palisade; f., funicle; h.gl., hour-glass cell; mes., mesophyll; pal., palisade cells; t.b., tracheid bar. (×135)

Corner (1951) has shown the origin of the 'Linea lucida' as a continuous boundary line between mucilage stratum and palisade cells. Earlier Hamly (1932, 1934) and Reeve (1946*a*, *b*) have described the same structure as a 'light line'. We did not find a light line in many species. Only *Lathyrus sativus*, *Lens culinaris* and *Pisum sativum* have shown it. The presence of a light line has been associated with the absorbency of the testa. We are not convinced that this is so. We are inclined to believe that it is probably due to a 'simple phenomenon of light refraction' (Marlière, 1897).

While studying the longitudinal sections of the testa, we came across a 'mucilage stratum' only in *Cicer arietinum*, but the seeds of this species did not produce any mucilage after soaking for several days. On the other hand, *Cyamopsis tetragonolobus* seeds produced considerable mucilage after soaking, although no mucilage stratum was found in them.

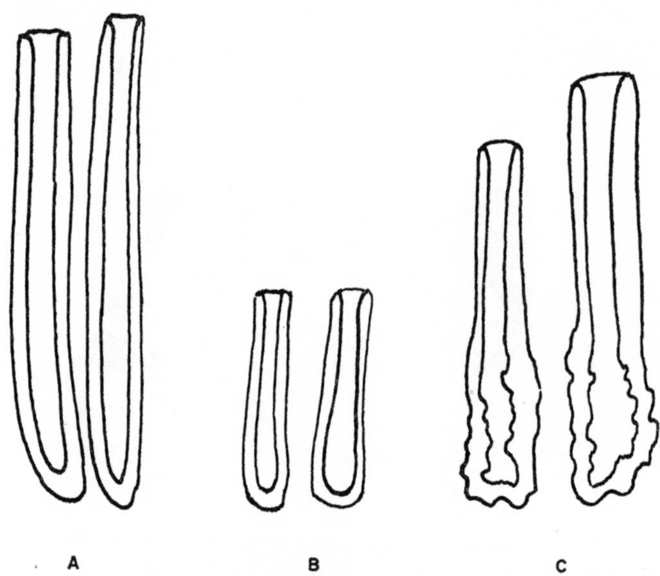

FIGURE 3. Palisade cells. **A.** Type I, *Cyamopsis tetragonolobus*. **B.** Type II, *Phaseolus mungo*. **C.** Type III, *Lathyrus sativus*. (×480)

As mentioned earlier, maceration of seed coat was made with a view to measuring accurately the height of palisade cells. The three different parts, namely near the hilum, 3 mm away from the hilum and an entire piece of seed coat, were macerated, and the mean height and its standard deviation of palisade cells in each portion were determined. Our results show that the height of palisade cells found in the entire pieces of seed coat is most reliable and therefore useful for our purpose of classification.

In *Cicer arietinum*, the surface of the seed coat is undulating. Where there are ridges, the palisade cells are tall and where there are depressions, the palisade cells are comparatively short. This has brought about a wide range of variation in height of the palisade cells in the macerated material measured. That is why in the key, instead of giving mean height and standard deviation, we have given the range of variation found in the species.

Hour-glass cells

These cells are situated between the palisade cells and the mesophyll tissue. Their shape is often characteristically like hour-glasses but not always. Occasionally, the upper and the lower ends are thick, somewhat resembling I-girders. The shape of the hour-glass cells varies from species to species and even within a species. Although longitudinal sections give some idea of the structure of these cells yet they are often not very clear in outline. To have an accurate idea of the cell shape one must macerate the material and study the individual cells.

The size varies to a great extent. The biggest hour-glass cell has been found in *Vicia faba* (Fig. 2**B**) and the smallest in *Vigna catjang*. The thickness of the wall of the cells also varies. The thickest wall has been found in *Lathyrus sativus* and *Pisum sativum*. In *Cicer arietinum*, the walls of the cells are rather thin.

Most species examined have a single layer of hour-glass cells. We have, however, come across layers of hour-glass cells in *Cajanus cajan*, *Dolichos lablab* and *D. biflorus*. The cells of the second and third layers are usually shorter than those of the first layer. The third layer is also often crushed along with the mesophyll cells.

Mesophyll

The remainder of the testa is composed mainly of mesophyll cells. Their size and shape vary, depending on where they occur. Near the hilum, these cells are usually small and thick-walled; away from palisade cells they are large, thin-walled or thick-walled. They often contain deposits, which fill the entire cell cavity (Fig. 2**A**).

A KEY FOR IDENTIFICATION OF INDIAN PULSES WITH THE AID OF SEED STRUCTURE AND SEED COAT ANATOMY

1. Maximum axial length of seed more than 11 mm 2
- **2.** Height of palisade cells more than 125 μm; hilum 2·5 × 2 mm; cuticle smooth . . . . *Dolichos lablab*
- **2.** Height of palisade cells less than 75 μm; hilum 3 × 2 mm; cuticle rough . . . . *Vigna sinensis*

1. Maximum axial length of seed less than 9 mm 3
- **3.** Maximum axial length of seed less than 9 mm but more than 7 mm 4
  - **4.** Seed coat prominently blistered; hilum 2·3 × 1·6 mm in a sunken pouch below the level of seed surface; cuticle slightly rough; height of palisade cells between 110–165 μm *Cicer arietinum*
  - **4.** Seed coat smooth; cuticle slightly rough or rough . . 5
    - **5.** Height of palisade cells more than 160 μm; hilum 4·6 × 2 mm, level with seed surface . . . . *Vicia faba*
    - **5.** Height of palisade cells less than 100 μm . . . 6
      - **6.** Hilum 4·6 × 2 mm, below the level of seed coat, completely covered with corky tissue . *Vigna catjang*

6. Hilum 3 × 3 mm, level with seed coat . *Pisum sativum*

3. Maximum axial length of seed less than 7 mm . . . 7

7. Maximum axial length of seed more than 6 mm . . 8

8. Height of palisade cells more than 75 μm; hilum 3 × 2 mm, covered partially with hard corky tissue *Cajanus cajan*

8. Height of palisade cells less than 75 μm; hilum 1·8 × 0·5 mm . . . . *Dolichos biflorus*

7. Maximum axial length of seed less than 6 mm . . . 9

9. Seed surface lightly blistered; palisade cells of type I; cuticle smooth; hilum 1 × 0·5 mm, below the level of seed surface, partially covered with whitish tissue *Cyamopsis tetragonolobus*

9. Seed coat smooth 10

10. Seed coat many-coloured 11

11. Palisade cells of type III, height more than 85 μm; cuticle dentate; hilum 2 × 1 mm *Lathyrus sativus*

11. Palisade cells of type II, height less than 65 μm 12

12. Cuticle dentate; hilum 1·5 × 0·5 mm; seed round and flat . . . *Lens culinaris*

12. Cuticle not dentate; hilum 3 × 2 mm; seed oblong . . . *Phaseolus mungo*

10. Seed coat one-coloured 13

13. Cuticle with papillae or papilla-like outgrowths *Phaseolus aureus*

13. Cuticle smooth, without papillae *Phaseolus aconitifolius*

ACKNOWLEDGEMENTS

We are indebted to Dr H. K. Jain of Indian Agricultural Research Institute, New Delhi, Dr S. S. Saxena of Agricultural College, Kanpur, and Dr T. R. Veeraswamy of Agricultural Research Institute, Coimbatore, for the supply of authentically named seeds from their collection. Our grateful thanks are also due to Professor E. J. H. Corner, Botany School, Cambridge, for many helpful suggestions during this investigation. We thank Dr S. M. Ali of the Statistics Department of this University for helping us with analysis of data.

REFERENCES

Backer, C. A. & Bakhuizen van den Brink, R. C., 1963. *Flora of Java*. **I**: 642. Groningen: Noordhoff.

Backer, C. A. & Bakhuizen van den Brink, R. C., 1968. *Flora of Java*. **III**: 651 (Addenda et Corrigenda). Groningen: Noordhoff.

Coe, H. S. & Martin, J. N., 1920. Sweet-clover seed. II. Structure and chemical nature of the seed coat and its relation to impermeable seeds of sweet clover. *U.S. Dept. Agric. Bull.*, **844**: 26–39.

Corner, E. J. H., 1951. The leguminous seed. *Phytomorphology*, **1**: 117–150.

HAMLY, D. C., 1932. Softening of the seeds of *Melilotus alba*. *Bot. Gaz.*, **93**: 345–375.
HAMLY, D. C., 1934. The light line in *Melilotus alba*. *Bot. Gaz.*, **96**: 755–757.
HYDE, E. O. C., 1954. The function of the hilum in some Papilionaceae in relation to the ripening of the seed and the permeability of the testa. *Ann. Bot.*, *N.S.*, **18**: 241–256.
KONDO, M., 1913. [Seeds of *Glycine*, *Vigna*, *Dolichos*, *Canavalia*, *Lathyrus* and *Cicer*.] *Z. Unters. Nahr.-u. Genussmittel*, **25**: 1.
MAHESHWARI, P. & SINGH, U., 1965. *Dictionary of economic plants in India*. New Delhi: ICAR publ.
MARLIÈRE, H., 1897. Sur la graine et specialement sur l'endosperme dur de *Ceratonia siliqua*. *Cellule*, **13**: 5–59.
MARTIN, A. C., 1946. The comparative internal morphology of seeds. *Am. Midl. Nat.*, **36**: 513–660.
MARTIN, A. C. & BARKLEY, W. D., 1961. *Seed identification manual*. University of California Press. Also Calcutta: Oxford and IBH Publishing Co.
MARTIN, J. N. & WATT, J. R., 1944. The strophiole and other seed structures associated with hardness in *Melilotus alba* L. and *M. officinalis* Willd. *Iowa St. Coll. J. Sci.*, **18**: 457–469.
MUSIL, A. F., 1963. *Identification of crop and weed seeds*. Agricultural Handbook No. 219. Washington, D.C.: U.S. Dept. Agric.
NETOLITZKY, F., 1926. Anatomie der Angiospermen-Samen. In Linsbauer, K. (ed.), *Handbuch der Pflanzenanatomie Bd.* II. 2, **10**. Berlin: Borntraeger.
OTT, A. C. & BALL, C. D., 1943. Some components of the seed coats of the common bean, *Phaseolus vulgaris* and their relationship to water retension. *Arch. Biochem.*, **3**: 189–192.
PAMMEL, L. H., 1899. Anatomical characteristics of the seeds of Leguminosae, chiefly genera of Gray's Manual. *Trans. Acad. Sci. St Louis*, **9**: 91–275.
REEVE, R. M., 1946*a*. Structural composition of the sclereids in the integument of *Pisum sativum* L. *Am. J. Bot.*, **33**: 191–204.
REEVE, R. M., 1946*b*. Ontogeny of sclereids in the integument of *Pisum sativum* L. *Am. J. Bot.*, **33**: 806–816.
SANTAPAU, H., 1967. *Rev. Bot. Surv. India*, 3rd ed., **16** (1): 70.
SINGH, B., 1964. Development and structure of angiosperm seed. I. Review of the Indian work. *Bull. nat. bot. Gdns, Lucknow*. No. 89.
STERLING, C., 1954. Development of the seed coat of lima bean (*Phaseolus lunatus* L.). *Bull. Torrey bot. Club*, **81**: 271–287.
ZIMMERMAN, K., 1936. Zur physiologischen Anatomie der Leguminosentesta. *Landwn Vers. Stnen*, **127**: 1–56.

Wood anatomy of Hawaiian, Macaronesian, and other species of *Euphorbia*

SHERWIN CARLQUIST

Claremont Graduate School and Rancho Santa Ana Botanic Garden, Claremont, California, U.S.A.

Qualitative and quantitative data on wood anatomy are presented for three groups of the genus *Euphorbia*: the Hawaiian species (section *Anisophyllum* or *Chamaesyce*), species of the Canary Islands and Madeira (section *Tithymalus*), and cactoid African species of the section *Euphorbia*. Longer vessel elements are correlated with wetter habitats in the Hawaiian species. The cactoid euphorbias have notably long vessel elements, while the Macaronesian species, which are leafy stem succulents, are intermediate in this respect. Scalariform lateral wall pitting of vessels is striking in the cactoid species. Width of vessels is in inverse ratio to abundance. Libriform fibres (fibre-tracheids in some species) are often gelatinous. Axial parenchyma is abundant in the cactoid species, much less so in the Hawaiian species. Ray histology is related to vessel-element length: erect ray cells are most abundant in those Hawaiian species with longer vessel elements. Summaries of differences among the three species groups studied are given. To explain the nature of wood anatomy in the various taxa, concepts of juvenilism (paedomorphosis) and evolution of wood features with relation to ecology are invoked.

CONTENTS

INTRODUCTION

This study was originally undertaken in order to investigate relationship between wood anatomy of the Hawaiian species of *Euphorbia* and the highly diverse ecological sites in which they grow. In the process of assembling these data, I realized that comparisons with non-Hawaiian species might offer further dimensions. As a basically herbaceous genus, several species groups of which have become woody under various circumstances, *Euphorbia* offers ideal material for study of phenomena termed paedomorphosis by the writer (1962) and for study of woodiness under insular and island-like conditions.

The Hawaiian species of *Euphorbia* belong to the section *Chamaesyce* (section *Anisophyllum* of some authors). The system of Sherff (1938) has been followed for identification of the Hawaiian taxa, although alterations in this monograph may be desirable in the light of recent studies on the nature of the Hawaiian flora. The stock basic to the Hawaiian euphorbias is probably represented by the wide-ranging *E. atoto* of maritime Pacific regions. This stock reached the Hawaiian islands, where it has diversified into various coastal species, such as *E. degeneri* and *E. skottsbergii.* These are low mat-like perennials with little wood accumulation. Larger, shrub-like mats are formed by *E. celastroides* var. *amplectans* of dry lowland Lanai, Maui, Molokai and Hawaii. *Euphorbia celastroides* var. *lorifolia* is a true small shrub of the *Acacia—Styphelia* woodlands of low montane Kauai. *Euphorbia celastroides* var. *mauiensis* is a small fountain-shaped tree of the dry *Diospyros* forest of Maui, while *E. celastroides* var. *haupuana* grows on cliffs and lowlands of Kauai. In moderately moist forest of Kauai is a tree species, *E. atrococca*, and on intermittently wet cliffs of Oahu is a shrub, *E. multiformis*. These seem clearly allied to the *E. celastroides* group.

The above species have glaucous, relatively thin leaves. These differ from a series of wet-forest euphorbias in the Hawaiian Islands. These species have thick, shiny glabrous leaves. Included in these species are: *E. clusiifolia*, a small shrub of Oahu; *E. halemanui*, a scandent sub-lianoid shrub of Kauai; *E. remyi* var. *molesta*, a dwarf, little-branched bog plant; *E. remyi* var. *pteropoda*, from rain forest on the Kauai plateau; *E. hillebrandii*, a small shrub of Oahu forests; and *E. rockii*, a tree of very wet Oahu forests. Thus the range of habits and habitats represented by the Hawaiian euphorbias is exceptional and offers an interesting example of adaptive radiation. These species have been illustrated by Carlquist (1970*a*). There is reason to believe that arborescence has evolved in Hawaiian euphorbias from mat-like plants in which the main shoot always aborts in the seedling stage (Degener, 1937).

Euphorbias of the Macaronesian islands (Canary Is., Madeira, Azores, etc.) belong, with the exception of some cactoid species, to the section *Tithymalus*. The *Tithymalus* euphorbias are derived from a growth form in which lanceolate, thin glaucous leaves and semisucculent stems are present. The *Tithymalus* euphorbias become fountain-shaped or irregularly branched shrubs. Of these, *E. balsamifera* grows nearest the coast, in very dry localities. *Euphorbia regis-jubae* and *E. atropurpurea* are plants of the coastal to inland scrub. On Madeira, *E. piscatoria* occurs in seaward-facing bluffs, but also inland in the wet laurel forest. Wet-forest individuals of this species can be called small trees, whereas the coastal individuals are only small shrubs.

Succulent cactoid euphorbias belong to the section *Euphorbia*. Many of these are native to Africa. Of the cactoid euphorbias, only a few have been utilized in the present study. One of these is a shrub of the lowland zones of the Canary Islands, *E. canariensis*. Another, *E. candelabrum*, occurs on the high plateau of Kenya, and is a large tree. *Euphorbia lactea* is an African species, although the sample of wood came from Haiti, where the plant has become naturalized.

For illustrations and descriptions of the Canarian euphorbias, the reader is referred to Schenck (1907) and Burchard (1929). Succulent species of *Euphorbia* as a whole have been illustrated by Jacobsen (1960), and by White, Dyer & Sloane (1941). The cactoid species studied here all belong to the 'Group 24' of section *Euphorbia* in Jacobsen's treatment, and in White, Dyer & Sloane's monograph are all placed in the subsections Trigonae and Polygonae of section *Euphorbia*.

Although the representation of Hawaiian and Macaronesian species here is relatively good, only a few of the cactoid euphorbias have been included. Those studied seem relatively stereotyped; additional species may or may not prove to follow this pattern, and would probably be worth studying. Most of these have relatively limited xylem accumulation, and study of such xylem might be rather like study of earlier-formed xylem in the more arborescent cactoid species.

Surprisingly little has been reported about wood anatomy of *Euphorbia*, although woody species of the Euphorbiaceae as a whole are better known. Lack of studies on wood of *Euphorbia* probably relates to the fact that this genus is not a source of timber. Studies of succulent euphorbias (Trumke, 1913; White, Dyer & Sloane, 1941) deal with aspects other than wood anatomy. A few anatomical features are mentioned for woods of *Euphorbia* species by Heimsch (1942) and Metcalfe & Chalk (1950).

MATERIALS AND METHODS

Most of the samples were collected by the writer and associates in the field. Because woods of *Euphorbia* are very soft and, when collected, contain much moisture, cortex and bark must be removed to aid drying if dried samples are desired. Retardation of decay was achieved by packing wood samples with paraformaldehyde powder when they were shipped from collecting areas to Claremont. They dried without going mouldy in transit, or drying was completed in the relatively dry climate of Claremont. The softness of woods of some *Euphorbia* species would make preservation in liquid a preferable procedure in many instances. One would like to know if fibres remained nucleated for long periods of time, for example. Also, microtoming of paraffin-embedded species would probably be preferable for species with very soft wood, such as *E. canariensis* (Plate 2**A, B**) or *E. balsamifera*. All woods in the present study were sectioned on a sliding microtome and stained with safranin and haematoxylin according to the usual techniques. In addition to permanent slides of sections, permanent stained slides of macerations were also prepared. Quantitative data shown in Table 1 are based on 50 or more measurements of a feature for each collection. Excessive reliance should not be placed on quantitative data on wood anatomy in this study or, for that matter, other studies on wood anatomy. These features are too variable, a fact realized when one takes samples from various portions of a single plant or even when one takes measurements from the same slide on two different occasions. However, when marked differences occur, they are probably significant. The features summarized in the table are those in which significant differences were observed among the species studied. In the case of *E. rockii*, data from two portions of a single stem are given: from near the pith, and from near the periphery of a stem about three inches in diameter. Because significant differences in this respect have been found in some families (Carlquist, 1970*b*), analysis of different portions was undertaken. The figures

Table 1. Anatomical data for *Euphorbia* species studied

Name	Collection	Vessel element average length (μm)	Longest vessel element (μm)	Average vessel diameter (μm)	Widest vessel (μm)	Vessels per group	Average diameter pits (μm)	Vessel pits mostly circular (C) or elliptical (E)	Average length libriform fibres (μm)	Average diameter libriform fibres (μm)	Average thickness walls libriform fibres (μm)	Average height multiseriate rays (mm)	Average width multiseriate rays (cells)
Euphorbia atrococca Heller var. *kokeana* Sherff	Stern & Carlquist 1284 (US, RSA)	328	469	55	83	2·6	4	E	481	16	3	0·33	2·4
E. atropurpurea Brouss.	Carlquist 2481 (RSA)	414	663	67	125	2·2	8	E	631	28	5	0·79	2·1
E. balsamifera Ait.	Carlquist 2503 (RSA)	320	490	69	114	1·5	>10	E	483	33	2	0·52	2·1
E. canariensis L.	Carlquist 2544 (RSA)	458	714	104	195	2·4	>10	E	640	33	5	1·62	2·2
E. candelabrum Trém. ex Kotschy	Carlquist 2671 (RSA)	637	1255	115	182	1·5	>10	E	1288	38	6	0·70	2·0
E. celastroides Boiss. var. *amplectans* Sherff	Carlquist 2023 (RSA)	282	469	42	62	3·0	4	C	436	14	5	0·36	2·5
E. celastroides Boiss. var. *haupuana* Sherff	Stern & Carlquist 1258 (US, RSA)	313	439	38	49	2·3	4	C	460	14	4	0·43	2·6
E. celastroides Boiss. var. *lorifolia* (Gray) Sherff	Carlquist 1998 (RSA)	248	333	44	62	2·1	5	C	377	13	4	0·26	2·9
E. celastroides Boiss. var. *mauiensis* Sherff	Carlquist 1949 (RSA)	327	469	56	91	3·1	5	C	470	15	5	0·52	3·4
E. clusiifolia H. & A.	Carlquist 2371 (RSA)	423	969	39	52	1·6	5	C	634	17	4	0·61	2·6
E. degeneri Sherff	Carlquist 1805 (RSA)	206	367	32	44	3·9	5	C	298	13	4	0·21	2·7
E. halemanui Sherff	Stern & Carlquist 1275 (US, RSA)	406	581	64	117	1·9	5	C	710	11	4	0·46	2·1
	Carlquist 1796 (RSA)	444	581	75	146	2·1	7	C	669	20	4	0·42	2·0
E. hillebrandii Lévl.	Carlquist 1905 (RSA)	352	510	40	57	1·3	5	C	557	15	3	0·37	2·0
E. lactea Haw.	USw-1820	472	673	83	104	2·9	11	E	983	34	4	0·56	2·0
E. multiformis H. & A. var.	Carlquist 2212 (RSA)	314	490	45	73	3·3	4	C	517	15	4	0·28	2·0
microphylla Boiss.	Carlquist 2419 (RSA)	330	551	36	57	3·0	5	C	517	13	3	0·30	2·0
E. piscatoria Ait.	Carlquist 2646 (RSA) (Inland)	452	704	62	91	3·1	9	E	789	22	4	0·49	2·6
	Carlquist 2626 (RSA) (Coastal)	321	469	60	107	3·1	8	E	500	31	2	0·45	2·1
E. regis-jubae Webb & Berth.	Carlquist 2428 (RSA)	390	530	74	130	2·9	10	E	690	20	3	0·52	2·0
E. remyi A. Gray var. *molesta* Sherff	Carlquist 1982 (RSA)	420	694	30	44	2·3	6	C	661	16	4	?	2·0
E. remyi A. Gray var. *pteropoda* Sherff	Stern & Carlquist 1244 (RSA)	356	541	53	106	2·2	6	E	534	18	4	0·64	2·0
E. rockii Forbes	Carlquist 2196 (RSA) (Stem centre)	607	918	54	75	1·4	5	C	840	22	3	0·82	2·0
	Carlquist 2196 (RSA) (Periphery)	624	1064	61	88	1·5	6	C	889	22	3	0·74	2·0

for *E. rockii* are very similar for both portions of the stem, a fact in itself significant, because it suggests little alteration of the wood pattern during ontogeny, or a form of juvenilism.

There are many who have aided my field work or provided technical assistance. For their help, I should like to thank Miss Jane Benjamin, Mr Ralph Daehler, Dr and Mrs George W. Gillett, Mr Gunther Kunkel, Mr Timothy O. Magee, Dr and Mrs Ernest Ross, Dr Harold St John, Dr A. C. Smith, Dr William L. Stern and Dr Warren H. Wagner, Jr.

ANATOMICAL DATA

Vessels

For a specialized group of basically herbaceous dicotyledons, most species of *Euphorbia* have rather long vessel elements. Several factors seem to be responsible. With respect to the Hawaiian species, there is an almost perfect correlation between habitat and vessel-element length. The most important ecological variable involved is rainfall. Of the species studied, *E. degeneri*, with vessel elements averaging 206 μm* long, grows in the driest locality, a coastal zone of Oahu where only about 15 inches of rain fall per year. The other extreme is represented by *E. rockii* (Plate **6D**), with vessel elements averaging more than 600 μm in length. This species is confined to the wettest area of Oahu, the upper Punaluu Valley of the Koolau Mts, where rainfall averages about 250 inches per year. Between these extremes, the species range approximately as expected. One could, in fact, estimate rainfall with reasonable accuracy from vessel element length. *Euphorbia remyi* var. *molesta* (Plate 5**B**), with vessel elements averaging 420 μm, grows in Wahiawa Bog on southern Kauai. *Euphorbia halemanui*, with vessel elements about the same length, grows in the Kokee rain forest of Kauai. *Euphorbia clusiifolia*, in which vessel elements average 423 μm, is native to Koolau rain forest of Oahu, where about 100 inches of rain per year fall. The various taxa within *Euphorbia celastroides* reflect their relatively dry forest habitats in their relatively short vessel element lengths. Close relationship between vessel element length and rainfall would be expected on the basis of considerations formulated in Asteraceae by the writer (Carlquist, 1966).

Vessel element length in the two other groups of *Euphorbia* species shows interesting correlations. The cactoid species of section *Euphorbia* all have exceptionally long vessels, averaging 637 μm in *E. candelabrum* (Plate **2D**), 458 μm in *E. canariensis* (Plate **2B**), and 472 μm in *E. lactea*. The large plant size of *E. candelabrum* might be related to the exceptionally long vessel elements, for in an evolutionarily plastic group, longer vessel elements occur in trees, shorter ones in shrubs (Carlquist, 1966). One can consider long vessel elements in these *Euphorbia* species to be related to a mesic *internal* environment. In other words, the xylem of the succulent euphorbias does not supply actively-transpiring leaves, but supplies a water-rich parenchyma from which transpiration is minimal. Cactoid euphorbias accumulate secondary xylem slowly, so that a form of juvenilism might be involved.

The Macaronesian species of section *Tithymalus* show some interesting correlations

* The symbol μm is used for micrometer, 10^{-6}m (instead of μ, micron, 10^{-3}mm), in accordance with revised international standards.

in vessel element length, suggesting the patterns of both the Hawaiian and the cactoid euphorbias. Notable in this regard is the difference in element length between the coastal and inland populations of *Euphorbia piscatoria*. The *Tithymalus* species have vessel elements mostly longer than those of Hawaiian species but shorter than those of the cactoid species. The *Tithymalus* species could be called leafy stem succulents, so this intermediacy in vessels might be expected to match intermediacy in habit.

In diameter, vessel elements of euphorbias show interesting distinctions. The Hawaiian species have relatively narrow vessels (e.g. Plates 3**A, C**; 4**A**; 5**A** and 6**A, C**). Notably wide vessels occur in the sub-lianoid *E. halemanui* (Plate 4**C**), a correlation one would expect on the basis of the wide vessels possessed by most vine-like dicotyledons. Exceptionally wide vessels characterize the cactoid euphorbias (Plate 2**A, C**). As with vessel element length, the semisucculent *Tithymalus* species (Plate 1**C**) fall into an intermediate vessel-diameter category. Wider vessels characterize mesic Asteraceae, narrow vessels xeric Asteraceae (Carlquist, 1966). If this correlation operates in *Euphorbia* also, as it appears to in a rough way, then the cactoid euphorbias must be regarded as 'internally mesic' on this indication also.

Vessel diameter in *Euphorbia* should not be viewed without reference to vessel abundance, however. As can be seen by examining the photographs of transverse sections, the species with very wide vessels (Plate 2**A, C**) have, in general, few vessels per unit area of transverse section, whereas species with narrow vessels (Plate 3**A, C** etc.) have more numerous vessels. A figure for vessels per unit area would have been informative, although this figure overlaps somewhat with the figures for vessel diameter and vessels per group—these express phases of similar phenomena.

The pattern of vessel grouping in all euphorbias studied can be described chiefly as radial chains. This is quite conspicuous in *E. celastroides* var. *amplectans* (Plate 3**A**), *E. celastroides* var. *lorifolia* (Plate 3**C**), *E. piscatoria* (Plate 1**A**) or *E. regis-jubae* (Plate 1**C**). One might expect an inverse relationship between degree of vessel grouping and vessel abundance. This proves true in many cases, although not in others. The low figure for vessels per group in the cactoid euphorbias would be understandable on this basis. Likewise, the relatively high number of vessels per group in *E. celastroides* and *E. degeneri* relates to narrowness of vessels. Species which do not fit this pattern closely include *E. clusiifolia*, *E. halemanui* and *E. hillebrandii*. These have a low figure for vessels per group. These are all wet-forest species of the Hawaiian *Anisophyllum* group, and a lower figure for vessels per group would be expected for mesic species, if the tendencies established in Asteraceae (Carlquist, 1966) are applicable here. In this case, the low figure for vessels per group in the cactoid euphorbias could be explained as a sort of mesic phenomenon, related to succulence.

All perforation plates observed in *Euphorbia* woods were simple. No aberrant bars or other irregularities were seen. In lateral-wall pitting, however (Plate 8**A–F**), the *Euphorbia* species studied exhibited a wide variety, ranging from scalariform to alternate, large to small. Scalariform pitting was previously observed for *Euphorbia* vessels by Metcalfe & Chalk (1950).

The cactoid euphorbias have a very clear representation of scalariform lateral-wall pitting. Such pitting is illustrated here for *E. lactea* (Plate 8**C**), although similar photographs could have been shown for *E. canariensis* and *E. candelabrum*.

The species of section *Tithymalus* show near-scalariform pitting, as shown for *E. regis-jubae* (Plate 8**B**). The large diameter of pits, as given in the table, for cactoid species and for species of section *Tithymalus* is due to the laterally elongate shape of pits. The only exception to scalariform pitting in vessels of cactoid and Macaronesian euphorbias is in the inland population of *E. piscatoria* (Plate 8**A**). The coastal plants have typical scalariform pitting. This exception might suggest that scalariform pitting characterizes stem succulents. The significant correlation here might be with juvenilism, as in other families (Carlquist, 1962). There are scalariform-like pits in *E. clusiifolia* (Plate 8**D**), *E. remyi* var. *molesta* and *E. atrococca*. This type of pitting typifies taxa in typically herbaceous families, taxa in which woodiness seems likely to be secondary rather than an ancestral feature of the group.

As can be shown by the comparisons in Plate 8**A**–**F**, pits in vessels (disregarding lateral widening, elliptical shapes) are larger in the cactoid and Macaronesian species (Plate 8**A**–**C**) than they are in the Hawaiian species (Plate 8**D**–**F**). Larger pits do appear to characterize woods of stem succulents (Carlquist, 1962).

Libriform fibres and fibre-tracheids

There are no tracheids in woods of *Euphorbia* species examined. Observation of cells that appear to be libriform fibres reveals no borders on pits in most species. However, two species, *E. clusiifolia* and *E. lactea*, were observed to have vestigial borders on pits of elements which must thereby be termed fibre-tracheids. The gelatinous texture of libriform fibres (or fibre-tracheids) might obscure the occurrence of borders in some species.

Libriform fibre walls often appear as an outer non-shrinking wall surrounding an inner gelatinous wall portion shrunken in permanent slides. This is shown for *E. piscatoria* in Plate 7**A**. Sometimes fibres with no shrinkage patterns occur in the same section as do normal-appearing fibres. This is apparent in *E. remyi* var. *pteropoda* (compare Plate 7**E** and **F**). This might reflect occurrence of a sort of tension wood in some species. However, others, such as *E. canariensis*, show prominent gelatinous shrinkage patterns in all libriform fibres. Conspicuous shrinkage patterns are observed in *E. atrococca*, *E. celastroides* var. *lorifolia*, *E. celastroides* var. *mauiensis*, *E. hillebrandii*, *E. piscatoria* (especially coastal plants), *E. remyi* var. *molesta* and *E. remyi* var. *pteropoda*. In other species, shrinkage patterns are not observed: *E. regis-jubae* (Plate 7**B**), *E. candelabrum* (Plate 7**C**), and *E. celastroides* var. *amplectans* (Plate 7**D**), for example. Even where shrinkage patterns are not apparent, gelatinous nature could perhaps be said to be present by virtue of staining and refraction qualities. In some species (notably *E. balsamifera*), lack of staining of fibres suggests minimal lignification; the grey colour in the transverse section of *E. canariensis* (Plate 2**A**), as opposed to the dark colour of vessels, is suggestive of this.

As the figures in the table show clearly, the three groups of *Euphorbia* species differ in fibre diameter. The cactoid euphorbias have quite wide fibres, as shown for *E. candelabrum* (Plate 7**C**). Next in fibre diameter are the *Tithymalus* species; fibres of *E. piscatoria* (Plate 7**A**) and *E. regis-jubae* (Plate 7**B**) illustrate this. The Hawaiian section *Anisophyllum* species have fibres relatively small in diameter; the range is illustrated in Plate 7**D**–**F**.

A number of species have fibres that appear markedly widened tangentially as viewed in a transverse section. These are seen most clearly in *E. atropurpurea* and *E. piscatoria* (Plate 7**A**). Thickness of fibre walls varies widely within *Euphorbia.* Notably thin-walled fibres are seen in *E. balsamifera* and *E. regis-jubae* (Plate 7**B**), whereas thicker-walled fibres characterize *E. celastroides* var. *amplectans* (Plate 7**D**) and other Hawaiian species.

Axial parenchyma

The Hawaiian species of *Euphorbia* have only a small number of parenchyma cells. A few of these are distributed in a diffuse apotracheal pattern (Plate 7**E**). However, most of the axial parenchyma cells in the Hawaiian species are paratracheal in occurrence, and cells with contents to the left of vessels in Plate 7**D** offer a typical example. In no species could axial parenchyma be described as more than scanty. However, somewhat more numerous axial parenchyma cells are seen in *E. celastroides* var. *mauiensis* than in the other Hawaiian taxa. Axial parenchyma in strands of two to four cells, chiefly four, is seen in these species.

The Macaronesian and cactoid euphorbias, in contrast to the Hawaiian taxa, have relatively abundant axial parenchyma. These cells are primarily apotracheal in distribution. In a wood with relatively abundant apotracheal parenchyma, of course, at least a few cells are going to be adjacent to vessels. Axial parenchyma cells, identifiable by thin walls and contents in transverse section (and by occurrence as strands in tangential sections) are visible in Plates 6**D** and 7**A**–**C**. Axial parenchyma cells in the Macaronesian and cactoid species occur in strands of four, with few exceptions.

Vascular rays

Uniseriate rays are abundant in woods of euphorbias. As examination of the tangential sections shown here indicates, uniseriate rays are as abundant or more abundant than multiseriate rays in most taxa (Plates 1**B, D**; 2**B, D**; 4**D**; 5**B, D** and 6**B, D**). However, in some of the Hawaiian euphorbias, particularly those of drier lowland localities, multiseriate rays more than three cells in width are relatively abundant (Plates 3**B, D** and 4**B**).

With respect to vertical height of rays, there is a rather close correlation with length of the axial elements: vessel elements and libriform fibres (or fibre-tracheids). Because some species have many, some few multiseriate rays, the measurements in Table 1, based on multiseriate rays only, may not provide an ideal measure. However, one notes that, of the species with multiseriate rays above 0·7 mm in height, all have vessel elements averaging more than 400 μm in length, all but one of these with vessels averaging more than 600 μm.

Similarly, ray histology seems related to length of tracheary elements. Five of the taxa studied had few or no procumbent ray cells: *E. clusiifolia* (Plate 6**B**), *E. halemanui* (Plate 4**D**), *E. hillebrandii*, *E. remyi* var. *molesta* (Plate 5**B**) and *E. rockii* (Plate 6**D**).

These species all have relatively long vessel elements. This condition seems to apply to the Hawaiian species only. The other *Euphorbia* species with relatively long vessel elements, however, do approach this pattern in that erect ray cells seem relatively more frequent in them than in species with shorter vessel elements.

An interesting feature in ray histology is the strong tendency of ray cells to be radially elongate when seen in transverse section. This is particularly true of the Macaronesian species (Plate 1**A, C**) and the cactoid species (Plate 2**A, C**).

As comparison of photographs here shows, ray cells are relatively large in the cactoid euphorbias (Plate 2**A–D**), large to intermediate in the Macaronesian species (Plate 1**A–D**), and intermediate to small in the Hawaiian species (Plates 3 to 6).

Latex and laticifers

The narrowness of rays in most euphorbias would make occurrence of laticifers in rays seem unlikely. This expectation does appear to be justified, for only in species with wide multiseriate rays were laticifers observed: *E. celastroides* var. *amplectans* (Plate 3**B**) and *E. celastroides* var. *mauiensis* (Plate 8**G**).

Latex in the form of droplets or otherwise congealed deposits was almost universally observed in parenchyma cells, both axial and ray, of *Euphorbia* woods. Such latex formations, in fact, aided in identification of parenchyma cells. More abundant latex deposits, occluding some cells, were seen in a few species, such as *E. celastroides* var. *amplectans* (Plate 3**A, B**) and *E. clusiifolia* (Plate 6**A**).

Growth rings

Growth rings are not conspicuous except in some of the Hawaiian euphorbias. Growth rings are shown here for *E. celastroides* var. *amplectans* (Plate 3**A**, slight differences in vessel diameter and fibre-wall thickness); *E. halemanui* (Plate 4**C**: wider, more numerous vessels above); *E. remyi* var. *molesta* (Plate 5**A**: slight differences in fibre-wall thickness); and *E. remyi* var. *pteropoda* (Plate 5**C**: vessels larger and more numerous, libriform fibres thicker-walled in upper portion of photograph).

Lack of growth rings in the Macaronesian and cactoid species of *Euphorbia* is noteworthy and might be explained if one supposes that succulence and leaflessness prevent marked seasonal differences in transpiration from occurring and thus affecting wood anatomy.

SYSTEMATIC CONSIDERATIONS

The euphorbias of the present study represent three groups, and these emerge clearly from data on wood anatomy. The differences can be summarized as follows:

Hawaiian species (section *Anisophyllum*): Vessels relatively narrow, of varied length; lateral-wall pitting of vessels mostly alternate, rarely elliptical, pits small; vessels often many per group, in radial chains; libriform fibres narrow. Axial parenchyma very scanty, mostly paratracheal with very few apotracheal cells; ray cells small; rays uniseriate and narrow to wide multiseriate; growth rings often present.

Macaronesian species (section *Tithymalus*): Vessels of intermediate diameter, intermediate length; lateral-wall pitting of vessels mostly scalariform or elliptical, pits large; vessels of intermediate abundance and degree of grouping, often in radial chains; libriform fibres wide, thin-walled. Axial parenchyma relatively abundant, diffuse apotracheal; rays uniseriate and narrow multiseriate; ray cells large; growth rings lacking.

Cactoid species (section *Euphorbia*): Vessels long, wide, with scalariform lateral-wall pitting; pits large; vessels few per unit area of transverse section, few per group; libriform fibres long, wide. Axial parenchyma abundant, diffuse apotracheal; ray cells large, rays uniseriate and narrow multiseriate; growth rings lacking.

The differences cited may bear some relation to ecology, as for example in the absence of growth rings and large parenchyma-cell size in the succulent and semisucculent species. However, some of the differences (e.g. presence of laticifers in wide multiseriate rays; scanty paratracheal versus abundant diffuse apotracheal parenchyma) do not seem directly related to differences in ecology. At any rate, the fact that differences among the three groups are so consistent (despite the wide ecological gamut of the Hawaiian species) clearly suggests that the three groups have developed woodiness independently from herbaceous ancestors. Certainly this would not be improbable in such a large and evolutionarily plastic genus as *Euphorbia*. In fact, some other sections of the genus *Euphorbia* not included in the present study would very likely furnish additional instances of this: the section *Poinsettia*, for example, has species, such as *E. pulcherrima*, that can become notably woody.

JUVENILISM, ECOLOGY AND OTHER FACTORS

An herbaceous group in which certain phylads are in the process of becoming more woody may be expected to have wood anatomy different from that of woody phylads stemming from a woody ancestry. The euphorbias of the present study exemplify this well.

Several criteria were cited as indicators of juvenilism, or paedomorphosis, in dicotyledonous woods by the writer (Carlquist, 1962). Although *Euphorbia* species were not included in that paper, they excellently illustrate principles related to it. Delay in transverse subdivision of fusiform and ray initials of the cambium, resulting in long vessel elements and libriform fibres and tall rays, characterizes many of the euphorbias studied here. Simple perforation plates, universal in *Euphorbia* woods, are a good indicator of the specialized nature, basically, of *Euphorbia* wood anatomy. For scalariform lateral-wall pitting to occur in vessels of such specialized wood seems best explained as a juvenilism.

However, one can press the issue further and ask why such a feature as scalariform pitting should occur in the particular *Euphorbia* species where it exists, yet be lacking in others. Its presence in succulent and semisucculent species of this genus is reminiscent of the condition in lobelioids, where it is also present in woods, also apparently as a juvenilism, in species that could be called stem succulents: *Brighamia insignis* and *Delissea undulata*, for example (Carlquist, 1970*b*). Wide pit areas, like those of metaxylem vessels, might reflect a growth pattern during which the plant is responding to mesic conditions and also typically does not build marked mechanical support. These features obtain in metaxylem, presumably because growth does not occur until sufficient water to permit growth is present (and thus a mesic condition could be said to exist), and a stem capable of elongation does not have mechanical elements of strong rigidity. One might view succulent and semisucculent species as experiencing internally mesic conditions, protected from marked variations in transpiration. Succulents as a growth form do not seem to exhibit features related to marked mechanical support.

Paucity of vessels, broad, thin-walled libriform fibres or fibre-tracheids, and relatively abundant apotracheal parenchyma cells, cells which are large in size, in the succulent and semisucculent euphorbias may reflect lack of selective pressure for mechanical strength. If these features represent persistence of primary xylem features into the secondary xylem, they can be said to persist because of lack of pressure toward a mechanically strong pattern of construction.

One would expect long elements in mesic plants, short elements in xeric ones (Carlquist, 1966). Long vessel elements in the succulent euphorbias do suggest this, as do wider vessels, and fewer vessels per group—the latter two also criteria of mesomorphy in Asteraceae (Carlquist, 1966). The tendency of rays to be taller in mesic Asteraceae as opposed to xeric ones is also paralleled in *Euphorbia* woods. Taller rays may be a side-effect of the slowness in transverse subdivision of cambial initials, and this suggests that in analyzing such concepts as mesomorphy, succulence and paedomorphosis, we find difficulty in separating such concepts.

The Hawaiian species of *Euphorbia* show little or no well-marked adaptations of a succulent nature, but their evolution from dry coastal to wet montane situations seems to have featured the following changes: longer vessel elements and libriform fibres (or fibre-tracheids); fewer vessel elements per unit transverse section; wider vessels (with some exceptions); fewer vessels per group; taller rays; higher proportion of erect ray cells; and thinner-walled fibres. This phylesis from xeric to mesic can be said to have involved paedomorphosis.

ACKNOWLEDGEMENT

This study was aided by two grants from the National Science Foundation, GB-4977x and GB-14092.

REFERENCES

BURCHARD, O., 1929. Beiträge zur Ökologie und Biologie der Kanarenpflanzen. *Biblthca bot.*, **98**: 1–262.

CARLQUIST, S., 1962. A theory of paedomorphosis in dicotyledonous woods. *Phytomorphology*, **12**: 30–45.

CARLQUIST, S., 1966. Wood anatomy of Compositae: a summary, with comments on factors controlling wood evolution. *Aliso*, **6** (2): 25–44.

CARLQUIST, S., 1970*a*. *Hawaii: a natural history*. New York: Natural History Press.

CARLQUIST, S., 1970*b*. Wood anatomy of Lobelioideae (Campanulaceae). *Biotropica*, **1**: 47–72.

DEGENER, O., 1937. Euphorbiaceae. In O. Degener, *Flora Hawaiiensis*. (Parts published by the author at various intervals.)

HEIMSCH, C., 1942. Comparative anatomy of the secondary xylem in the 'Gruinales' and 'Terebinthales' of Wettstein, with reference to taxonomic grouping. *Lilloa*, **8**: 93–198.

JACOBSEN, H., 1960. *A handbook of succulent plants*, Vol. I. London: Blandford Press.

METCALFE, C. R. & CHALK, L., 1950. *Anatomy of the dicotyledons*. Oxford: Clarendon Press.

SCHENCK, H., 1907. Beiträge zur Kenntnis der Vegetation der Canarischen Inseln. In C. Chun (ed.), *Wissenschaftliche Ergebnisse der deutschen Tiefsee-Expedition auf dem Dampfer 'Valdivia' 1898–1899*, **2** (1.2): 225–406. Jena: Fischer.

SHERFF, E. E., 1938. Revision of the Hawaiian species of *Euphorbia* L. *Ann. Missouri bot. Gdn*, **25**: 1–94.

TRUMKE, H., 1913. *Beiträge zur Anatomie der sukkulenten Euphorbien.* Thesis, Breslau Univ.

WHITE, A., DYER, R. A. & SLOANE, B. L., 1941. *The succulent Euphorbieae* (*Southern Africa*). Pasadena: Abbey Garden Press.

EXPLANATION OF PLATES

PLATE 1

Woods of Macaronesian species (sect. *Tithymalus*) of *Euphorbia*.

A, B. *E. piscatoria*, Carlquist 2646 (inland): **A,** T.S., showing veins in radial chains; **B,** tangential section; fibres wide, difficult to differentiate superficially from rays.

C, D. *E. regis-jubae*, Carlquist 2428: **C,** T.S. (cambial side at left); ray cells prominently radially elongated; **D,** tangential section; most rays are uniseriate.

A photomicrograph of a stage micrometer, enlarged at the same scale as wood sections and showing 1·3 mm is shown above **A.** This scale applies to Plates 1 to 6 inclusive. T.S., Transverse section.

PLATE 2

Woods of cactoid species (sect. *Euphorbia*) of *Euphorbia*.

A, B. *E. canariensis*, Carlquist 2544: **A,** T.S., cambial side at left; **B,** tangential section, note large ray cells.

C, D. *E. candelabrum*, Carlquist 2671: **C,** T.S., cambial side above, note sparsity of vessels; **D,** tangential section, most rays are uniseriate; apotracheal parenchyma cells are abundant.

PLATE 3

Woods of Hawaiian species (sect. *Anisophyllum*) of *Euphorbia*.

A, B. *E. celastroides* var. *amplectans*, Carlquist 2023: **A,** T.S., prominent radial chains of vessels are visible; **B,** tangential section, laticifers visible in two rays; rays dark because of latex deposits.

C, D. *E. celastroides* var. *lorifolia*, Carlquist 1998: **C,** T.S., vessels relatively narrow; **D,** tangential section, multiseriate rays relatively abundant.

PLATE 4

Woods of Hawaiian species (sect. *Anisophyllum*) of *Euphorbia*.

A, B. *E. atrococca* var. *kokeana*, Stern & Carlquist 1284: **A,** T.S., thicker-walled fibres indicate a growth ring at top of photograph; **B,** tangential section, multiseriate rays few but tall.

C, D. *E. halemanui*, Carlquist 1796: **C,** T.S., outer portion of stem has wider vessels, suggesting lianoid wood anatomy; **D,** tangential section, very few multiseriate rays are present.

PLATE 5

Woods of Hawaiian species (sect. *Anisophyllum*) of *Euphorbia*.

A, B. *E. remyi* var. *molesta*, Carlquist 1982: **A,** T.S., cambial edge at left, note small diameter of vessels; **B,** tangential section, erect ray cells make rays difficult to detect.

C, D. *E. remyi* var. *pteropoda*, Stern & Carlquist 1244: **C,** T.S., note growth ring phenomena; **D,** tangential section; this section is taken from a zone of wider vessels.

PLATE 6

Woods of Hawaiian species (sect. *Anisophyllum*) of *Euphorbia*.

A, B. *E. clusiifolia*, Carlquist 2371: **A,** T.S., vessels are notably narrow; **B,** tangential section, rays are almost exclusively uniseriate.

C, D. *E. rockii*, Carlquist 2196: **C,** T.S., many of the libriform fibres are tangentially widened; **D,** tangential section, ray cells are all erect, so that rays are difficult to see; rays are mostly uniseriate.

PLATE 7

Portions of *Euphorbia* woods in T.S. to show libriform fibres. Further discussion in text.

A. *E. piscatoria*, Carlquist 2646. **B.** *E. regis-jubae*, Carlquist 2428. **C.** *E. candelabrum*, Carlquist 2671. **D.** *E. celastroides* var. *amplectans*, Carlquist 2023. **E, F.** *E. remyi* var. *pteropoda*, Stern & Carlquist 1244: **E,** portion of section to show fibres lacking shrinkage patterns; **F,** portion of same section as in **E** showing shrinkage patterns in fibres. Scale of magnification shown above **F** (each division of photographed stage micrometer is 10 μm).

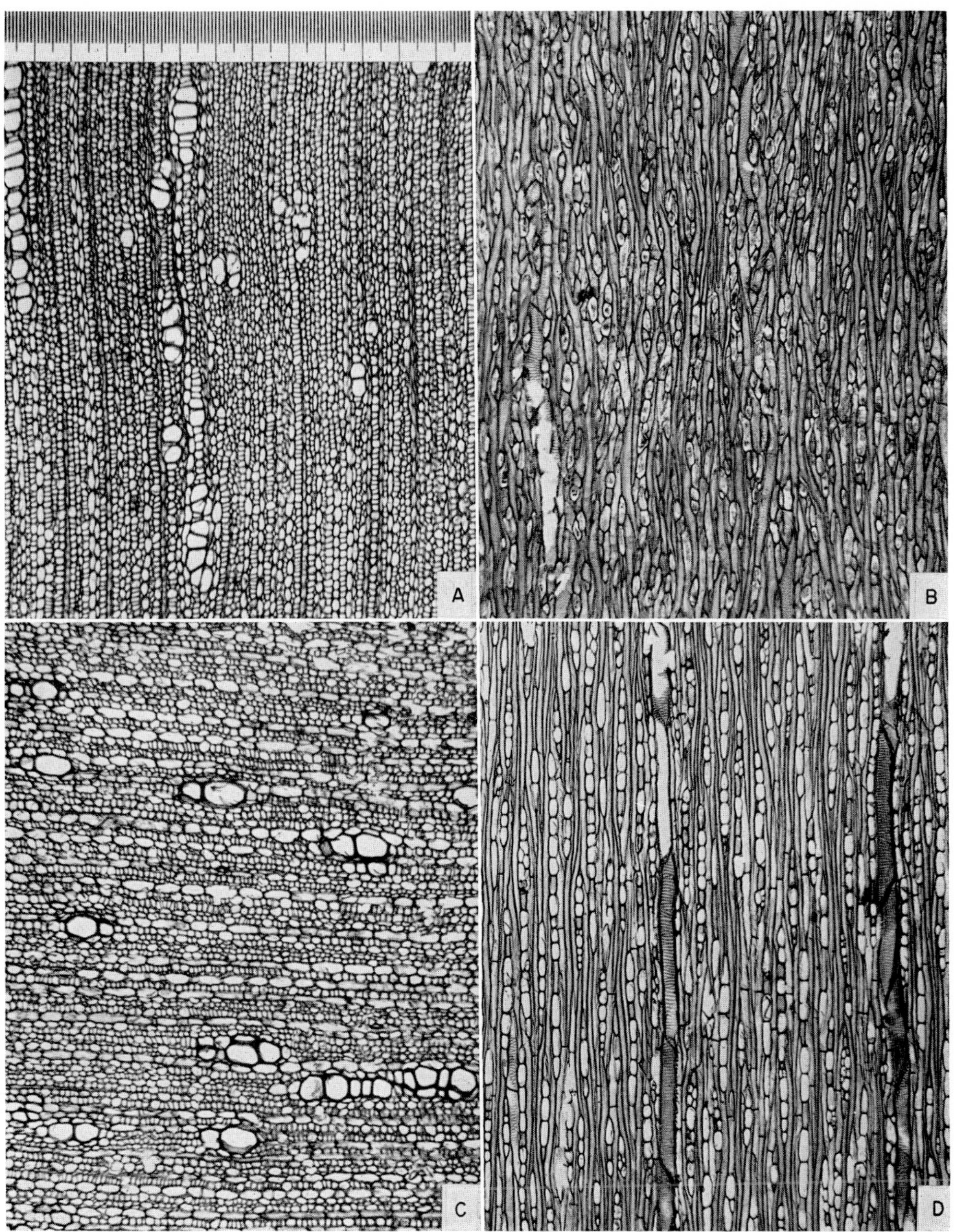
A
B
C
D

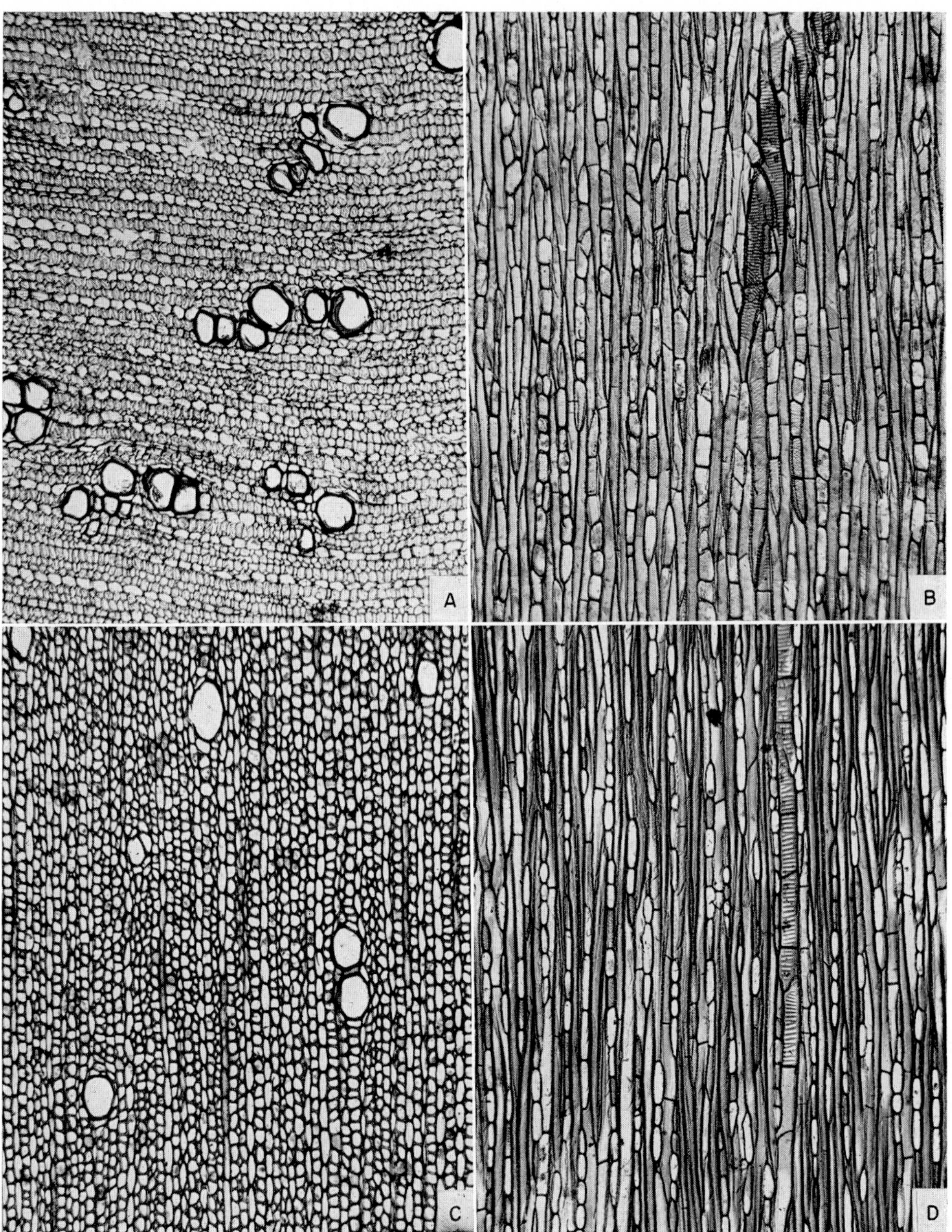

S. CARLQUIST

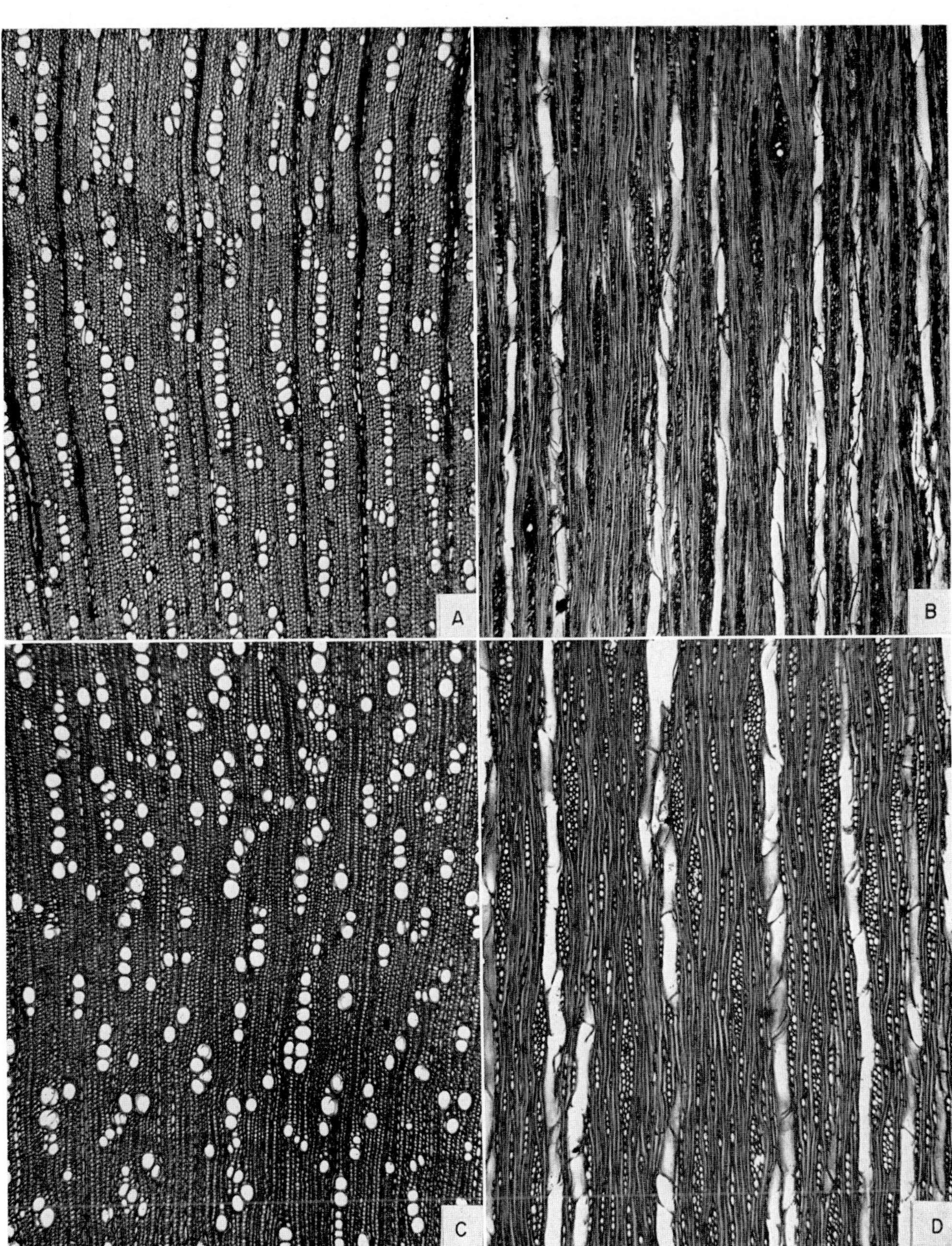

S. CARLQUIST

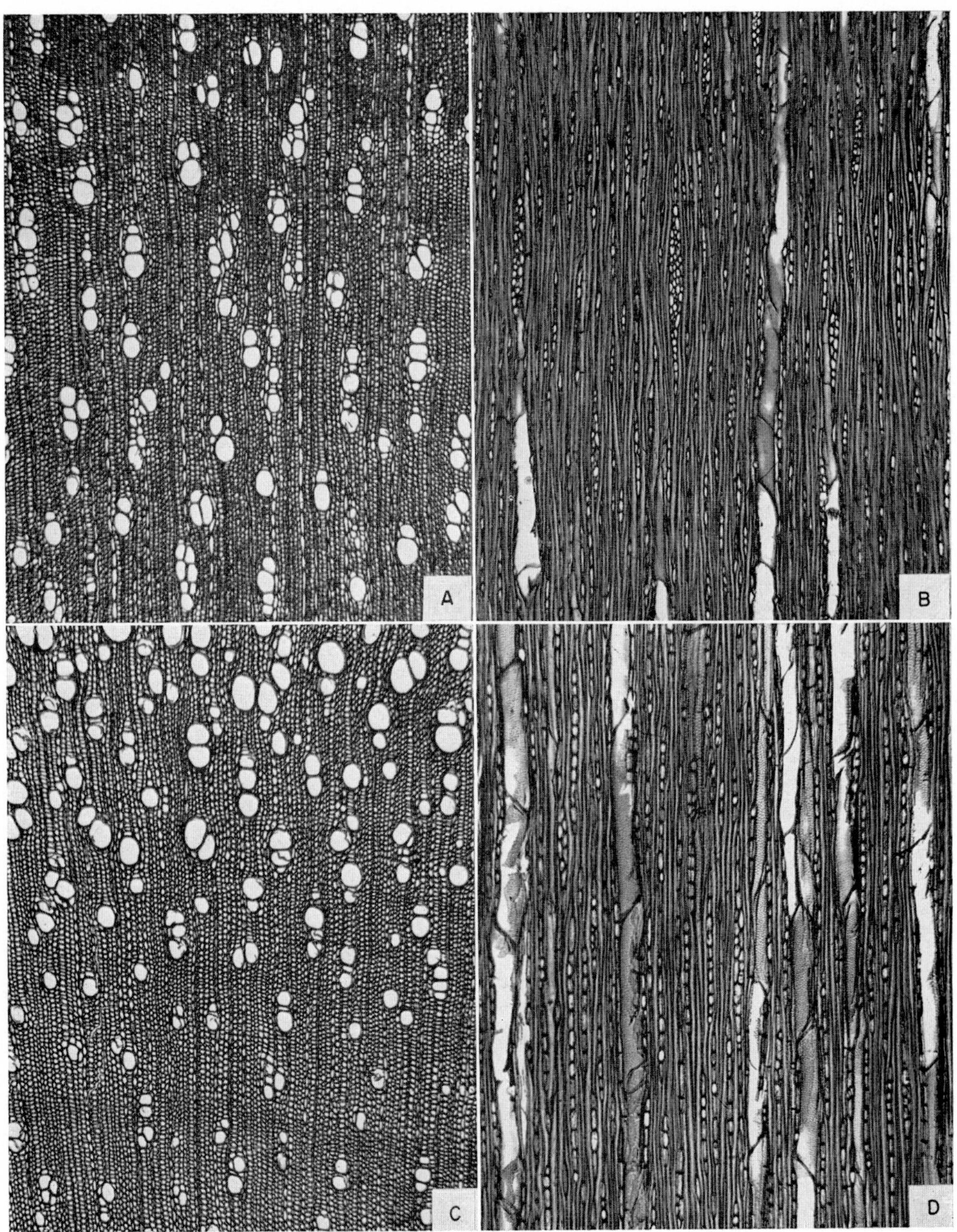

S. CARLQUIST

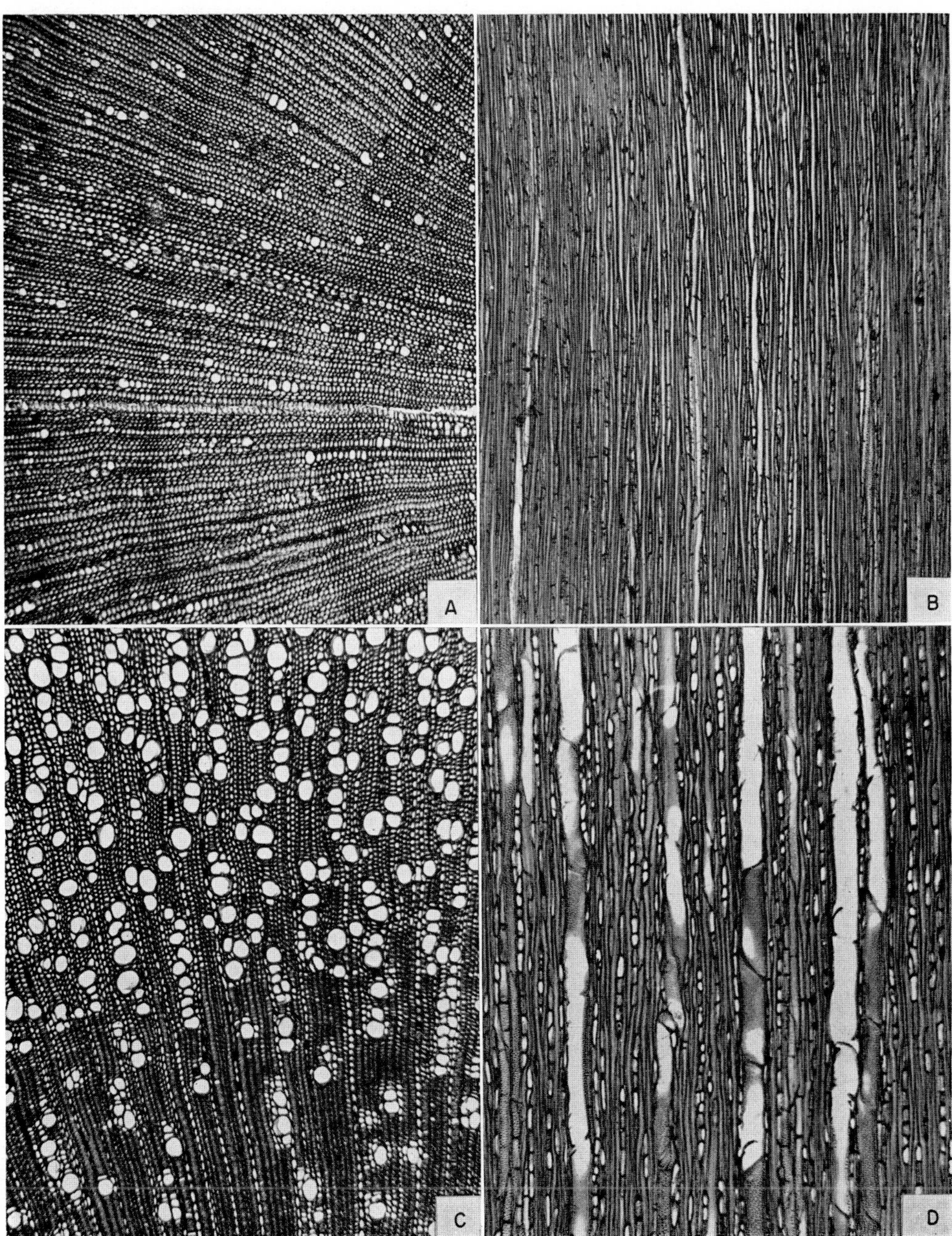

S. CARLQUIST

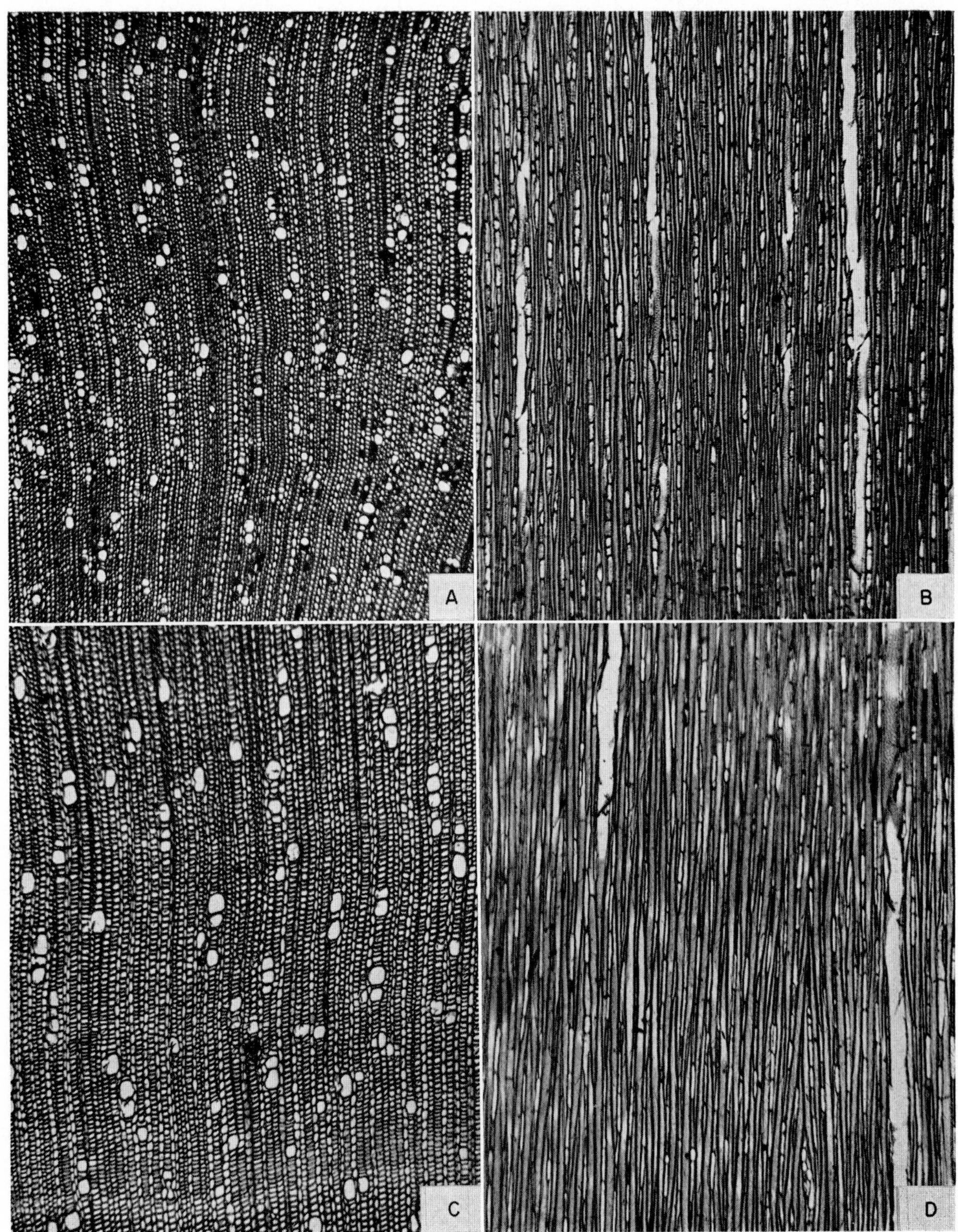

S. CARLQUIST

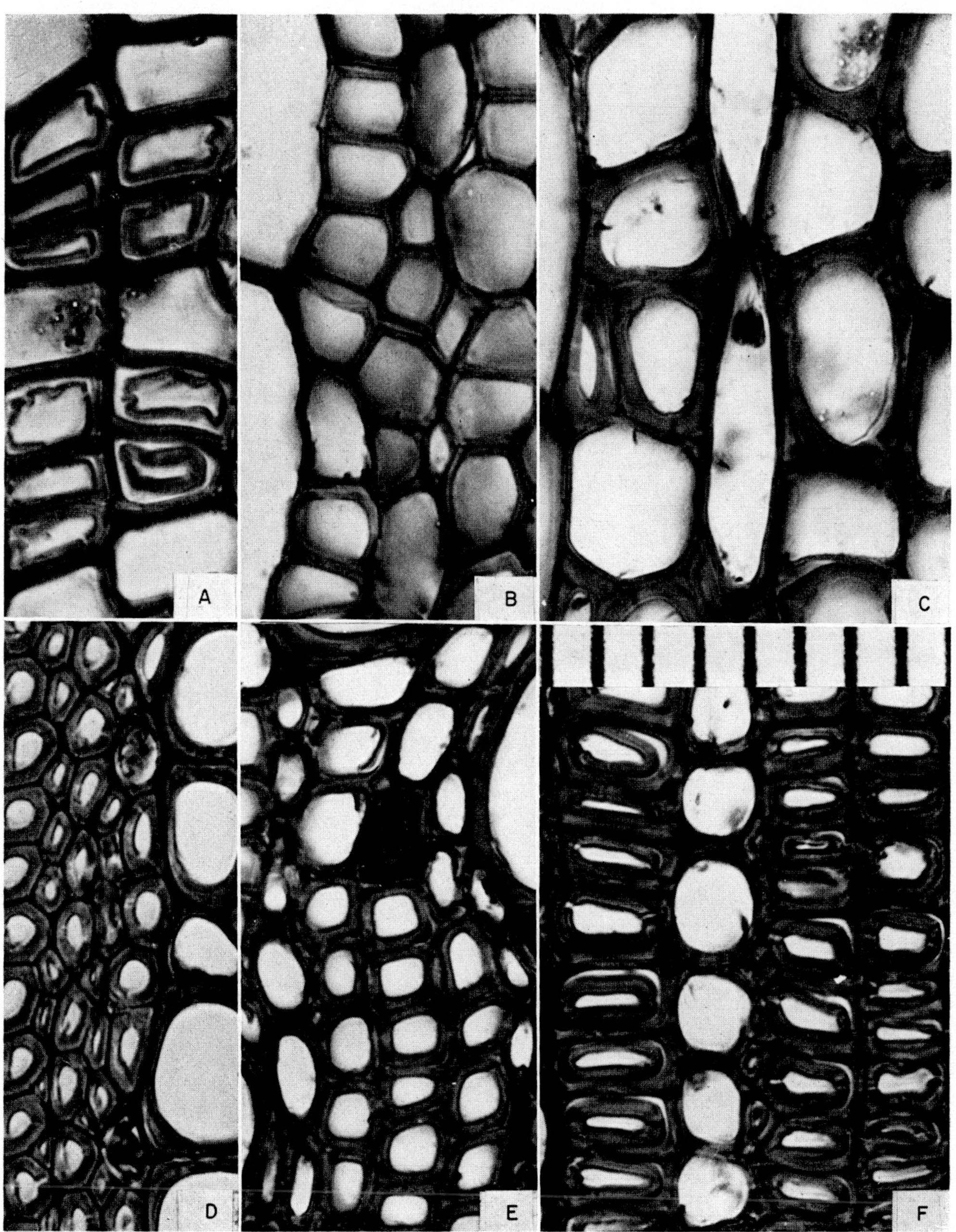

S. CARLQUIST

S. CARLQUIST

PLATE 8

A–F. Portions of walls of vessels, from tangential sections of *Euphorbia* woods, to show lateral-wall pitting. Further discussion in text: **A,** *E. piscatoria*, Carlquist 2646; **B,** *E. regis-jubae*, Carlquist 2428; **C,** *E. lactea*, USw-1820; **D,** *E. clusiifolia*, Carlquist 2371; **E,** *E. rockii*, Carlquist 2196; **F,** *E.celastroides* var. *mauiensis*, Carlquist 1949.

G. *E. celastroides* var. *mauiensis*, portion of tangential section to show laticifer in centre of ray.

Scale of magnification (each division of photographed stage micrometer is 10 μm) for **A–F** above **F**; for **G** above **G**.

Some observations on the nodes of woody plants with special reference to the problem of the 'split-lateral' versus the 'common gap'

RICHARD A. HOWARD

Arnold Arboretum of Harvard University, Jamaica Plain, Massachusetts, U.S.A.

The term node in botanical literature is ambiguous. It is applied to joints or swollen portions of a stem or a fruit; to locations on the stem where one, two or more leaves may be attached; to the internal structure of the stem involving the association of vascular tissue from one leaf or several leaves with that of the stele. Sinnott established a classification of three nodal types, unilacunar, trilacunar and multilacunar, which the majority of workers have followed. Marsden and Bailey proposed a fourth type of nodal anatomy. This classification is unsatisfactory and further consideration must be given to the anatomical variation between the internodal stele and the petiole.

Opposite leaves may have lateral traces which enter and girdle the stem in the cortex and join with the stele at a point midway between the leaf bases. Bailey *et al.* have regarded these as double traces departing from a common gap. Other workers have referred to the traces as a 'composite trace' or 'compound trace'. For descriptive use the vascular bundles can be referred to as 'split-laterals'. These are now reported to occur in the following families: Caprifoliaceae, Chloranthaceae, Compositae, Gentianaceae, Gesneriaceae, Rhizophoraceae, Rubiaceae and Zygophyllaceae. *Alloplectus ambiguus* (Gesneriaceae), when studied in the field, showed for opposite leaves two median traces and two split-lateral traces at the stele. When grown in a greenhouse the stelar pattern varied from regular split-laterals, i.e. from four lacunae, to two trilacunar arrangements from six lacunae.

Split-laterals are usually associated with the presence of stipules. This is another type of nodal anatomy with taxonomic value.

CONTENTS

INTRODUCTION

The node is defined in most textbooks and floras in relation to the external morphology or the appearance of the stem. It is either a joint or a place where a leaf or whorl of leaves may be borne. Variations of this definition may indicate a diversity of opinion on the direction of development, i.e. the node is a point on the stem *from* which a leaf develops, or, in contrast, the zone of the stem where the leaves are *inserted.* The word node has also been associated with the internal structure of the stem, where the term indicates the relationship of the vascular supply of the leaf base or the petiole with that of the stele. Regrettably, a simplistic anatomical classification of three types of nodes was established by Sinnott in 1914, and this has affected adversely much study and interest in the nodal region of the stem ever since. The majority of botanical textbooks today still refer to the three distinctive types of nodal anatomy recognized by Sinnott, and a few add the 'fourth' type of nodal anatomy proposed by Marsden & Bailey (1950), implying that these comprise a survey of the nodal types in the Dicotyledoneae.

The present paper, indicating still another type of nodal anatomy, will avoid the designation of a 'fifth' type by calling attention to the basic problem and indicating some of the types of variation that have been found in our broad survey of the Dicotyledoneae, some that have been described by other workers, and those that remain to be considered in an overall classification of nodal types.

There is no easy way to avoid the multiple meaning associated in botany with the word node. Certainly the node may be a swelling or joint where a leaf is borne, as in the stem of the Polygonaceae. The adjective nodose is often used to describe terete fruits which possess serial constrictions as with Cruciferae, Capparaceae and Leguminosae. By their very presence the leaves may indicate superficially an unenlarged region of the stem where leaves are arranged in an alternate, opposite or whorled pattern. If the leaf abscises, the petiole scar (leaf scar) formed by the zone of abscission is equally indicative of where a leaf was borne on the stem. However, leaf scars enlarge with age in some plants and may often show on mature tree trunks as exaggerated in size as those of *Paulownia*, where the scar may be 24 inches wide and 12 inches high.

The use of the term node for the internal structure of the stem can be applied to a simple union of leaf and stele vascular bundles or to a complex pattern of truly amazing diversity. The term node has been used when leaves are single and alternate or the word may be applied with equal frequency to the stem area where two or more leaves occur at one level. The stem and the leaf may be said to have a continuous vascular supply. The units of conducting tissue which enter the stem from the leaf, or conversely depart from the stem and enter the leaf, have been termed bundles or traces, and their association with the total vascular supply of the stem is such that a lacuna or gap generally filled with parenchymatous tissue is present. Sinnott's general grouping of taxa on the characterization of nodes as unilacunar, trilacunar or multilacunar is the one most widely referred to. Although each lacuna normally has but one trace, he recognized that two, three or more groups of vascular strands could be associated with a given lacuna. Marsden & Bailey (1950), after a study of *Clerodendrum trichotomum* Thunb., proposed a fourth type of nodal anatomy where two traces departed from the stele and entered the leaf together but were independent in origin from separate parts of the eustele and remained separate to varying degrees in the petiole and the blade. The relationship of this type of double trace to the condition found in cotyledons was described, and phylogenetic relationships to other groups of vascular plants were indicated.

Subsequent to Sinnott's publication, many individual workers have called attention to variations, not only to the data for the families and orders that Sinnott studied (e.g. Bailey & Howard, 1941; Money, Bailey & Swamy, 1950), but in the number of gaps on a given stem and the size and number of bundles which may be associated with each (Kato, 1966, 1967). Several workers have suggested that the data used for comparison should be derived only from studies of the primary tissues (Post, 1958), or from comparable parts of the plant (Philipson & Philipson, 1968), or presented only in relation to the traces of leaf-node associations above and below the one under examination (Bailey, 1956). Sufficient evidence is now available to acknowledge that the arrangement of appendages and the internal structural pattern at the cotyledonary node is different from the arrangement of leaves and the structural pattern of the first leaf-nodes of the stem, but that after a few leaves are developed in the stem of the seedling a pattern of great stability is obtained in succeeding nodes and this pattern continues until altered near the inflorescence or at the development of cataphylls.

NODAL PATTERNS

The complicated associations of leaf-traces within the primary body have been indicated as distinctive patterns most recently in the work of Philipson & Balfour (1963), Benzing (1967), Jensen (1968) and others. These patterns cannot be determined readily in most stems and the same authors have described the difficulties of working with these tissues before secondary tissues are formed. Mature stems, however, are the ones available for general observations in most readily identifiable living plants or for comparative studies of most herbarium material. These cannot be excluded from study or from practical use by the anatomist or taxonomist. The need remains for a better knowledge of the variation to be expected in the mature node, and there are characteristics which have not been used fully. The size of the bundle, the attitude of the bundle approaching the stele, the relative position of each bundle in joining the stele, the role of cortical bundles, the nature of stipule vascularization, and the role of medullary bundles represent some of the possible variations found in mature stems which should be represented in a general classification of nodal types in the stems of dicotyledons. Except for the following brief discussion, the ultimate classification of nodal types will be left for a later paper.

For the unilacunar node Sinnott indicated relatively little variation. However, there can be the significant difference as to whether the bundles are collateral or bicollateral, quickly limiting the identification to a relatively few families. The bundle may be nearly isodiametric in transverse section or can be described as oblong in outline, elliptic, or curved. A single gap may show one, two, three or more units of vascular tissue closely associated as an apparently single bundle or seemingly unassociated. Bailey has pointed out the presence of a double trace as characteristic of certain families, as well as the fact that the multiple traces of a unilacunar node may commingle with the vascular supply of an axillary bud or of the inflorescence. Plants with a unilacunar node may possess stipules which receive their vascular supply from the margins of the single trace. The single trace, or the supply to the stipules, may enter the stele within the area represented by the leaf-scar, or the vascular tissue may run down the stem in the cortex and enter the stele at a level well below the base of the leaf-scar or the pulvinus.

In a trilacunar node the bundles may be collateral or bicollateral. The central trace, usually called the median, may be larger or smaller than the lateral traces. The median trace may be a single bundle in appearance or may appear to consist of several units of vascular tissue. The lateral traces probably supply the stipules if they are present, either forming the complete vascular supply of the stipule or giving rise to smaller bundles which vascularize the stipules. Lateral traces may arise near or at some tangential distance from the median trace. Lateral traces often appear in a horizontal path around the stele as a girdling trace contributing one or more bundles to the stipules, or none if the stipules are wanting. The three traces of a trilacunar node may all enter the stele at approximately the same level. The median trace may enter precociously with reference to the laterals or may be retarded in its entry to the stele. The laterals may enter at the same level of the stem, or each may enter at a different level. Any or all of the traces may descend in the cortex for varying distances, occasionally many internodes, before entering the stele.

Cortical bundles may be found in stems which have a trilacunar condition. The cortical bundles may contribute the entire vascular tissues of the lateral bundles which enter the petiole, or only a portion of them. Bundles of the cortical system may be interconnected at the node yet exist as separate bundles in the ridges of the stem in the internodal regions.

Medullary bundles are usually associated with stems possessing a trilacunar nodal condition. The medullary bundles may be complete, having both xylem and phloem, or be incomplete, lacking xylem tissue and possessing only phloem tissue. Medullary bundles may be collateral or bicollateral. Much has been written of their position relative to the stele in one or several series, inverted or of normal orientation. All medullary bundles may contribute to the vascular supply of each leaf, or only a few of the medullary bundles may depart through the stele into the leaf at each node. Medullary bundles at one internodal level of the stem may become stelar or even cortical bundles at other levels. Medullary stem bundles may be medullary petiole bundles within the leaf or incorporated within the eustele proper of the petiole or the midrib.

Multilacunar nodes are found in a relatively few families of the Dicotyledoneae. The median bundle of a multilacunar vascular-pattern is usually larger than the laterals and is easily distinguished. The laterals may be equal in number on either side of the median trace, or the leaf-base may have an asymmetrical distribution of bundles. The lateral traces may arise at varying distances around the stem from the median trace and in some cases may completely encircle the stem. The 'lateral' trace that is on the opposite side of the stem from the median trace, may be independent of the leaf and only supply vascular tissue to the stipule. The other laterals may follow a girdling course, fusing during their horizontal course and commonly giving off bundles which vascularize stipules or sheathing stipules. In the cortex or in the leaf base the many traces of a multilacunar node commonly form a complicated matrix of interwoven and anastomosing strands before becoming reorganized into a special vascular pattern in the petiole.

Philipson & Philipson (1968) stress that comparable material must be used in describing the nodal pattern of woody plants. The significance of the variations which can be described within the individual nodal pattern must be determined by the examination of a large number of specimens or even those of related taxa. Philipson & Philipson (1968) have recently recognized five variations of nodal pattern within the genus *Rhododendron* after investigating 264 species of the genus. *Rhododendron* was originally reported to possess simple unilacunar nodes, but an examination of a larger number of species has revealed the existence of trilacunar nodes and four recognizable and consistent patterns of variation in the unilacunar node. The trilacunar node was found to be present in some species of five different sections of the genus, in one species of still another section, but in all species examined of the subsection *Grandia*. Dormer (1945*b*) was able to determine by careful study that the so-called multilacunar node of the Epacridaceae was in most cases a misinterpretation of the branching pattern of a single trace from a unilacunar node. Saha (1952) found trilacunar, bilacunar and unilacunar nodes in various plants of *Citrus grandis* Hassk. Kato (1967) has shown variations possible on the same plant of *Lindera obtusiloba* Bl., where one, two or three traces may depart from a single gap. In *Rubus palmatus* Thunb. (= *R. microphyllus* L.), which normally has trilacunar nodes, he found tetralacunar and pentalacunar nodes on the same branches. *Poncirus trifoliatus* (L.) Raf. was reported to have variations of unilacunar, bilacunar and trilacunar nodes on the same branch, and material grown at the Arnold Arboretum showed a similar variation. Our material of *Sorbus conmixta* Hedlund, however, had uniform pentalacunar nodes, while Kato reported variations in successive nodes of 3–5–5–5–3–4–3–5 gaps. Kato also reported a great variation in the number of traces from the multilacunar nodes of species of *Ficus*. Canright (1955) reported diversity in multilacunar nodes of the various genera of the Magnoliaceae, to contrast with the stable patterns reported by Ozenda (1947). Swamy (1949) reported that the regular nodal vascular pattern in mature stems of *Degeneria* was pentalacunar but that the first few seedlings were vascularized with only three traces, and leaves of vigorously growing saplings may receive more than five traces. Swamy & Bailey (1949) found in *Cercidiphyllum* that the seedlings and the regular long shoots were trilacunar at the node but that the leaves of the short shoots differed markedly in having three strands that were related to unilacunar nodes. This observation for short shoots has not been verified with additional material from the same plant. Pant & Mehra (1964) reconsidered the evidence Bailey had used in stressing the phylogenetic significance of the double trace and disagreed with Bailey's conclusions. However, they proposed that nodes of various vascular plants be termed (1) alacunar if there is no gap in the stele, (2) unilacunar if there is a single gap in the stele and (3) multilacunar where the number of gaps is more than one, thus putting less emphasis on the number of gaps. They further proposed a subclassification on the basis of the number of traces which supply a leaf, e.g. unilacunar, one-trace; unilacunar, two-trace; etc.

The nodal condition in herbaceous plants is even less well known. Philipson & Philipson (1968) advise caution in comparing nodal types in herbaceous plants.

Müller (1944) found that the mature foliage leaves of *Helleborus foetidus* L. possessed 7 to 11 traces but that this number diminished in higher leaves and only a single trace was found in the bracts of the inflorescence. Post (1958), in examining successive nodes of *Frasera* spp. and *Swertia perennis* L., found what he termed four types of nodal anatomy in *Frasera* and a fifth type in *Swertia*. Each type depended either on the consistency of the pattern or the variation of the pattern in the number of traces to successive leaves as the stem was examined from the basal leaves to the inflorescence. Since *Helleborus* and *Frasera* have strong underground rhizomes, the upright shoot system examined might well be considered as an inflorescence axis, and the variations comparable to those found in other inflorescences. Plants of *Lisianthus laxiflorus* Urb., also of the Gentianaceae, but variously considered to be woody herbs or shrubs and without horizontal rhizomes, were examined in the field in Puerto Rico where abundant material could be sectioned. No seedlings could be located to determine the pattern in the cotyledonary and first seedling leaves, but other plants ranged in size from those possessing only five nodes to a plant where 75 consecutive nodes could be sectioned. In all, the nodal pattern was consistently trilacunar. Husson (1965) examined herbaceous Polygonaceae with multilacunar traces and found in them a lower number of traces and gaps in the lower nodes, a higher number in the middle nodes of the stem, and a reduction to primarily trilacunar nodes immediately below the inflorescence.

The presence of two or more leaves at the same level in mature stems is recognized as a stable characteristic often associated with certain families. Examples of opposite as well as whorled leaves associated with unilacunar and trilacunar nodes are known. There are examples of multilacunar nodes associated with opposite leaves but few verified reports of multilacunar condition for whorled leaves.

THE 'SPLIT-LATERAL' VERSUS THE 'COMMON GAP'

There is a general tendency to regard the vascular supply of a leaf as distinct from that of its neighbour. The work on the vascularization of the primary body by Dormer (1945*a*), Balfour & Philipson (1962), Philipson & Balfour (1963), Fahn & Broido (1963), Ezelarab & Dormer (1963), Benzing (1967) and Jensen (1968) has shown that many types of bundle association can be recognized. Open, closed and intermediate vascular systems are recognized where adjacent sympodia contribute to the vascular supply of a given leaf. A single sympodium may be associated with traces to a given leaf, or bundles from adjacent sympodia may supply a portion of the vascular supply to leaves in successive or superposed nodes. In a few plants the vascular supply of a single leaf may be derived from completely independent stelar and cortical systems, as in *Chimonanthus* (Balfour & Philipson, 1962).

In mature stems the association of a single bundle from the stele with more than one leaf has received little attention and the interpretation offered is conflicting. The earliest reference to a trace dividing with branches entering adjacent leaves, or conversely lateral traces from two leaves fusing and entering the stele as a single bundle, is in the illustration of *Sambucus nigra* L. published by Nägeli in 1858. Here a single trace midway between the bases of opposite leaves, is shown forking, with a portion going to each of the opposite leaves. Esau (1945) reported this condition in her study of the primary body of *Sambucus* but did not indicate that the condition could be found in mature stems as well. In 1927 Cunningham described a similar branching in the vascular structure of the stem of *Zygophyllum fabago* Thunb. as follows: 'Tracing the vascular system from below upwards, the first change, as the node is approached, is the departure of a narrow part of the vascular cylinder from the middle of the convex side. When this strand has ascended about one or two millimetres and has moved outward from the main cylinder, it divides into two forks, which extend laterally to form horizontal girdles, each of which passes round about a quarter of the stem. Meanwhile a similar change has commenced on the opposite flattened side of the stem and the central cylinder soon resolves itself into eight components. Two of these, placed opposite each other in the transverse plane, constitute the petiolar supply. As each petiolar strand passes into the leaf-base it is joined by the girdle components, first on the rounded side and later, at a slightly higher level, on the flattened side.' Carlquist (1957) described and illustrated the vascular pattern of spirally arranged leaves of *Argyroxiphium sandwichense* DC. (Compositae). He stated, 'The node is pentalacunar with three unbranched central traces. Each of the marginal traces, however, branches shortly above its departure from the vascular cylinder of the stem. Half of the veins derived from a lateral trace enter the margin of one leaf, the other half enter the margin of the adjacent leaf.' If one accepts the acropetalous development of xylem within the procambial strand from the stem into the leaf, this condition can be called a split-lateral trace.

Swamy & Bailey (1950), in the first of their studies of the Chloranthaceae, called attention to the vesselless wood structure of the genus *Sarcandra* and in the same paper to an unusual type of nodal vascular pattern in the primitive taxon. In *Sarcandra*, and also *Chloranthus* as later reported, Swamy & Bailey referred to the node as a modified unilacunar type, for the marginal strands of opposite leaves originated from two 'common gaps'. They state that in these genera the lateral traces of the pairs of leaves traverse the node in the cortex before entering the stele through the same gap. Bailey & Swamy emphasize their opinion in several papers that this is not a dichotomy of a single strand but the approximation or fusion of two strands. They suggest that at lower levels in the stem the bundles appear to separate and join different parts of the eustele. Swamy (1953) stated, 'the four marginal veins (two for each of the oppositely arranged leaves) arise from two 'common gaps' in contrast to the typically trilacunar nodal situation where the four marginal strands of two leaves confront a corresponding number of independent gaps. Therefore the type of nodal structure in *Chloranthus* and *Sarcandra* cannot be assigned either to a strictly unilacunar type or to a typically trilacunar one. However, it appears reasonable to consider the nodal structure of *Chloranthus*, *Sarcandra* and the like as being a modification of the unilacunar type on account of their sporadic occurrence, always among representatives that possess predominantly unilacunar nodal anatomy.' Bailey and Swamy were seeking evidence for the proper placement of the family Chloranthaceae, and the similarity of unilacunar nodes was striking evidence. They did not consider the possibility of recognizing both unilacunar and trilacunar nodes within the family.

One further example of a single lateral trace supplying a portion of the vascular supply of a pair of opposite leaves is shown by Mitra & Majumdar (1952) in an illustration of the node of *Ixora parviflora* Vahl. They refer to this bundle as a 'Composite trace', recognizing its origin from a single gap and the division into two girdling traces with its ultimate participation in the vascular supply of the opposite leaves.

Whether this should be called a 'composite trace', or an approximation of two traces entering a 'common gap', or a single trace which divides, becoming a 'split-lateral', cannot be settled here. The condition is found more frequently than had been assumed, and examples of its occurrence have been reported in the Caprifoliaceae (Nägeli), Chloranthaceae (Swamy), Compositae (Carlquist), Rubiaceae (Mitra & Majumdar), and Zygophyllaceae (Cunningham). Additional examples have been encountered in these families and in the Gentianaceae, Gesneriaceae and Rhizophoraceae as well.

CAPRIFOLIACEAE

Nägeli's illustration of the split-lateral trace in *Sambucus nigra* L. is believed to be the earliest reference to this condition. *Sambucus* is often cited as being aberrant in the Caprifoliaceae in the possession of pentalacunar nodes with opposite, exstipulate leaves which are compound. Sinnott (1914), in discussing the Rubiales, stated 'The Caprifoliaceae, Valerianaceae and Dipsacaceae are entirely trilacunar except for a few instances (*Sambucus* and others) where they [sic] may be five bundles and gaps.' In sectioning living and herbarium material of various species of *Sambucus* in the Arnold Arboretum, we determined that *Sambucus gaudichaudiana* DC. and *Sambucus racemosa* L. forma *flavescens* (Sweet) Schwerin are consistently trilacunar with independent gaps for each trace. *Sambucus nigra* L., as grown in the Arnold Arboretum, is pentalacunar with independent gaps for each trace. Material of *Sambucus canadensis* L. (Fig. 1**A**), however, showed variation in nodal pattern from plant to plant and within the same plant. The split-lateral illustrated by Nägeli was found at either or both lateral positions for the opposite leaves at some nodes, while nodes immediately above or below had all traces entering independent gaps. The remainder of genera examined within the Caprifoliaceae had trilacunar nodes and independent lateral traces. Airy Shaw, in a revision of Willis' work (Willis, 1966), considers *Sambucus* to be the only genus in the Sambucaceae differing from the Caprifoliaceae in the pinnate stipulate leaves, among other characters. Schwerin (1920) concluded that the genus *Sambucus* differed from other genera of the Caprifoliaceae in the possession of five vascular bundles in the petiole.

CHLORANTHACEAE

Swamy (1953), in his summary paper on the Chloranthaceae, recognizes four genera within the family, two of which, *Ascarina* and *Hedyosmum*, have 'nodes typically of the unilacunar type', and two, *Chloranthus* and *Sarcandra*, have 'nodes of a modified unilacunar type' (Fig. 1**C**). Swamy groups the Chloranthaceae in 'Category I: Nodes unilacunar' in the division of the ranalian families proposed by Money, Bailey & Swamy (1950), concluding that the addition of marginal strands originating from two common gaps represents only 'a modification of the unilacunar type on account of their sporadic occurrence, always among representatives that possess predominately unilacunar nodes'.

Rousseau (1927) had referred earlier to these bundles of *Chloranthus inconspicuus* Blanco (*Chloranthus spicatus* (Thunb.) Makino) as 'foliaires composés' and noted that the laterals split with a portion entering each of the opposite leaves. Additional sections

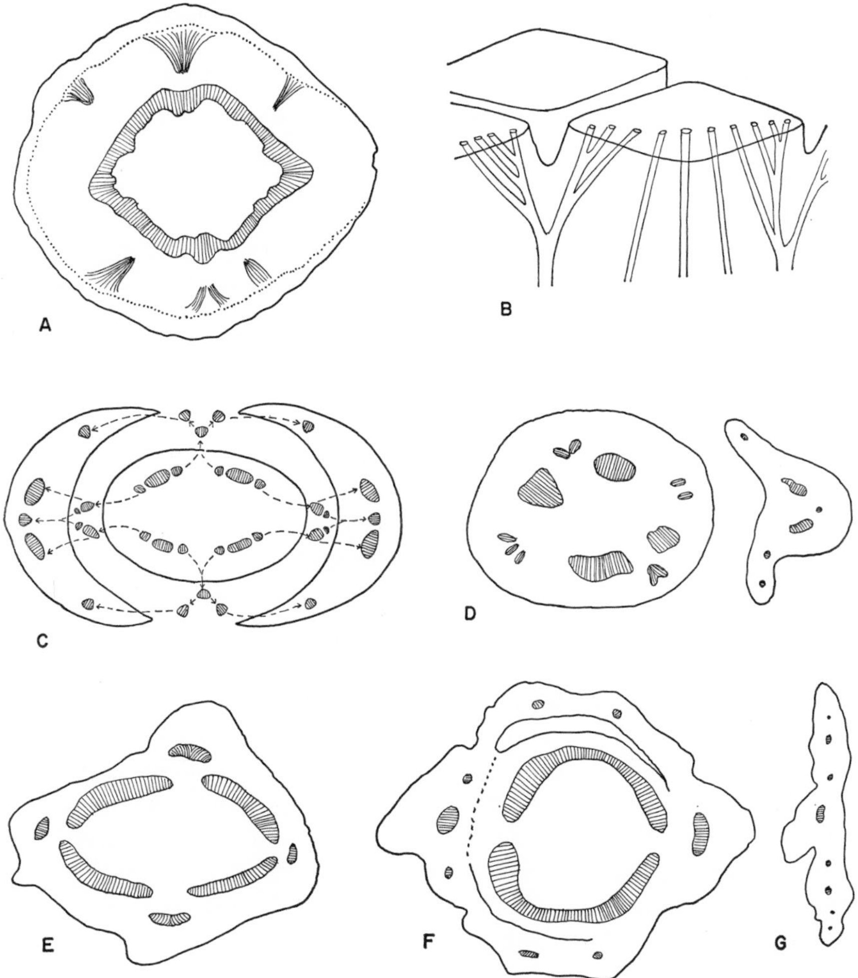

FIGURE 1. **A.** Node of *Sambucus canadensis*. Median traces are not shown. Two split-laterals are indicated. **B.** *Argyroxiphium sandwichense*. Redrawn from Carlquist (1957). Lateral traces bifurcate before entering adjacent leaves and then branch. **C.** *Sarcandra glabra*. Nodal vascular pattern redrawn from Swamy & Bailey (1950). The authors describe the association of two vascular bundles from the eustele departing from a common gap and separating to follow a girdling path before entering the leaves as lateral traces. **D.** Diagram of the node of *Chloranthus brachystachys*. The division of the lateral traces is shown. Within the leaf base the lower lateral is shown divided. **E, F, G.** *Enicostema verticillatum*. Camera lucida drawings of the node. **E,** lateral traces are undivided. **F,** median trace of the leaf on the left has divided to produce two lateral bundles. A space is shown on the upper side of the node suggesting the development of a stipule sheath. The split-laterals have each divided. **G.** The lower portion of the petiole showing further divisions of the products of the split-lateral traces.

of the herbarium material which Swamy studied failed to show any variation on the pattern he described, nor did ample living material of a single plant of *Chloranthus brachystachys* Bl. obtained from Hawaii (Fig. 1**D**).

Swamy is incorrect in his assumption that the split-lateral condition is found only in families which are predominantly unilacunar.

Both *Sarcandra* and *Chloranthus* (Fig. 6**C**) possess stipule-like processes.

COMPOSITAE

The split-lateral traces of *Argyroxiphium sandwichense* DC. as illustrated by Carlquist are seen in the drawing reproduced from Carlquist's paper (1957) (Fig. 1**B**).

GENTIANACEAE

Sinnott reports the Gentianaceae to have one and many traces in the node. He states, 'All genera examined of the five families included by Engler in the Contortae are unilacunar save *Menyanthes*, which (with another genus) composes the subfamily Menyanthoideae of the Gentianaceae. In this case three or five bundles enter the base of each leaf from as many gaps. Either *Menyanthes* should not be included under the Gentianaceae or else we must believe that its nodal structure has been so modified by its aquatic habitat, which has caused its leaves to become sheathing, that evidence from this region should be disregarded.' Sinnott does not state what material he has examined under the various family entries in his publication. The genus *Menyanthes* has been segregated by many authors as the family Menyanthaceae by many characters including the alternate leaves. Within the Gentianaceae *sensu strictu*, the leaves are opposite and without stipules. Post (1958) has reported that the nodal structure of *Frasera* spp. and *Swertia perennis* L. have variable nodes with one to seven traces from an equal number of gaps. In our admittedly incomplete survey of the family, of 24 genera examined *Enicostema verticillatum* (L.) Engl. (Fig. 1**E–G**), *Faroa salutaris* Welw. and five species of *Lisianthus* (*L. axillaris* Perkins, *L. capitatus* Urb., *L. cordifolius* L., *L. latifolius* Sw. and *L. laxiflorus* Urb.) have been determined to have opposite leaves with split-lateral traces supplying a part of the vascularization of the leaves. *Enicostema* and *Faroa* can be regarded as herbaceous plants. Some species of *Lisianthus* are stout, almost shrubby, and clearly woody plants. The median trace in all species of *Lisianthus* divides quickly and is triple in the leaf base before it is joined by the portions of the split-laterals.

GESNERIACEAE

Sinnott noted that, of the many genera of the Tubiflorae which he investigated, only *Cyrtandra* of the Gesneriaceae displayed other than a single-gap condition. Three or five strands he regarded as typical of that genus.

During the course of field studies in an elfin forest on a mountain-top of Puerto Rico (Howard, 1968, 1969), anatomical studies were made of the component taxa. Of the two members of the Gesneriaceae present in that flora, material of *Alloplectus ambiguus* Urb. presented the unusual pattern we referred to as the split-lateral trace. *Alloplectus ambiguus* (Fig. 2**A**) is a semi-woody epiphyte which, in that wet environment, may also be found on the ground. The semi-fleshy stems ascend or are pendant and are characterized by noticeable heterophylly, that is, leaves of the opposite pairs are of strikingly different sizes. In the course of the study material was sectioned for

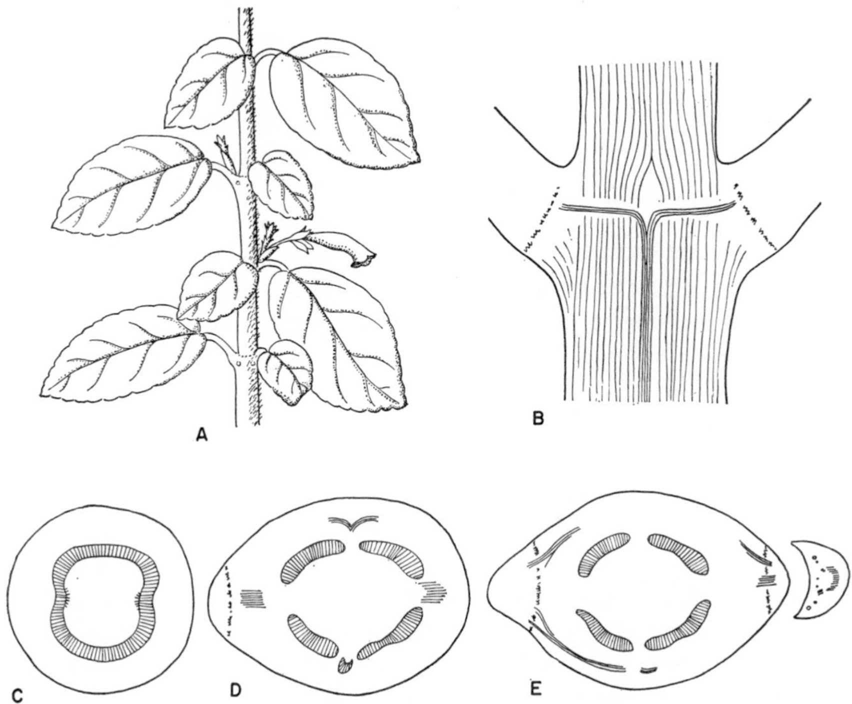

FIGURE 2. *Alloplectus ambiguus* (Gesneriaceae). **A.** Habit of *Alloplectus ambiguus* showing the conspicuous heterophylly of the mature shoot. **B.** Diagrammatic representation of the split-lateral trace within the cortical area. The girdling path is indicated as is a single lacuna. **C.** Diagram of the internodal stele. **D.** Two split-lateral traces are shown together with two median traces. **E.** Diagram to show the girdling path of the split-lateral derivatives and their relative position in the pulvinus.

nodal studies from over 20 individual plants, and all nodes studied showed that of the opposite leaves, in spite of differences in size, each had a single strong median trace and two smaller lateral traces entering the petiole (Fig. 2**C–E**). The lateral traces of opposite leaves extended horizontally in the cortex to a midpoint between the leaves where the traces came together and entered the stem (Fig. 2**B**). Each leaf, therefore, was supplied with a median trace, and one half of each of two lateral traces. The cross section of the stem at the point of attachment of the paired leaves showed four gaps.

This example appeared to be so useful for student laboratory exercises that cuttings and seeds were returned to the Arnold Arboretum and plants were grown to be used in subsequent laboratory exercises. It was surprising, therefore, on the occasion of first use of this material, to discover that the nodal vascular patterns were not those seen in sections made in Puerto Rico. In fact, the examination of a large number of stems and nodes showed a considerable variation, from opposite leaves each being supplied with traces from three independent gaps to the expected condition of six traces from four gaps for the opposite leaves. Stems were sectioned, and nine patterns as illustrated were found as variations (Fig. 3**A–I**).

We have re-examined material collected in Puerto Rico which was preserved in alcohol and studied additional stems propagated from plants grown in greenhouses in the Boston area. The variable pattern persists in the greenhouse material and the stable pattern of split-lateral traces is all we can find in the field-grown material. This is the only case we know in published literature of striking changes in nodal vascular patterns of plants grown under cultivation.

Studies were also made of plants grown from seeds, but these have been limited to material grown in the Arnold Arboretum greenhouses (Fig. 3**Q**). No seedlings of *Alloplectus ambiguus* have been located in the wild. The two small cotyledons are opposite in the seedlings and are vascularized by a single trace to each cotyledon. No double strand condition has been found to the present. The leaves of successive nodes above the cotyledons are arranged in opposite and decussate pairs. Heterophylly is apparent in the seedlings upon measurement but is not noticeable to the eye, as the condition is in mature plants. The leaves of the first two leaf-pairs above the cotyledonary node all have but a single trace. The first appearance of the lateral traces in the seedlings examined was at the third node above the cotyledons when a single thin lateral trace appeared. These have been irregular in location in reference to the larger or smaller leaf or to the side of the stem. At the fourth node both lateral gaps were evident, with the lateral traces numbering one, two, three or four in various combinations (Fig. 3**J–Q**). Here too, there is no correlation in the development of lateral traces which can be associated with either the large or small leaf of the pair or in the side of the stem relative to the position of the plant within the flat of seedlings. By the fifth node above the cotyledons the laterals of seedlings were regular in occurrence and the several patterns previously described were found.

In other seedlings the lateral traces were delayed in appearance as late as the seventh node above the cotyledons.

We have found several examples of seedlings in which the lateral traces are present in the petiole at the third node above the cotyledon, yet these have not established a connection with the stele. An example is illustrated diagrammatically as Fig. 3**P**, where one lateral trace joins the median trace before it enters the stele while a second trace enters and ends in the cortex. Such observations are based on cleared seedlings and we cannot determine from these preparations if undifferentiated tissue completes the path of the lateral trace to the stele. From this scanty material it appears that the lateral traces originate in the petiole and are completed by subsequent differentiation through the cortex and into the stele.

A trilacunar node, or the occurrence of split-laterals, has been found in other genera of the Tribe Columneeae of the Gesneriaceae by Hans Weihler, who is working on this problem at Cornell University and whose work will be published later. We have found the split-lateral plus a median trace in material of *Alloplectus grisebachianus* (Kuntze) Urb. and *Drymonia spectabilis* Mart. collected in Jamaica and Panama, respectively, and in herbarium materials of *Didissandra aspera* Drake and *Hemiboea follicularis* C. B. Clarke. Material of unidentified species of *Cyrtandra* collected in

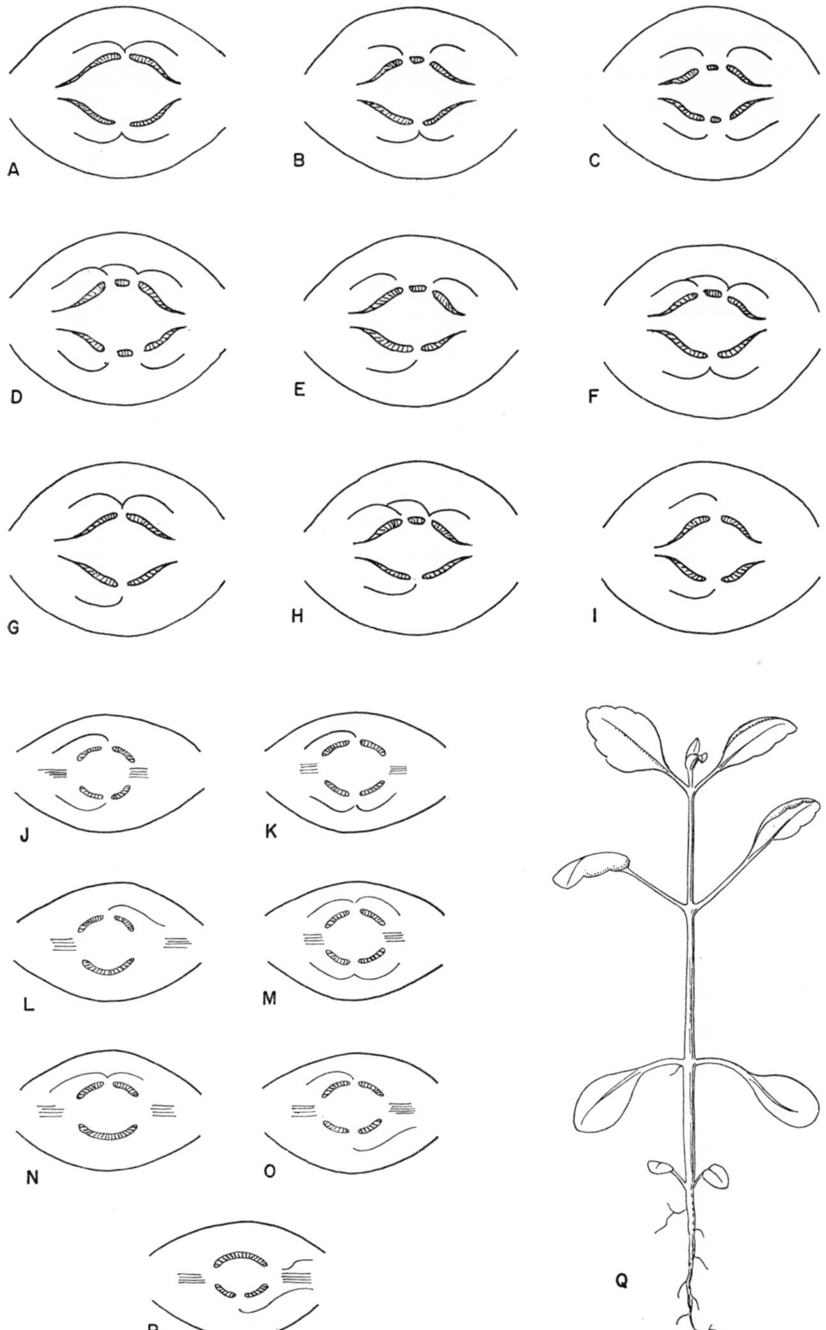

FIGURE 3. *Alloplectus ambiguus* (Gesneriaceae). **A–I.** The variation of nodal vascular pattern found in material of *Alloplectus ambiguus* grown in a greenhouse from cuttings made in Puerto Rico. **J–P.** Variations in the development of lateral or split-lateral traces found in the third node above the cotyledons in seedlings grown from seed under greenhouse conditions at the Arnold Arboretum. **P.** An anomalous vascular pattern showing one lateral trace of the leaf entering the same stelar gap as the median bundle while the other lateral trace girdles the node in the cortex and approaches the stele midway between the two leaves. **Q.** A seedling of *Alloplectus ambiguus*. The cotyledons did not show a double trace in any plant examined. The first two pairs of leaves above the cotyledons had only median traces in all material examined.

Hawaii had opposite leaves with pentalacunar nodes where the extreme lateral traces of each node entered a common gap. Boldt (1897) illustrates a split-lateral in his paper on *Didymocarpus* (*Chirita*) *hamosa* Wall.

RHIZOPHORACEAE

Sinnott reported the Rhizophoraceae to have trilacunar nodes. He noted that this family, along with the Nyssaceae and Alangiaceae, stood apart in the Myrtiflorae on this characteristic. He stated, 'Nodal anatomy also indicates that the Rhizophoraceae should not be placed in the Myrtiflorae which, aside from these three families, seems to be a natural order.'

In a survey of the nodal pattern in fresh material of several taxa of this family we have found patterns of three, five or seven bundles departing to each leaf in which the extreme laterals of each of the opposite leaves represent split-laterals or depart from a common gap. Material of *Cassipourea elliptica* (Sw.) Poir. (Fig. 4**B**) from the West Indies and from Panama, *Cassipourea gummifera* Tul. from South Africa, and *Cassipourea elliottii* Alston (Fig. 4**A**) from Kenya all show a single large median bundle entering the leaf base, accompanied by two lateral traces which are derived from a single lateral which departed from the stele from midway between the leaf bases. *Bruguiera sexangula* (Lour.) Poir. from South Africa has five traces to each leaf plus the bundles from split-laterals (Fig. 4**C, D**). *Rhizophora stylosa* Griff. and *Rhizophora mucronata* Lam. (Fig. 4**F**) have a comparable nodal pattern of five traces from five gaps plus the branched trace from a common gap contributing a small extreme lateral bundle to each leaf. Material of *Rhizophora mangle* L. from Florida and five places in the West Indies, however, revealed a nodal pattern of seven traces from seven gaps for each of the opposite leaves (Fig. 4**E**). Perhaps with the examination of a larger number of specimens, the same condition of completely separate vascular systems of the split-lateral condition might be found in other taxa of *Rhizophora*.

Fused stipules occur regularly in this family and examples are illustrated for *Cassipourea elliptica* (Fig. 6**B**) and *Bruguiera sexangula* (Fig. 6**G**).

RUBIACEAE

Sinnott (1914) noted that the Rubiaceae are 'overwhelmingly unilacunar, the only exception being the genus *Sarcocephalus*, where the presumably ancient trilacunar condition persists'.

In our survey of genera trilacunar nodes have been found in 12 other genera representing eight tribes as follows: Condamineae: *Portlandia*; Naucleeae: *Nauclea*, *Sarcocephalus*; Mussaendeae: *Schradera*, *Gouldia*; Gardenieae: *Casasia*, *Genipa*, *Gardenia*; Alberteae: *Alberta*; Vanguerieae: *Vangueria*; Ixoreae: *Ixora*; Coussareae: *Faramea*, *Coussarea*.

The leaves of the Rubiaceae are opposite or whorled, and each leaf possesses a pair of stipules. In many genera and species the stipules may be free or are fused between the pairs of leaves. Fused stipules may show a bifurcation at the apex or be united into a single collar with an apiculate apex (Fig. 6**D**). In some unilacunar nodal types, branches from the single trace may be retrorse in direction entering the stipule and remain independent or may fuse, as Varossieau (1940) has shown for *Coffea arabica* L. (Fig. 6**E**). In trilacunar nodal types the lateral traces also supply the vascular tissue for the stipules. The vascular tissue to the stipules may also be slightly retrorse in direction if the lateral traces arise a distance apart. In a few specimens we have examined, the lateral traces arise from a single gap and the lateral splits with a portion entering each of the pair of leaves. In its girdling path at the node the lateral trace

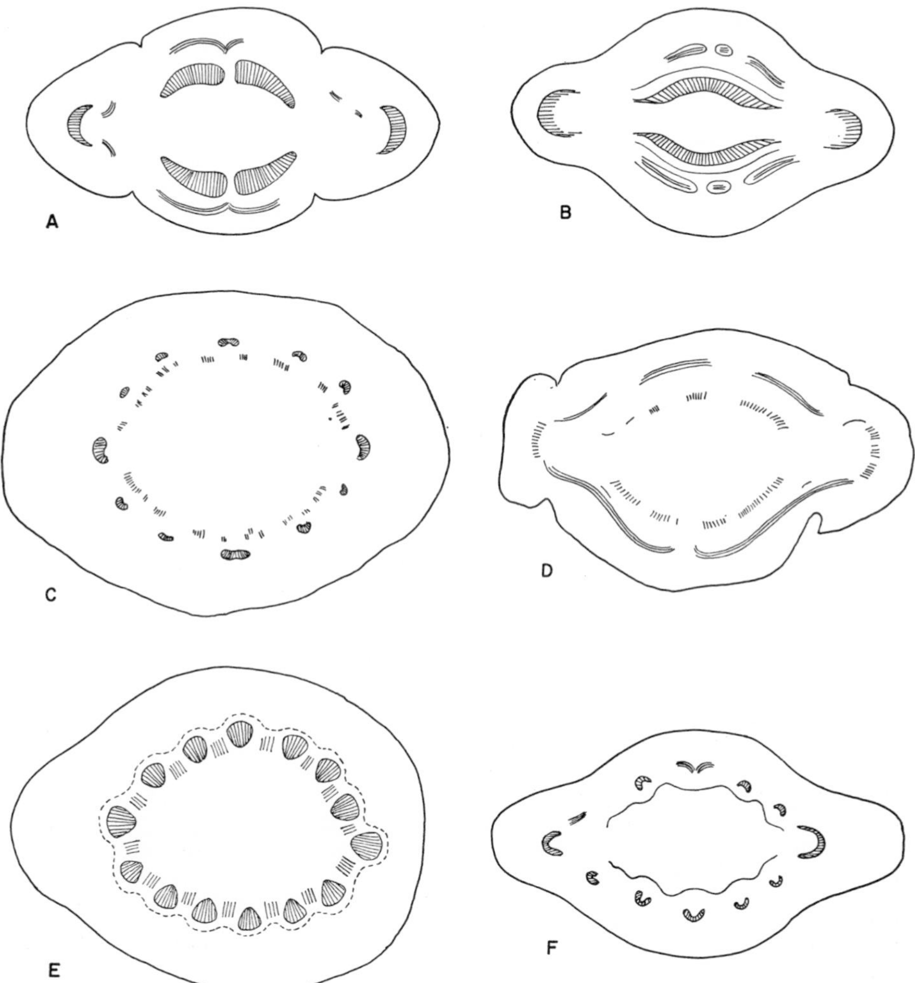

FIGURE 4. **A.** The nodal vascular pattern of *Cassipourea elliottii* (Rhizophoraceae) showing the split-lateral traces. **B.** The nodal pattern of *Cassipourea elliptica* (Rhizophoraceae). The girdling path of the split-lateral traces is shown. Branches of one or both divisions of the split-lateral may enter the stipule immediately above the lacuna within the stele. **C.** The nodal vascular pattern of *Bruguiera sexangula* (Rhizophoraceae) showing for each leaf a median plus two pairs of lateral traces plus the derivatives of a split-lateral trace. **D.** *Bruguiera sexangula* (Rhizophoraceae). A diagram showing the girdling path of the split-lateral traces. The other traces had already entered the base of the leaf. **E.** *Rhizophora mangle* (Rhizophoraceae). A diagram of the nodal area showing a median and three pairs of lateral traces from individual lacunae for each leaf. **F.** *Rhizophora mucronata* (Rhizophoraceae). A diagram of the nodal area showing split-lateral traces in addition to the median and two pairs of laterals for each leaf.

supplies vascular tissue to the stipules before entering the leaf base. Mitra & Majumdar (1952) illustrated this condition of split-lateral and girdling trace, referring to the

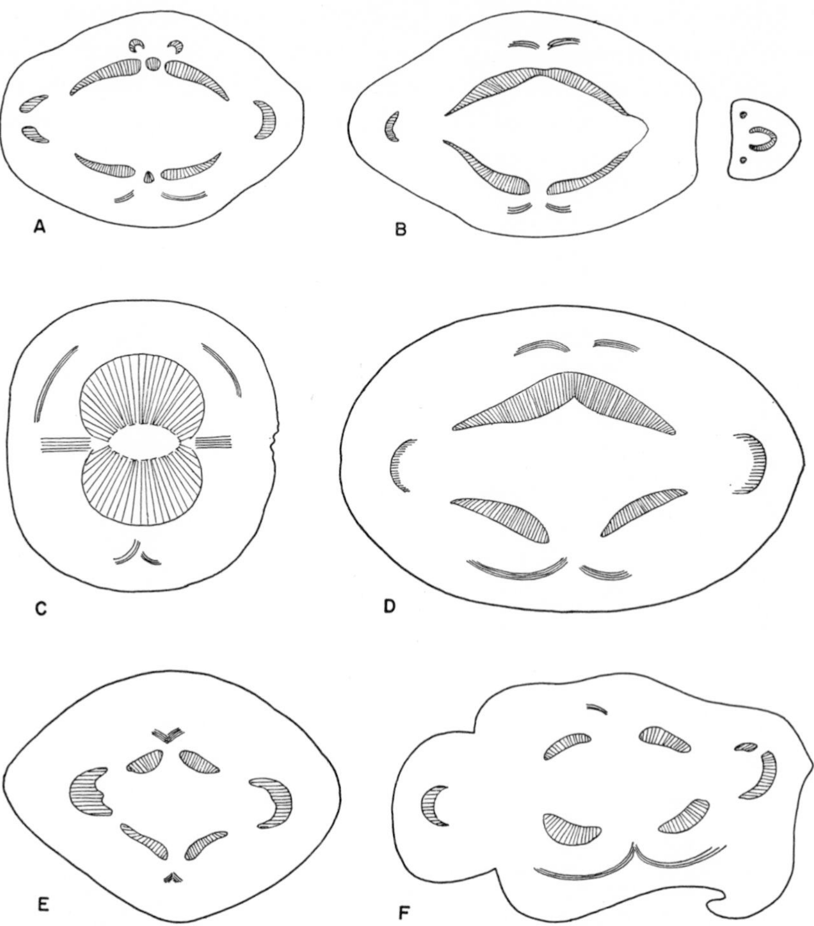

FIGURE 5. **A.** A nodal diagram of *Faramea occidentalis* (Rubiaceae) showing three bundles departing from individual lacunae for each leaf. The gaps for the lateral traces are very close to each other. **B.** A nodal diagram of *Faramea luteovirens* (Rubiaceae) showing the 'common gap' for the split-lateral traces. **C.** A diagram of the node of *Porlieria angustifolia* (Zygophyllaceae). The derivatives of the upper split-lateral are shown in girdling position in the cortex. The lower trace shows the division. **D.** *Zygophyllum stapfii* (Zygophyllaceae). A diagram of the nodal pattern showing the split-lateral traces. **E.** *Larrea divaricata* (Zygophyllaceae). A diagram showing the split-lateral traces at the node of opposite leaves. **F.** *Bulnesia arborea* (Zygophyllaceae). This plant has very large leaves in comparison to other members of the family, yet the diagram shows the simple nodal pattern of a median and the products of two split-lateral traces for each leaf.

FIGURE 6. The nature of the stipules associated with split-lateral traces. **A.** Stipules of *Porlieria angustifolia* (Zygophyllaceae). **B.** Stipules of *Cassipourea elliptica* (Rhizophoraceae). **C.** Stipules of *Chloranthus brachystachys* (Chloranthaceae). **D.** Stipules of *Faramea luteovirens* (Rubiaceae). **E.** A diagram of the node of opposite leaves of *Coffea arabica* (Rubiaceae) redrawn from Varossieau (1940). A single trace to each leaf produces basal branches which divide. One portion of each branch continues in the petiole and the other portion girdles the stem in the sheathing stipules before joining that of the opposing leaf. **F, H.** Stipules of *Zygophyllum stapfii* (Zygophyllaceae). **G.** Stipules of *Bruguiera sexangula* (Rhizophoraceae).

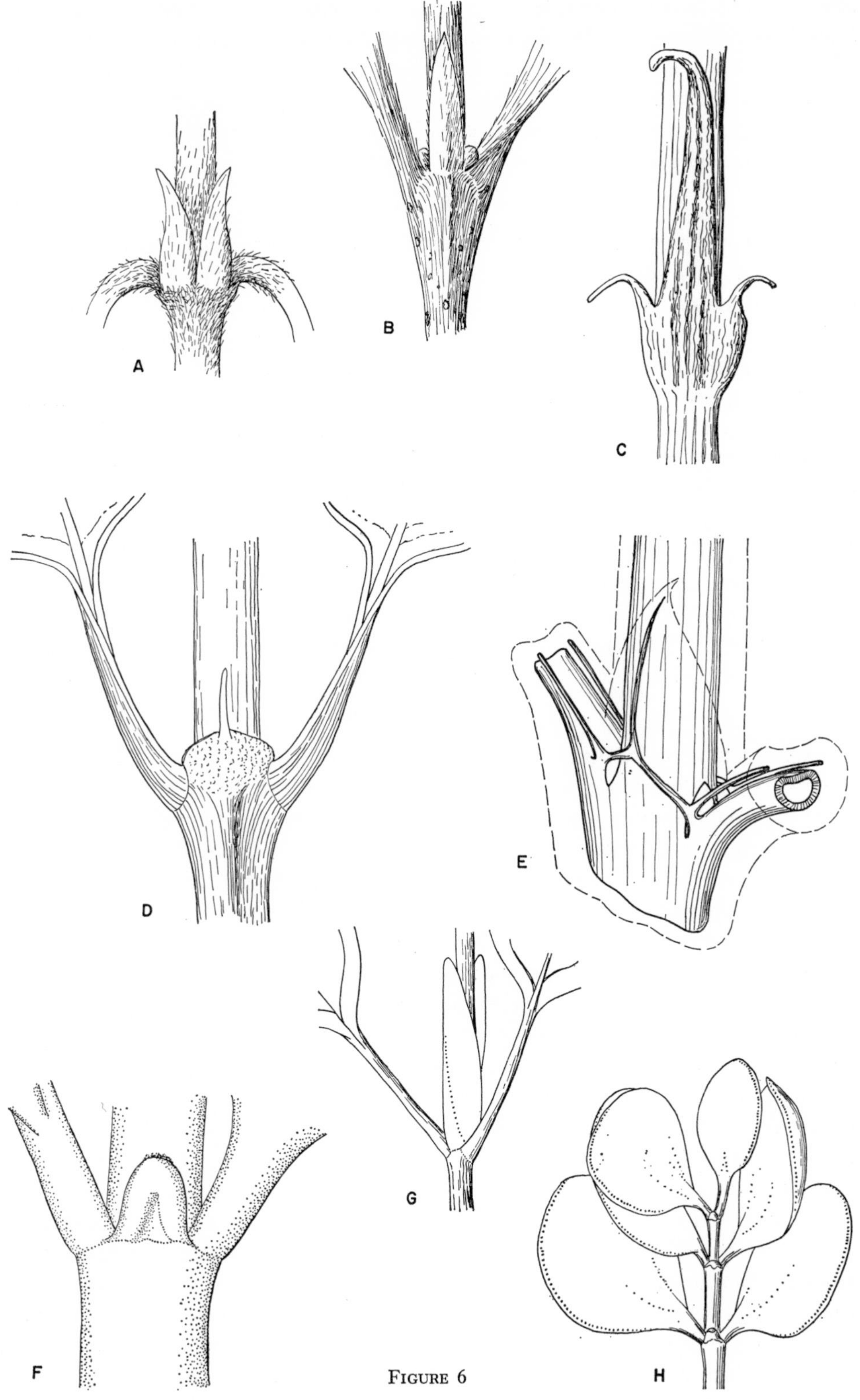

FIGURE 6

single bundle as a 'Composite trace'. We have found this condition in *Faramea luteovirens* Standl. (Fig. **5B**) from Panama without any variation, but material of *Faramea occidentalis* (L.) A. Rich. from the same area shows two lateral gaps on either side of the median in very close proximity (Fig. 5**A**). *Coussarea impetiolaris* Donn. Sm. from Panama shows the split-lateral at each node without variation. Both *Coussarea* and *Faramea* are considered to be within the tribe Coussareae.

ZYGOPHYLLACEAE

Sinnott found that most of the families of the Geraniales possessed trilacunar nodes. In the process of sectioning nodes of various species of the family Zygophyllaceae collected in nature, we have found that the three traces which enter the base of the leaf are derived as a median and two split-laterals in *Bulnesia arborea* (Jacq.) Engl. (Fig. 5**F**), *Fagonia glutinosa* Del., *Larrea divaricata* Cav. (Fig. 5**E**), *Porlieria angustifolia* (Engl.) Gray (Figs 5**C** and 6**A**) and several species of *Zygophyllum* including *Z. foetidum* Scrad. & Wendl., *Z. flexuosum* Eckl. & Zeyh., *Z. sessilifolium* L., *Z. spinosum* L., and *Z. stapfii* Schinz. (Fig. 5**D**). Stipules are commonly fused between the leaves as illustrated for *Z. stapfii* (Fig. 6**F, H**).

Cunningham's report (1927), referred to earlier, of the split-lateral trace of *Zygophyllum fabago* L. was encountered in a search of the literature.

The genus *Balanites*, of which material was collected in Africa, differs from the genera of the Zygophyllaceae in the absence of stipules and by having alternate leaves. The genus is often separated as a family distinct from the Zygophyllaceae. Material sectioned revealed a typical trilacunar node.

CONCLUSION

Swamy felt that the fusion of lateral traces from pairs of opposite leaves was a modification of the unilacunar nodal type and not a nodal type worthy of recognition. It is clear from the numerous occurrences of the split-lateral or common gap associated with trilacunar and multilacunar nodes that this is not a modification of a single median trace. Our knowledge of differentiation within the procambial tissue leading to the development of one or more traces per leaf and node is still scanty. Mitra & Majumdar have referred to the cylinder of procambial tissue within which the vascular tissue differentiates. If this is the pattern within the dicotyledonous shoot apex, then the impetus for differentiation and the determination of the path of differentiation as one or more traces remains the key to the problem of interpretation. At present we can only count the number of traces which differentiate from the stele into the leaf and acknowledge that opposite leaves can receive a fixed number of bundles and that the lateral traces can depart from individual separate loci or that intervening vascular tissue can fail to differentiate or be absent. The latter situation gives rise to one trace which divides, or what may be a pair of closely associated traces which depart from a common gap as a split-lateral trace. In *Alloplectus ambiguus* we have found in one plant a great deal of variation in the nature of the path of the lateral traces from the leaf base into the stele. We have some evidence that the pattern shifted from a stable one of split-lateral traces to completely independent trilacunar nodes when the plant was taken into cultivation. We will attempt to return the cultivated plants to the field environment to determine if the pattern can be reversed.

With the exception of the example discovered by Carlquist in the Compositae, all other plants found to possess split-lateral traces in the mature stem also have stipules or stipule-like processes. The Gentianaceae are commonly reported to lack stipules but do possess a transverse line connecting the opposite leaf bases. The stipules may or may not possess conspicuous vascular tissue and they may or may not be fused into a single median stipule. The nature or the origin of the stipule and its phylogenetic relationship to the leaf remains one of the enigmas of morphology and is not solved here.

The frequency of occurrence of the split-lateral indicates that it is another of the many types of nodal anatomy worthy of recognition. It appears to have value as a taxonomic character.

ACKNOWLEDGEMENTS

It is a pleasure to submit this paper for publication in a collection honouring C. Russell Metcalfe. I have also accepted the great honour of contributing further information on the node of dicotyledons and the vascular structure of the petiole to the proposed new edition of the *Anatomy of the Dicotyledons*.

Most of the information used in this paper has been obtained from a study of living material collected personally in the wild or occasionally obtained from living greenhouse collections at various botanical gardens. Some data have been obtained from herbarium collections, which of necessity has placed a limit on the number of nodes which can be taken for study. I am grateful to the directors and curators of the collections from which material has been obtained. Many technicians have aided me in the long-term study, which is nearing completion, in which nodes of plants of as many families as possible have been examined. It is impractical to name each technician individually, but I do express my gratitude to Mrs Helen Roca-Garcia, who is currently involved in this study and who has prepared the illustrations for the present paper.

REFERENCES

Bailey, I. W., 1956. Nodal anatomy in retrospect. *J. Arnold Arbor.*, **37**: 269–287.

Bailey, I. W. & Howard, R. A., 1941. The comparative morphology of the Icacinaceae. I. Anatomy of the node and internode. *J. Arnold Arbor.*, **22**: 125–132.

Balfour, E. E. & Philipson, W. R., 1962. The development of the primary vascular system of certain dicotyledons. *Phytomorphology*, **12**: 110–143.

Benzing, D. H., 1967*a*. Developmental patterns in stem primary xylem of woody Ranales. I. Species with unilacunar nodes. *Am. J. Bot.*, **54**: 805–813.

Benzing, D. H., 1967*b*. Developmental patterns in stem primary xylem of woody Ranales. II. Species with trilacunar and multilacunar nodes. *Am. J. Bot.*, **54**: 813–820.

Boldt, C. E., 1897. Om epifylla blommor hos *Chirita hamosa* R. Br. *Vidensk. Meddr dansk naturh. Foren.*, **10**: 332–355.

Canright, J. E., 1955. The comparative morphology and relationships of the Magnoliaceae. IV. Wood and nodal anatomy. *J. Arnold Arbor.*, **36**: 119–140.

Carlquist, S., 1957. Leaf anatomy and ontogeny in *Argyroxiphium* and *Wilkesia* (Compositae). *Am. J. Bot.*, **44**: 696–705.

Cunningham, L. M., 1927. Observations on the structure of *Zygophyllum fabago* L. *Trans. bot. Soc. Edinb.*, **29**: 353–361.

Dormer, K. J., 1945*a*. An investigation of the taxonomic value of shoot structure in Angiosperms with special reference to Leguminosae. *Ann. Bot.*, N.S. **9**: 141–153.

DORMER, K. J., 1945*b*. Morphology of the vegetative shoot in Epacridaceae. *New Phytol.*, **44**: 149–151.
ESAU, K., 1945. Vascularization in the vegetative shoots of *Helianthus* and *Sambucus*. *Am. J. Bot.*, **32**: 18–29.
EZELARAB, G. E. & DORMER, K. J., 1963. The organization of the primary vascular system in Ranunculaceae. *Ann. Bot.*, N.S. **27**: 23–38.
FAHN, A. & BROIDO, S., 1963. The primary vascularization of the stems and leaves of the genera *Salsola* and *Suaeda* (Chenopodiaceae). *Phytomorphology*, **13**: 156–165.
HOWARD, R. A., 1968. The ecology of an elfin forest in Puerto Rico. 1. Introduction and composition studies. *J. Arnold Arbor.*, **49**: 381–418.
HOWARD, R. A., 1969. The ecology of an elfin forest in Puerto Rico. 8. Studies of stem growth and form and of leaf structure. *J. Arnold Arbor.*, **50**: 225–262.
HUSSON, P., 1965. Structure nodale de quelques Polygonacées. *Bull. Soc. Hist. nat. Toulouse*, **100**: 96–100.
JENSEN, L. C. W., 1968. Primary stem vascular patterns in three sub-families of the Crassulaceae. *Am. J. Bot.*, **55**: 553–563.
KATO, N., 1966. On the variation of nodal types in the woody plants (1). *J. Jap. Bot.*, **41**: 101–107.
KATO, N., 1967. On the variation of nodal types in the woody plants (2). *J. Jap. Bot.*, **42**: 161–168.
MARSDEN, M. P. F. & BAILEY, I. W., 1950. A fourth type of nodal anatomy in dicotyledons, illustrated by *Clerodendron dichotomum*. *J. Arnold Arbor.*, **31**: 372–404.
MITRA, G. C. & MAJUMDAR, G. P., 1952. The leafbase and the internode, their true morphology. *Palaeobotanist*, **1**: 352–367.
MONEY, L. L., BAILEY, I. W. & SWAMY, B. G. L., 1950. The morphology and relationships of the Monimiaceae. *J. Arnold Arbor.*, **31**: 372–404.
MÜLLER, E., 1944. Die Nervatur der Nieder und Hochblätter. *Bot. Arch.*, **45**: 1–92.
NÄGELI, C., 1858. *Beitr. wiss. Bot.*, **1**: 117–118. Leipzig.
OZENDA, P., 1947. Structure du noeud foliaire des Magnoliacées et des Annonacées. *C. r. hebd. Séanc. Acad. Sci., Paris*, **224**: 1521–1523.
PANT, D. D. & MEHRA, B., 1964. Nodal anatomy in retrospect. *Phytomorphology*, **14**: 384–387.
PHILIPSON, W. R. & BALFOUR, E. E., 1963. Vascular patterns in dicotyledons. *Bot. Rev.*, **29**: 382–404.
PHILIPSON, W. R. & PHILIPSON, M. N., 1968. Diverse nodal types in *Rhododendron*. *J. Arnold Arbor.*, **49**: 193–217.
POST, D. M., 1958. Studies in Gentianaceae. I. Nodal anatomy of *Frasera* and *Swertia perennis*. *Bot. Gaz.*, **120**: 1–14.
ROUSSEAU, D., 1927. Contribution à l'anatomie comparée des Piperacées. *Mem. Acad. r. Belg. Cl. Sci., 8°*, **9**: 1–45.
SAHA, B., 1952. The phylogeny of the unilacunar node as illustrated by the nodal studies of three *Citrus* spp. and of *Phyllarthron commarense* DC. *Bull. bot. Soc. Bengal*, **6**: 89–94.
SCHWERIN, F. G. VON, 1920. Revisio generis *Sambucus*. *Mitt. dt. dendrol. Ges.*, no. 29, 194–231.
SINNOTT, E. W., 1914. The anatomy of the node as an aid in the classification of Angiosperms. *Am. J. Bot.*, **1**: 303–322.
SWAMY, B. G. L., 1949. Further contributions to the morphology of the Degeneriaceae. *J. Arnold Arbor.*, **30**: 10–38.
SWAMY, B. G. L., 1953. The morphology and relationships of the Chloranthaceae. *J. Arnold Arbor.*, **34**: 375–408.
SWAMY, B. G. L. & BAILEY, I. W., 1949. The morphology and relationships of *Cercidiphyllum*. *J. Arnold Arbor.*, **30**: 187–210.
SWAMY, B. G. L. & BAILEY, I. W., 1950. *Sarcandra*, a vesselless genus of the Chloranthaceae. *J. Arnold Arbor.*, **31**: 117–129.
VAROSSIEAU, W. W., 1940. *On the development of the stem and the formation of leaves in* Coffea *species*. Thesis, Leiden Univ.
WILLIS, J. C., 1966. *A dictionary of the flowering plants and ferns*, 7th edn, revised by H. K. Airy Shaw. Cambridge: University Press.

Comparative anatomy and systematics of woody Saxifragaceae. *Ribes*

WILLIAM L. STERN, EDWARD M. SWEITZER
AND ROBERT E. PHIPPS

Department of Botany, University of Maryland, College Park, Maryland, 20742, U.S.A.

Ribes is part of the saxifragaceous complex, the woody members of which are being studied from the anatomical and systematic points of view. Leaves of *Ribes* are characterized by a trimerous vascular structure: the node is trilacunar, three traces enter the petiole, three veins supply the lamina at its base, and three veins are associated with each hydathode. Hydathodes are invariably present in the leaves of all species as are unicellular, bulbous-based epidermal hairs, and druses in the mesophyll. The xylem is characterized by scalariform perforation plates in vessels, transitional to opposite intervascular pitting, absence of axial xylem parenchyma, imperforate tracheary elements with circular bordered pits, high heterocellular vascular rays, broad and narrow rays in each species, and by the presence of isolated radially uniseriate groups of vascular ray cells throughout the genus. Bona fide axial xylem parenchyma has been noted only in *R. americanum* and simple perforation plates as well as scalariform perforation plates occur in *R. americanum* and *R. aureum*. The genus cannot be validly separated on anatomical grounds into two taxa, *Ribes sensu stricto* and *Grossularia*.

CONTENTS

INTRODUCTION

Saxifragaceae, as considered by Engler (1928), are a large plant family containing both woody and herbaceous members. Recent studies by botanists have cast considerable doubt on the taxonomic homogeneity and relatedness among the taxa comprising the family. Some of the genera have been shown to belong to other families, e.g. *Kania* Schlechter to Myrtaceae and *Berenice* Tulasne to Campanulaceae (Erdtman & Metcalfe, 1963*a*, *b*), and taxonomists differ widely in their evaluation of the position of component taxa. Columelliaceae, a family long associated with Gesneriaceae, are probably allied to Saxifragaceae *sensu lato* (Stern, Brizicky & Eyde, 1969). Hutchinson (1959, 1967, 1969) divided the Englerian Saxifragaceae into two orders, Saxifragales and Cunoniales, clearly along the woody/herbaceous line. He,

Takhtajan (1969) and Cronquist (1968) elevated certain of Engler's infrafamilial taxa to familial level; on the other hand, Thorne (1968) followed Engler's treatment very closely.

This paper is the first in a series of anatomical studies of the vegetative structures of the woody plants which Engler included among his Saxifragaceae. Genera will form the taxonomic bases for study and the initial reports will be designed to establish a comprehensive anatomical foundation for subsequent comparisons and for evolutionary and phylogenetic discussions.

The edible currants and gooseberries of commerce, as well as numerous ornamental plants of considerable beauty, belong to the genus *Ribes* L. (Berger, 1924; Rehder, 1940). *Ribes* is also important as the alternate host in the life cycle of *Cronartium ribicola* J. R. Fischer, the causal organism of the white pine blister rust (Walker, 1969). This disease produces cankers on branches and trunks and destroys bark tissue in the five-needled pines (subgenus *Haploxylon* section *Cembra*).

Ribes, with about 150 species (Willis, 1966), is distributed throughout the North Temperate Zone in both America and Eurasia; in America, it extends southward along the Andes to Tierra del Fuego, and it occurs in Mediterranean north-west Africa. It is entirely absent from the remainder of the African continent, from Pacific Oceania, from Indomalaysia, from Madagascar and other Indian Ocean islands, and from Australia and New Zealand. Distribution maps appear in Pavlova (1927) and Hutchinson (1959), and Janczewski (1907) and Berger (1924) present detailed information on the geographical ranges of species.

Ribes is a genus of shrubs which are sometimes armed with spines. The leaves are alternate or clustered, simple, petiolate, plicate or convolute in the bud, and usually palmately lobed and always palmately veined. Stipules are absent; the dilated or clasping bases of petioles in some species, however, give the appearance of stipules. The flowers are bisexual, or unisexual by abortion and then the plants are dioecious. Inflorescences are racemose or they occur rarely in few-flowered sessile umbels. The calyx tube is adnate to the ovary but the 5 (4) lobes are produced beyond it. In some species the calyx lobes are large, brightly coloured and attractive. Petals are 5 (4), usually small and sometimes scale-like, and alternate with the calyx lobes. The 5 (4) stamens are opposite the calyx lobes and alternate with the petals. The ovary is inferior, unilocular and biplacental. Styles are 2 or completely connate. Ovules are usually many, anatropous and bitegmic. The fruit is a many-seeded, succulent berry. Seeds possess a rather hard testa with a gelatinous outer covering. The endosperm is fleshy and the embryo is very small. Cotyledons in the germinated seedling are foliaceous and epigeous. Detailed descriptions of the genus and of species may be seen in Janczewski (1907), Coville & Britton (1908), Berger (1924), Engler (1928) and Hutchinson (1967).

Most modern writers treat *Ribes* in the family Grossulariaceae, following the work of Lamarck & de Candolle (Grossulariae, 1805: 405). Richard (1823: 487) considered *Ribes* as the basis of Ribesieae and Dulac (1867: 263) placed *Ribes* in Pulpaceae. Coville & Britton (1908), Berger (1924) and others divide Grossulariaceae into two genera, *Ribes* L. and *Grossularia* Miller. The distinction is based on the predominant lack of spines, jointed pedicels and disarticulation of the fruit in *Ribes* versus the presence of spines, unjointed pedicels and non-disarticulation of the fruit in *Grossularia*. The most recent monographer of the currants and gooseberries, Janczewski (1907), does not make this distinction and considers all taxa under the genus *Ribes*. Most authors follow Janczewski and classify *Ribes* into six subgenera: *Ribesia* Berlandier, *Coreosma* Spach, *Grossularioides* Janczewski, *Grossularia* A. Richard, *Parilla* Janczewski and *Berisia* Spach.

MATERIALS AND METHODS

To achieve a broad taxonomic and geographic spectrum of specimens for study, an attempt was made to secure representative material from each of Janczewski's (1907) subgenera and sections. Recourse was made to the dried plant collections in the United States National Herbarium; to the wood collections of the Smithsonian Institution, Yale University and the Arnold Arboretum; and to living specimens from the field and from botanical gardens and arboreta, notably the Rancho Santa Ana Botanic Garden. We are grateful to the curators and directors of these institutional collections for making material available for our studies. In particular we would like to acknowledge with thanks the help of Dr Sherwin Carlquist, Dr José Cuatrecasas, Dr Kenton L. Chambers, Dr Frederick G. Meyer, Dr Arthur R. Kruckeberg and Dr James L. Reveal, for their kindness in gathering living collections of *Ribes* from the field and from botanical gardens. For performing tannin tests, we are pleased to note the help of Mr Donald Bissing; for her valuable assistance in performing translations from the Russian language, we thank Mrs Branka Popmijatov. Dr George H. M. Lawrence was helpful in securing rare bibliographic data for us. All specimens used in this research are documented in Table 1.

Xylem anatomy terminology for the most part is in agreement with that suggested by the Committee on Nomenclature of the International Association of Wood Anatomists (1957). Most additional terms conform to current usage. Terms which deviate from such usage are explained where employed in the text.

Standard techniques were used in preparing material for observation. Wood specimens preserved in formalin-acetic acid-alcohol (F.A.A.) were embedded in celloidin. Dried wood specimens were boiled in water and then placed in a 50:50 solution of 95% ethyl alcohol and glycerin for storage. Twigs smaller than 1 cm in diameter were frozen to the stage of a sliding microtome for sectioning; larger specimens were sectioned with the sliding microtome in the ordinary manner. Transverse, radial and tangential sections were stained with Heidenhain's iron-alum haematoxylin and counterstained with safranin. Macerations of wood were prepared using Jeffrey's fluid, washed, stained with safranin and dehydrated. Sections and macerations were mounted on microscope slides with Canada balsam.

Both fluid-preserved (F.A.A.) and dried leaves were available for study. Wherever possible, fluid-preserved leaves were used for sectioning in preference to dried leaves from herbarium specimens. Leaves were cleared using Arnott's (1959) technique, employing 5% NaOH followed by washing and further clearing with a saturated aqueous solution of chloral hydrate. Leaves were washed with water, stained with safranin, dehydrated and mounted on slides with Canada balsam. Transverse and

Table 1. Specimens of *Ribes* examined

Species[a]	Collector	Locality	Herbarium[b]	Xylarium[c]	Garden	Leaf	Wood	Dried	Preserved
Subgenus *Ribesia* Berlandier									
multiflorum Kit.	Krummel 1137		US		Bot. Gart. Tech. Hochs., Braunschweig	+		+	
rubrum L.		Poland		Yw 23158[d]			+	+	
rubrum	Iltis 3029	France	US			+		+	
Subgenus *Coreosma* Spach									
affine Kunth	J. N. Rose, Painter & J. S. Rose 8785	Mexico	US			+	+	+	
americanum Mill.	Stern 2801	Virginia	MARY				+		+
aureum Pursh					RSA *s. n.*	+	+		+
aureum	Proctor 70	Idaho	Y[d]	Yw 40499			+	+	
X bethmontii Jancz.					Univ. Washington Arbor. 1474–45		+		+
bracteosum Douglas	Stern 2806	Washington	MARY				+		+
bracteosum	Stern 1657	Oregon	US	USw 30128			+	+	
canthariformis Wiggins		California			RSA 6541	+	+		+
cereum Douglas	Stern & Chambers 1417	Oregon	US	USw 24628			+	+	
ciliatum Humb. & Bonpl. ex Roem. & Schult.	Balls 4166	Mexico	US				+	+	
coloradense Coville	Eggleston 11937	Colorado	US				+	+	
erythrocarpum Coville & Leiberg	Coville & Applegate 172	Oregon	US			+	+	+	
fragrans Pallas	Sukatchew & Poplaws 314lb	USSR	US			+	+	+	
indecorum Eastw'd (= *malvaceum* var. *indecorum* Jancz.)					RSA 6759	+	+		+
inebrians Lindl.	Bailey 663	New Mexico	US			+	+	+	
laxiflorum Pursh	Coville & Kearney 593	Alaska	US				+	+	
laxiflorum	Coville & Kearney 2554	Alaska	US				+	+	
malvaceum Sm.					RSA 6544	+	+		+
neglectum Rose	Palmer 190	Mexico	US			+	+	+	
nevadense Kellogg	Quick 22	California		Yw 26994			+	+	
nigrum L.	Blocki 613	Galicia	US			+	+	+	
odoratum[e]	Hicock *s. n.*	Connecticut (cult.)		Yw 39333			+	+	
petiolare Douglas (= *hudsonianum* Richardson var. *petiolare* Jancz.)	Proctor 87	Idaho	Y	Yw 26968			+	+	
prostratum L'Hérit.	McKery *s. n.*	Alaska	US			+	+	+	
sanguineum Pursh					RSA 9966	+	+		+
tortuosum Benth.	Ferris 8548	Mexico	US			+	+	+	
viburnifolium A. Gray					RSA 11861	+	+		+
viscosissimum Pursh	Stern 2808	Washington	MARY				+		+
viscosissimum	Proctor 48	Idaho	Y	Yw 40480			+	+	
viscosissimum	Proctor 112	Idaho	Y	Yw 40532			+	+	

Species[a]	Collector	Locality	Herbarium[b]	Xylarium[c]	Garden	Leaf	Wood	Dried	Preserved
Subgenus *Grossularioides* Jancz.									
lacustre Poiret	Vogel *s. n.*	Montana	US			+	+	+	
lacustre	Proctor 50	Idaho		Yw 40482			+	+	
montigenum McClatchie	Tidestrom 198	Utah	US			+	+	+	
Subgenus *Grossularia* A. Richard									
alpestre Dcne.	Rock 8347	China	US			+	+	+	
buriense Schmidt	Tang 1011	China	US			+		+	
californicum Hooker & Arnott (=*menziesii*?)					RSA 6034	+	+		+
cruentum Greene (=*amictum* Greene var. *cruentum* Jancz.)	Chambers 2808	Oregon	OSC			+	+		+
cynosbati L.	Biltmore Herbarium 3252d	North Carolina	US			+	+	+	
divaricatum Douglas		British Columbia		Yw 17152			+	+	
divaricatum	Stern 2810	Washington	MARY				+		+
formosanum Hayata	Wilson 10930	Formosa	US				+	+	
grossularia L.	Fiori 2877	Italy	US			+		+	
grossularia	Opdyke *s. n.*	Ohio (cult.)	Akron Mus. Nat. Hist.	USw 20826			+	+	
leptanthum A. Gray	Quick 3	California		Yw 26993			+	+	
lobbii A. Gray	Stern 2807	Washington	MARY			+	+		+
lobbii	Constance & Rollins 2864	California	US			+	+	+	
menziesii Pursh	Eastwood 13	California	US			+	+	+	
menziesii	Chambers 1823	Oregon	US, OSC	USw 30612			+	+	
rotundifolium Michaux	Ebinger 4130	Virginia	US	USw 32383			+	+	
rotundifolium	Stern 2802	Virginia	MARY			+			+
speciosum Pursh					RSA 7316	+	+		+
speciosum	Purer 6512	California	US			+	+	+	
Subgenus *Parilla* Jancz.									
andicola Jancz.	Ariste-Joseph A81	Colombia	US			+	+	+	
bogotanum Jancz.	Smith & Idrobo 1363	Colombia	US			+	+	+	
catamarcanum Jancz.	Venturi 4620	Argentina	US			+	+	+	
densiflorum Philippi		Patagonia		Yw 1746			+	+	
fasciculatum Sieb. & Zucc.	Lee 6004	China	US			+	+	+	
glandulosum Ruiz & Pavon	Bernath *s. n.*	Chile		Yw 34048			+	+	
hirtum Humb. & Bonpl.	Cazalet & Pennington 5702	Ecuador	US			+		+	
hirtum	Rimbach 154	Ecuador	Y	Yw 24093			+	+	
leptostachyum Benth.	Cuatrecasas 27681	Colombia	US				+		+

Table 1.—*Cont.*

Species[a]	Collector	Locality	Herbarium[b]	Xylarium[c]	Garden	Leaf	Wood	Dried	Preserved
leptostachyum	Cuatrecasas 27428	Colombia	US				+		+
magellanicum Poiret	Sleumer 1047	Argentina	US			+	+	+	
punctatum Ruiz & Pavon	Werdermann 8	Chile	US				+	+	
sardoum Martelli	Martelli *s. n.*	Sardinia	US			+	+	+	
valdivianum Philippi	Werdermann 1928	Chile	US			+		+	
valdivianum	Bullock 164	Chile	Y	USw 15364			+	+	
Subgenus *Berisia* Spach									
alpinum L.	Castella 758	Switzerland	US			+	+	+	
alpinum				Uw 9228			+	+	
diacantha Pallas	Roerich Exped. 701	Manchuria	US			+		+	
giraldii Jancz.	Ching 1158	China	US			+	+	+	
glaciale Wallich	Rock 5833	China	US			+	+	+	
humile Jancz.	Rock 18251	China	US			+		+	
laurifolium Jancz.					Univ. Washington Arbor. 606–55	+	+		+
orientale Desf.	Koelz 2817	Kashmir	US			+	+	+	

[a]Names used are those originally accompanying specimens; authorities for these and for synonyms were taken directly from Janczewski (1907) wherever possible.

[b]Abbreviations follow those recommended by Lanjouw & Stafleu (1964) in *Index Herbariorum.*

[c]Abbreviations follow those recommended by Stern & Chambers (1960) and Stern (1967) in *Index Xylariorum.*

[d]Yw and Y refer to the xylarium and herbarium collections of the Yale University School of Forestry, respectively. These collections comprise the Samuel James Record Memorial Collection. In December 1969, this collection was transferred to the U.S. Forest Products Laboratory, Madison, Wisconsin.

[e]Janczewski (1907) gives *R. ciliatum* Humboldt & Bonpland ex Roemer & Schultes and *R. aureum* Pursh as possible synonyms of *R. odoratum* Schlechtendal ex Hemsley and *R. Odoratum* Wendland f., respectively. No herbarium voucher exists for the wood specimen studied and it is therefore impossible to trace the nomenclature of this particular specimen.

paradermal sections of leaves were prepared from material embedded in paraffin. Sectioned material was stained with Heidenhain's iron-alum haematoxylin and safranin. However, some sections were left unstained because structural detail was obliterated by the strongly chromophilic pigmented deposits in some leaf cells. Unstained sections were also employed to study crystals because it was found that the ferric ammonium sulphate mordant [Sass (1958) formula containing sulphuric and acetic acids] used with the Heidenhain's haematoxylin dissolved these bodies in leaf cells. Some sections were stained only with haematoxylin to present another perspective for study.

Nodal and petiolar studies were carried out on dried specimens from the herbarium and on material which had been preserved in F.A.A. and stored in 70% ethyl alcohol. Nodes and petioles were examined using both hand sections and clearings. Successive sections of nodal regions were cut through at least two contiguous nodes. Vasculation of the petiole was studied from sections made at three points: the base where it joined the stem, the midpoint and distally at the base of the leaf blade. Sections of the petiole, young stem and nodal regions were treated with a supersaturated aqueous solution of phloroglucinol followed by hydrochloric acid to demonstrate lignified tissues in vascular strands. It was found that the reaction was more rapid and the colour more intense if concentrated hydrochloric acid was used rather than the 25% solution recommended by Johansen (1940).

Additional observations of petiolar vasculation and of nodal configuration were made after clearing the material in 5% NaOH. In some instances it was necessary to carry out further clearing using Stockwell's solution (Johansen, 1940) because of the dense deposits of dark-coloured materials which continued to obscure structure following treatment with NaOH. After the removal of all dark-coloured deposits, tissues were washed with water, stained with safranin, dehydrated and cleared in xylene or toluene prior to study.

Tests for tannins were carried out using fresh leaves. Three indicators were used: ferric chloride and nitroso reactions (Jensen, 1962) and protein-binding test (Swain, 1965).

Distributions of pores are based on percentages derived from counts of ten microscope fields. Tangential pore diameters, vessel element lengths and lengths of imperforate tracheary elements are based on 50 measurements each. Tangential pore diameters were measured only in the last completely formed growth ring. For all ring porous members, 25 tangential pore diameters were measured from the first-formed portion of the growth ring and 25 from the last-formed portion of the growth ring. Vessel elements were measured from tip to tip. Diameters of crystals in cells of leaves are based on ten measurements. Stomatal size is based on ten measurements, each across the short axis and the long axis of a pair of guard cells.

LEAF

Vascular skeleton and mesophyll

Leaves in *Ribes* are all 'marginal actinodromous' (von Ettingshausen, 1861), that is, the primary veins all arise from a point (Fig. 1, p) at the base of the blade and radiate from there in an almost straight course to the tips of the lobes. Leaves are 3-, 5- or 7-lobed. In some species, namely *R. laurifolium* and *R. viburnifolium*, leaves are unlobed; in *R. tortuosum*, leaves are unlobed or weakly 3- or 5-lobed. Lobation is generally constant within a species.

Nodes in *Ribes* are trilacunar and three traces enter the base of the petiole from the vascular cylinder of the stem (Fig. 1, a–c).

Regardless of lobation, the nervation of leaves throughout the genus is consistent. Three primary veins diverge from the distal end of the petiole and enter the base of the leaf blade. The middle of these (Fig. 1, a) runs directly to the leaf apex thereby dividing the leaf into two lateral halves. Each of the two lateral veins (Fig. 1, b, c), corresponding to the two outer traces which entered the base of the petiole, branches more or less immediately into three veins (Fig. 1, b^1, b^2, b^3 and c^1, c^2, c^3). Veins b^3 and c^3 may further divide to produce b^4 and c^4 (not shown in Fig. 1), which serve the most proximal portions of the leaf blade. Veins a, b^1, and c^1 are of about the same magnitude; b^2 and c^2 are of the same size but narrower than b^1 and c^1; veins b^3 and c^3 are of the same size but narrower than b^2 and c^2; and veins b^4 and c^4, where formed, are the narrowest of the veins noted. Each of the veins mentioned is branched further so that the penultimate vasculation forms islets in which veinlets, consisting of one or a few xylem elements, terminate blindly in the areole.

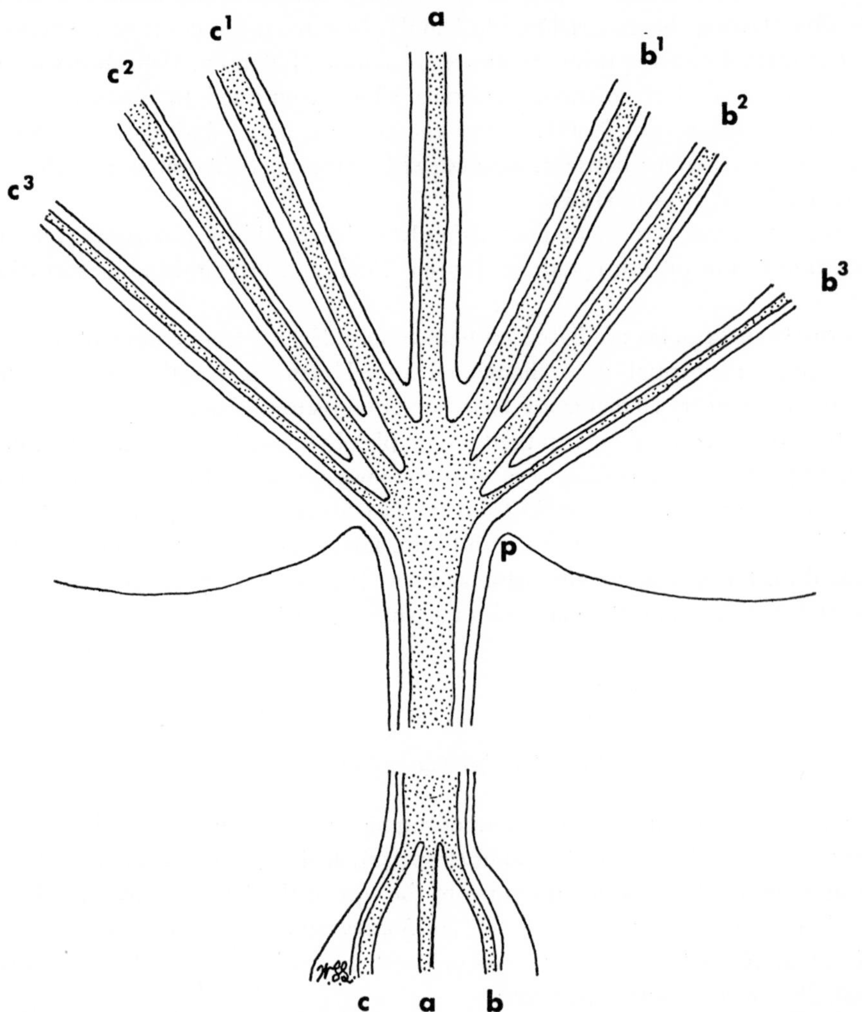

FIGURE 1. Idealized vascular pattern of the leaf in *Ribes*. Three traces, c, a and b, enter the petiole from the trilacunar node. These may remain fused throughout the length of the petiole, as shown, or they may remain separate. At the base of the blade, p, there are always three separate traces, two of which, c and b, divide immediately; a proceeds directly to the apex and becomes the mid-vein of the blade; c and b each divide into three veins, c^1, c^2, c^3 and b^1, b^2, b^3.

Leaves in *Ribes* are usually dorsiventral. The palisade mesophyll is usually 1- or 2-seriate but, at least in *R. californicum*, it may vary from 2- to 4-seriate. In some species, the palisade mesophyll is predominantly uniseriate with localized biseriate portions (e.g. *R. indecorum*, *R. malvaceum*, *R. speciosum*). In most species some cells of the palisade mesophyll are filled with a yellow deposit which becomes dark red-brown after staining with safranin. Somewhat less commonly, these deposits also occur in cells of the spongy mesophyll. These deposits contain tannins.

Crystals occur in mesophyll cells of all species of *Ribes*. These appear in both the palisade and spongy layers as well as in the adjacent adaxial collenchyma related to the larger veins. Crystals are usually solitary in cells but cells with two superposed crystals are not uncommon. In *R. erythrocarpum*, cells with up to four superposed crystals were noted. Crystalliferous cells of the palisade layers are fatter than adjacent chlorenchymatous cells; crystalliferous cells of the spongy region are morphologically similar to those of the adjacent chlorenchyma. In some species (e.g. *R. erythrocarpum*, *R. malvaceum*, *R. viburnifolium*), crystals occur in both regions of the mesophyll; in others (e.g. *R. affine*, *R. californicum*, *R. montigenum*), crystals occur only in the palisade layers; less commonly, crystals occur only in the spongy layers (e.g. *R. catamarcanum*). Generally, the palisade layers are better provided with crystals than the spongy layers. Where crystals occur in the spongy region, they are usually localized in cells near the lower epidermis. Leaves of all species examined contained druses except for *R. sardoum*, where all crystals were tabular. It is not uncommon to find a few tabular and irregularly-shaped crystals even in species where druses predominate.* Crystals vary in size within each leaf; the smallest druses encountered were only 6 μm in diameter (e.g. *R. affine*, *R. indecorum*, *R. lacustre*) and the largest were in excess of 35 μm (*R. canthariformis*).

Vascular bundles in leaves of *Ribes* are collateral, that is, the xylem is adaxial and the phloem abaxial. In larger bundles, noticeable secondary xylem and phloem are produced; small bundles include only primary vascular tissues. Vascular rays in secondary xylem and phloem are conspicuous in transverse section; the cells are enlarged, barrel-shaped, circular, or somewhat angular. In many instances they are completely or partially filled with dark-coloured tanniferous deposits. In the larger vascular bundles, subtending the primary phloem in the 'pericyclic' region, there is a zone of parenchymatous or collenchymatous cells.

The bundle sheath is a uniseriate layer of cells surrounding the vascular tissues of veins. It is usually conspicuous because some or all of the cells contain dark-staining tanniferous deposits. In all cases, a bundle sheath surrounds the larger veins and usually most of the smaller veins as well. Bundle sheaths were not seen around the smaller veins of *R. canthariformis* and *R. magellanicum*. Bundle sheath extensions occur on the adaxial side of the larger veins. Smaller veins lack these structures. A broad, collenchymatous layer of several cells in thickness commonly subtends the bundle sheath in the larger veins.

Petiole

Vascular bundles in petioles of *Ribes* are collateral; the xylem is adaxial and the phloem abaxial. At the node, petioles are provided with three distinct vascular bundles in all species. Depending upon the species,† these bundles remain discrete or they fuse as they pass distally through the length of the petiole into the base of the leaf blade. The distal end of the petiole always shows three separate bundles.

* It is possible that tabular and irregular crystals are incipient druses or fragments of druses injured during sectioning. The latter possibility stems from actual observation of crystals of all three types in adjacent cells.

† An exception among the species studied appears to be *R. speciosum*; in *Purer 6512*, the vascular strands in the central portion of the petiole are fused; in *RSA 7316*, there are three separate but contiguous strands at the mid-petiole.

In *R. alpestre*, *R. alpinum*, *R. californicum*, *R. diacantha*, *R. erythrocarpum*, *R. fasciculatum*, *R. fragrans*, *R. humile*, *R. lacustre* and *R. speciosum*, the mid-section of the petiole shows three discrete vascular bundles. These describe a more or less open arc; the bundles may be separated from one another by fundamental tissue, as in *R. erythrocarpum* and *R. humile*, or the bundles may be in direct contact, as in *R. fragrans* and *R. speciosum*. In the mid-section of the petiole of some species, the fused vascular tissue describes an almost complete circle, as in *R. affine* and *R. bogotanum*; in *R. prostratum*, a truly complete circle is produced. In most of the species studied, however, the mid-section of the petiole shows fused vascular tissue which in transverse section varies from broadly arcuate to shield-shaped with only a slight adaxial indentation. Table 2 lists the vascular condition at mid-petiole for selected species.

Table 2. Condition of vascular bundles at petiole mid-section in *Ribes*

	Vascular bundles				
Species	Separate	Fused at edges	Completely fused	*Ribes sensu stricto*	*Grossularia*
affine			+	+	
alpestre		+			+
aureum			+	+	
californicum		+			+
cruentum			+		+
cynosbati			+		+
erythrocarpum	+			+	
fragrans		+		+	
indecorum			+	+	
inebrians			+	+	
lacustre		+		+	
lobbii			+		+
malvaceum			+	+	
menziesii			+		+
montigenum		+		+	
multiflorum			+		+
prostratum			+	+	
sanguineum			+	+	
speciosum Purer 6512			+		+
speciosum RSA 7316		+			+
tortuosum			+	+	
viburnifolium			+	+	

Sclerenchyma is sometimes associated with petiolar vascular tissue. Most often it is abaxial and is only present in the mid-section of the petiole. However, in *R. bogotanum*, *R. catamarcanum* and *R. orientale*, sclerenchyma subtends each of the three vascular bundles as they enter the petiole from the stem. A complete cylinder of sclerenchyma surrounds the vascular tissue in the mid-section of the petiole in *R. affine*, *R. alpestre*, *R. bogotanum*, *R. catamarcanum*, *R. cynosbati*, *R. erythrocarpum*, *R. lacustre* and *R. sanguineum*. In some cases, abaxial sclerenchyma is present in the distal portion of the petiole where the vascular tissue enters the blade: *R. affine*, *R. aureum*, *R. erythrocarpum*, *R. magellanicum*, *R. montigenum*, *R. orientale* and *R. speciosum*. In *R. cruentum*, abaxial sclerenchyma was only encountered in the most distal portion of the petiole.

Epidermis

Both upper and lower epidermides in *Ribes* are uniseriate. Epidermal cell walls are thickened on their outer surfaces. Beneath major veins these walls are excessively thickened, so that the surface appears corrugated and the cells papillate when viewed in transverse section. In this region, the cuticle is often specially thickened as well.

Stomata (Fig. 2) are restricted to the lower epidermis. The stomatal apparatus* is anomocytic (*sensu* Metcalfe & Chalk, 1950), i.e. the guard cells are surrounded by cells

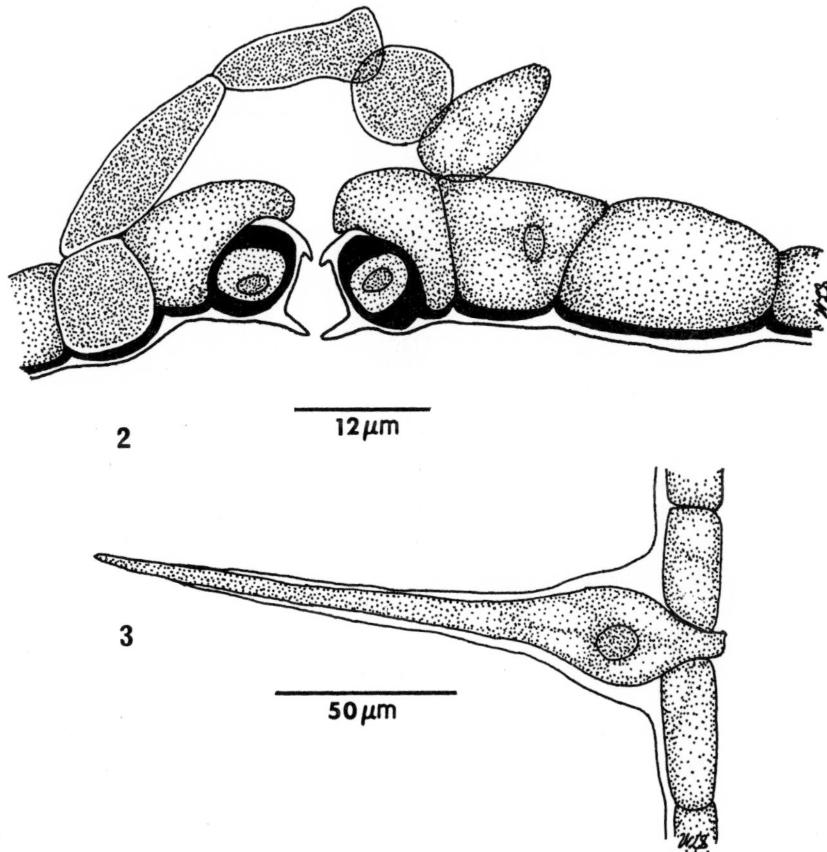

FIGURES 2 AND 3. 2. Transverse section of stomatal apparatus in *Ribes viburnifolium*. The cuticular horns are those described by Hryniewiecki (1913). All walls of the guard cells are thickened and the adjacent epidermal cells overlap the guard cells on the inner side.
3. Unicellular, bulbous-based hair found in all species of *Ribes*.

of varying number which are indistinguishable in form and position from the remainder of the epidermal cells. However, as seen in Fig. 2, the epidermal cells immediately adjacent to the guard cells are modified to overlap the guard cells on their inner surfaces. This modification is not obvious in paradermal view.

In transverse section, all walls of the guard cells are thickened, but only the lower and upper walls are specially thickened (Fig. 2). The cuticle is modified in the region

* See footnote, p. 48, in Stern, Brizicky & Eyde (1969) for discussion and use of this terminology.

of the stoma so that outer and inner horn-like projections appear in section. The cuticle extends around exposed surfaces of the guard cells through the stoma to where the guard cell surfaces are in contact with the overlapping adjacent epidermal cells.

The guard cells are typically reniform to elongate-reniform in paradermal view. The stomatal apparatus is longer than wide, an average for 17 species being 25 μm across the short axis and 30 μm along the long axis (Table 3).

Table 3. Dimensions of the stomatal apparatus in *Ribes* (averages in μm*)

Species	Short axis	Long axis
affine	19·7	22·1
alpestre	23·4	28·0
alpinum	17·6	25·0
californicum	18·7	24·4
canthariformis	25·1	27·9
catamarcanum	23·1	30·2
diacantha	23·4	27·0
erythrocarpum	18·5	21·1
fasciculatum	19·9	23·1
indecorum	22·1	25·7
lacustre	21·7	25·1
magellanicum	24·7	29·0
malvaceum	21·5	26·2
montigenum	20·9	27·9
sardoum	21·1	34·1
speciosum	22·4	26·5
viburnifolium	23·9	26·0

* The symbol μm is used for micrometer, 10^{-6}m (instead of μ, micron, 10^{-3}mm), in accordance with revised international standards.

Trichomes

Leaves of all species of *Ribes* are covered with a hairy indumentum consisting of unicellular and multicellular trichomes. It was noted that, in processing leaves for microscopic examination, trichomes are frequently removed, cut off or broken. Therefore, to gain a complete understanding of the hairy indumentum, it was necessary to study it from sections, clearings and in some cases from fluid-preserved or boiled dried leaves from herbarium specimens.

Unicellular, bulbous-based trichomes, with elongated 'necks', occur on the leaf blades and petioles of all species, on the upper and lower epidermis and on the margins of the blade (Fig. 3). These trichomes are more numerous on young leaves than on older leaves and toward the base of the blade than toward the apex. Unicellular trichomes also occur as outgrowths of the epidermis of the stipe on the multicellular, capitate trichomes of some species. The bulbous bases of unicellular trichomes are always intercalated among the regular epidermal cells. The lowermost end of the bulb is narrowed, as though pinched by the surrounding cells (Fig. 3). Unicellular trichomes are thin-walled and covered by a continuation of the cuticle; they are living and the nucleus occurs in the bulbous base.

All capitate trichomes in *Ribes* consist of a multiseriate, multicellular stipe and a multicellular capitulum. The stipe comprises several rows of cells arranged in a more or less columnar manner. It is narrowed at the apex beneath the capitulum and is supported by a more or less flared pedestal. The capitulum is usually globose or depressed globose. (In some species, e.g. *R. fragrans* and *R. viburnifolium*, the capitulum is flattened and the trichomes are truly peltate, Plate 1**A**.) The descriptions below refer to this kind of trichome, its distribution on the leaf and its modifications.

Capitate trichomes occur on the leaves of most species of *Ribes* at some stage of development. In general, they are more abundant and massive on the petiole and toward the base of the blade than toward the tip. They are more numerous and conspicuous along the major veins than in areas of the blade between the major veins. These trichomes are found on both abaxial and adaxial leaf surfaces in, e.g., *R. erythrocarpum, R. indecorum* and *R. malvaceum*. In *R. affine* and *R. montigenum*, for example, they occur only on the adaxial surface of the leaf; in *R. canthariformis*, they were noted only on the abaxial surface. Specimens of *R. californicum, R. diacantha* and *R. magellanicum* did not show any multicellular trichomes on the leaf blade.

Besides being situated on the petiole and on both surfaces of the leaf blade, capitate trichomes commonly occur along the margins of the leaf blade in *R. lacustre, R. montigenum* and *R. speciosum*, for example. In some species, e.g. *R. erythrocarpum, R. lacustre* and *R. montigenum*, massive, marginal trichomes are associated with hydathode-bearing cusps of the leaf blade. In *R. viburnifolium*, the abaxial, short-stipitate, peltate trichomes are always situated in an invagination of the leaf (Plate 1**A**).

Multicellular trichomes are usually long stipitate, as in *R. canthariformis, R. erythrocarpum, R. montigenum* and *R. speciosum*. Short stipitate trichomes occur in *R. alpestre, R. fragrans, R. lacustre* and *R. viburnifolium* and other species. In *R. catamarcanum*, the trichomes are virtually sessile. The capitate trichomes on the petioles of *R. viburnifolium* vary from long stipitate to almost sessile.

Trichome stipes are mostly simple in *Ribes*. However, as noted above, the multicellular, petiolar trichomes of *R. catamarcanum, R. fragrans, R. sardoum* and *R. viburnifolium* are often branched, the branches comprising the bulbous-based, unicellular trichomes (Plate 1**B**). Branched hairs in these species give the petioles a plumose appearance.

In a few instances, multicellular trichomes are not capitate (Plate 1**B**). This may represent an ontogenetic stage in the development of capitate trichomes. Short to long, conical, non-capitate trichomes occur along the veins on the abaxial surface of the blade in *R. fasciculatum*; petioles of *R. catamarcanum* bear elongate, non-capitate trichomes; leaf blades of *R. alpinum* have both capitate and non-capitate multicellular trichomes.

Hydathodes

Hydathodes are a constant feature of the leaves of all species of *Ribes*. In those species with lobed leaves, they occur in conjunction with the larger cusps along the margins of the blade. In the entire-leaved species, e.g. *R. viburnifolium*, they occur in association with the larger vein endings. Hydathode-bearing cusps of the blade contain a triple vein supply as seen in surface view: a central larger vein which characteristically flares distally and two smaller lateral veins, one on each side of the central vein (Plate 1**C**). Typically, the lateral veins approach the central vein toward its distal

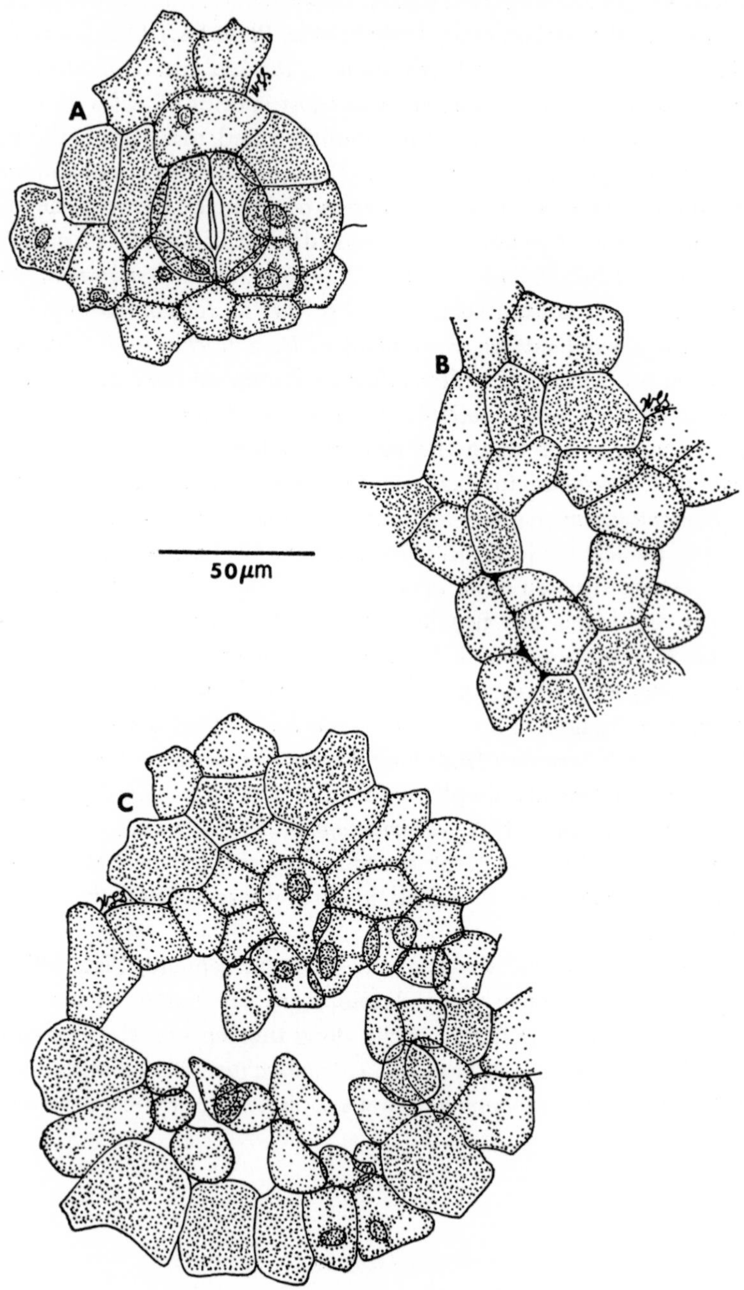

FIGURE 4. *Ribes viburnifolium.* Axial series of sections through hydathode. **A.** 'Water pore' and surrounding epidermal cells. **B.** Intercellular space directly beneath 'water pore'. **C.** Enlarging intercellular space and associated mesophyll cells.

end. These veins may fuse with the central vein; they may approach the central vein more or less closely without fusing; one lateral vein may approach more closely to the central vein than the other; one lateral vein may be of larger diameter than the other

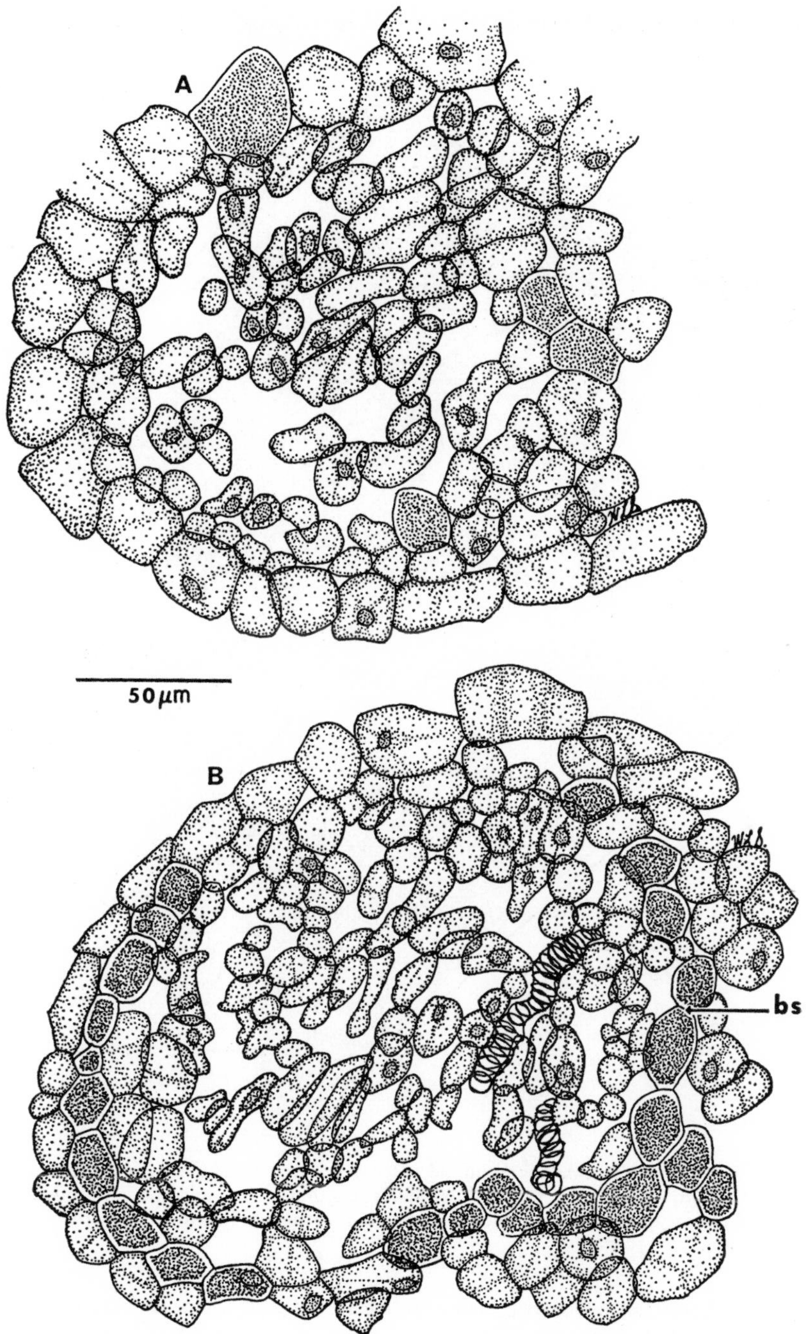

FIGURE 5. *Ribes viburnifolium.* Axial series of sections through hydathode. **A.** Intercellular space with beginning of epithem cells. **B.** Bundle sheath, bs, surrounding epithem and spiral elements of primary xylem.

or they may be equal in diameter. Lateral veins may narrow distally or they may flare, much as the central vein does. In entire-leaved species, a similar triple vein supply to the hydathode is also present.

The secretory tissues of the hydathode itself (Plate 1**C**; Figs 4, 5 and 6) are directly associated with the central vein of the three-vein complex described above. In a

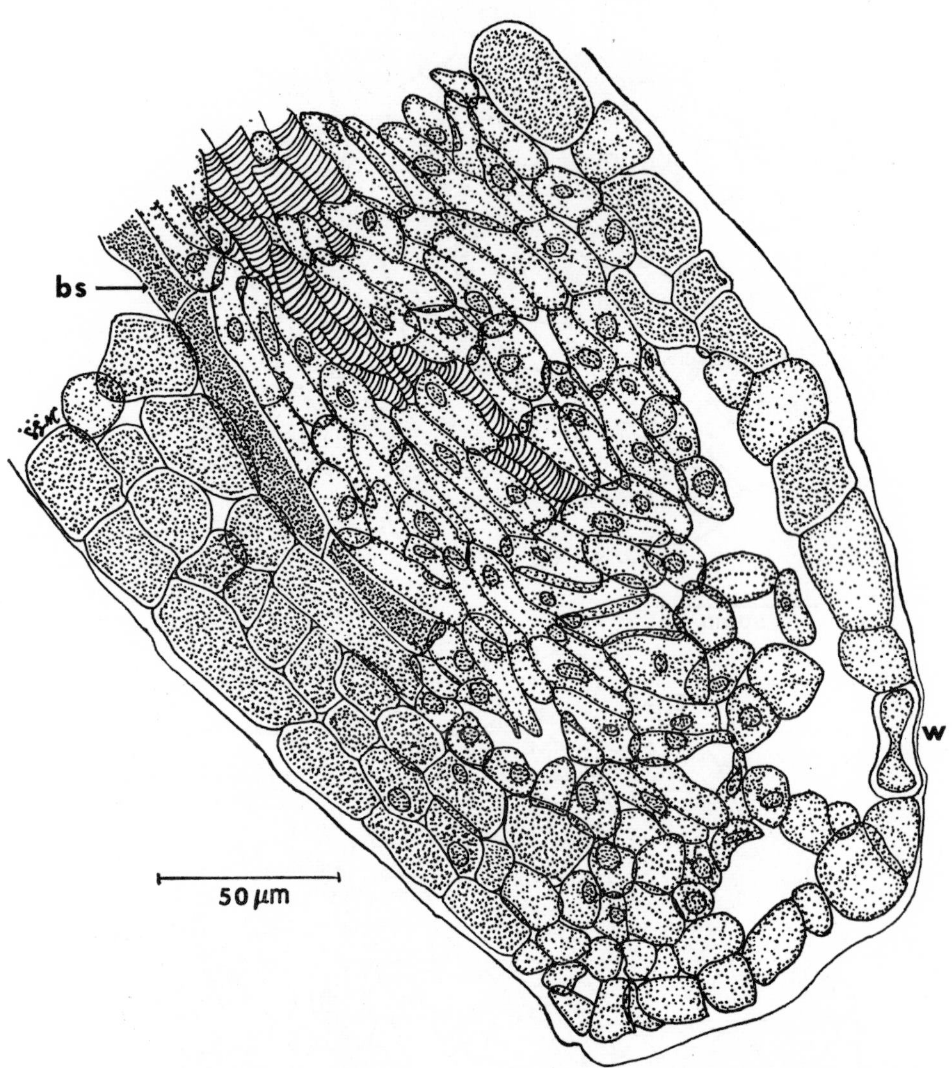

FIGURE 6. *Ribes viburnifolium.* Longitudinal section through hydathode. The 'water pore', w, is subtended by an irregular intercellular space. Epithem tissue surrounds terminal spiral elements of primary xylem. The bundle sheath, bs, is seen only beneath the epithem in this oblique section.

section of the hydathode-bearing leaf cusp cut parallel with the central vein, the epidermis is provided with a few (1–3) 'water pores' (Figs 4**A** and 6). Immediately subtending these is an intercellular space of variable size and shape (Figs 4**B**, **C**, 5**A** and 6). Proximally to the intercellular space, an epithem of loosely arranged, thin-walled, parenchymatous, chloroplast-free cells occurs (Figs 5 and 6). Extending into this mass of cells are the terminal elements of the central vein (Figs **5B** and 6). These elements are annular or spirally thickened tracheids. Surrounding the hydathode tissues at the proximal end are cells of the bundle sheath, which, in the species figured, are filled with tanniferous contents (Figs **5B** and 6). The structural sequence just described can be visualized best in a series of sections cut at right angles to the hydathode, beginning at the tip of the hydathode-bearing cusp and proceeding proximally (Figs 4 and 5).

'Water pores' consist of two modified guard cells enclosing a pore (Fig. **4A**). In surface view, 'water pores' are almost completely circular; they are somewhat larger than the functional stomatal apparatus in the same species. (In *R. erythrocarpum*, for example, the functional stomatal apparatus averages 18·5 μm wide and 21·1 μm long; circular 'water pores' average 27·5 μm in diameter.) In addition to 'water pores', tips of hydathode-bearing cusps frequently contain large, multicellular openings which appear at first to be regularly formed secretory ostioles. However, closer observation indicates that these openings are irregularly formed; the tissue is disorganized and the cells completely or partially disintegrated. These structures appear to be formed through a breakdown of tissues near hydathodes. No evidence of associated fungi has been found in the material under investigation, however, and it appears that the breakdown process is initiated by the plant itself.

Hydathode-bearing leaf cusps are hardly more bulky than adjacent cusps which are not associated with hydathodes. However, in *R. fragrans* and other species, the hydathodes are so massive as to produce teat-like callosities along the margin of the leaf blade. On the other hand, hydathodes in *R. sardoum* are poorly developed and the central vein is not broadly flared as in most other species.

WOOD

The wood of *Ribes* generally possesses well-defined growth rings, except in young material (Plate 2**A**, **B**). Some species of *Ribes* show diffuse-porous wood, e.g. some specimens of *R. aureum*, *R. menziesii*, *R. rubrum* and *R. speciosum*; and also *R. affine*, *R. californicum*, *R. cruentum*, *R. indecorum* (Plate 2**A**), *R. petiolare*, *R. sanguineum* and *R. viburnifolium*. In most of the species examined, however, pores are more concentrated in the early portion of the growth ring than in the later formed portion, giving rise to a definite pore zone (Plate 2**B**). In this early wood zone, the average diameter of pores may be somewhat wider, equal to, or somewhat narrower than that of pores in the late wood of the same species, but the pores are always more abundant here than in the late wood zone. To all intents and purposes more or less sharp rings are visible on the transverse surface, and we prefer to characterize this formation of pores and growth rings as ring-porous in conformity with the definition given in the 'Glossary of terms used in describing woods' (Committee on Nomenclature, International Association of Wood Anatomists, 1933). It seems to us unnecessarily restrictive to limit the term ring-porous only to those woods where the early wood pores are larger than those in the late wood as is done in the 'International glossary of terms used in wood anatomy' (Committee on Nomenclature, International Association of Wood Anatomists, 1957). Larger pores or more abundant pores are both probably a manifestation of rapid growth early in the growing season.

Pore distribution is mostly solitary with some radial multiples of two to six cells and some pore clusters of 3–13 cells (Plate 2**A**–**C**). Pores are angular in outline. Vessel wall thickness ranges from 0·5–2·5 μm. Perforation plates are exclusively scalariform (Plate 2**D**) in all but two species: *R. americanum* (*Stern 2801*) and *R. aureum* (*Proctor 70*). In addition to scalariform perforations, these species show simple perforations with vestigial bars and completely simple perforations (Plate 3**A**). Scalariform perforations are rare in *R. americanum* and only one or two bars per plate were noted. In *R. aureum*, scalariform perforations have one to four bars. In species with only scalariform perforation plates, bars range from 2–12 in number. Bars are sometimes branched to form a reticulate pattern. Openings in scalariform perforation plates vary from completely bordered, to bordered to the middle, to bordered to the ends. In some, borders are completely absent. End wall angles vary from 5°–58°.

Intervascular pitting is mostly transitional to opposite with some scalariform and alternate arrangements. Diameter of the pits ranges from 2–6 μm. Pits are circular to elongate in outline and uncrowded. It is not unusual to find unilaterally compound pit-pairs between vessel elements and cells of the vascular rays.

Vessel elements are ligulate. The range of the average lengths of vessel elements for all species is 255–594 μm; the total range of lengths for all species is 126–1380 μm.

Tyloses in *Ribes* are apparently traumatic in origin for they were accompanied by decay in the one specimen in which they were observed.

Imperforate tracheary elements include vasicentric tracheids (Plate 2**D**), non-septate fibre-tracheids and septate fibre-tracheids (Plate 3**B**). Wall thickness for imperforate elements ranges from 1·5–4·5 μm. The range of the average lengths of imperforate elements for all species is 366–714 μm; the total range of lengths for all species is 210–2250 μm.

Vascular rays include both homocellular and heterocellular types and are quite variable in height, both in numbers of cells and in microns. Homocellular rays are uniseriate and comprise upright cells (range, 1–30 cells; 300–1400 μm). Heterocellular rays range in width from 2–35 cells and 10–280 μm, and in height from 3–204 cells and 115–4775 μm. As seen in radial section, the cells making up the body of heterocellular rays vary in shape and they may be square, upright or roundish. Most of the ray cells are thick-walled and the lumina contain dark-staining deposits. Sheath cells are present in association with wider rays in most species. Uniseriate wings on some heterocellular rays are six to eight cells high. Because of the extensive vertical dimensions of these heterocellular rays, it is common to find two or more of them joined axially through uniseriate 'bridges'. Crystals were absent from cells of vascular rays except for *R. tortuosum* (*Ferris 8548*), where prismatic crystals were found.

Axial xylem parenchyma of the usual type is absent from all species studied except for *R. americanum* (*Stern 2801*), where diffuse and diffuse-in-aggregates parenchyma occurs (Plate 2**C**). Isolated strands of parenchyma are found throughout the ground mass of the wood in all species. These strands are mostly fusiform, two to four cells high and do not form part of an axial system as is ordinarily the case with axial parenchyma. Rather, these strands comprise short, solitary rows of cells, each axial row of cells being shorter than associated vessel elements (Plate 3**B**). Presumably, then, these parenchyma strands are not derivatives of the fusiform initials of the vascular cambium; rather, they have been produced intermittently by the ray initials to form modified vascular rays extending only one cell in the radial dimension. These give the impression of isolated strands of true axial parenchyma.

DISCUSSION

Considering its large size, floral diversity, and broad geographical distribution, it is remarkable that *Ribes* is basically such a homogeneous taxon from the anatomical point of view. The vascular skeleton of the leaf in all species is trimerous: three traces enter the petiole, three veins supply the lamina at its base, three veins are associated with each hydathode. The node is trilacunar. Leaves of all species are provided with hydathodes, and morphologically similar, unicellular, bulbous-based hairs are invariably present. Crystals occur in the mesophyll cells of leaves of all species and these are predominantly druses. The xylem is characterized by scalariform perforation plates in vessels, by transitional to opposite intervascular pitting, by the lack of axial parenchyma, by imperforate tracheary elements with circular bordered pits, by unusually high heterocellular vascular rays, by the occurrence of broad and narrow rays in each species, and by the presence of isolated radially uniseriate groups of vascular ray cells throughout the genus. There appear to be no consistent anatomical bases for the separation of *Ribes* into two genera: *Ribes sensu stricto* and *Grossularia*, as has been supported by Coville & Britton (1908), Berger (1924) and others.

The general vegetative anatomy of *Ribes* has been authoritatively covered through the original studies of Holle (1893), Kudelka (1907) and Janczewski (1907), and by the disorganized and questionable research of Bates (1933). The anatomy has been reviewed by both Solereder (1908) and Metcalfe & Chalk (1950). Specialized studies on the xylem have been reported by Tippo (1938), Record & Hess (1943), Yatsenko-Khmelevsky (1954) and Greguss (1959). However, there is no exhaustive investigation of the wood reported anywhere in the literature. The leaf has been described by Petit (1886), Hryniewiecki (1913), Morvillez (1918), Bugnon (1927), Brienne (1948) and Pyykkö (1966). In general the reports of these authors are corroborated by our own observations, except for the differences noted below.

Leaf

The often repeated observation that the stomatal apparatus in *Ribes* is circular in outline stems from the work of Holle (1893). For this reason a special attempt was made to measure these structures, and it can be appreciated from Table 3 that such is not the case. The stomatal apparatus in *Ribes* is always longer than it is wide.

Although in our investigations of *Ribes* leaves we have found stomata only in lower surfaces, Pyykkö (1966) reported stomata in both surfaces of leaves in *R. cucullatum*. Similarly, Kudelka (1907) mentioned the occurrence of stomata in both leaf surfaces in *R. cereum* and *R. inebrians*. He pointed to the work of Janczewski (1907), who correlated this phenomenon with the orientation of laminae in a vertical position under prolonged warmth in certain species. In this position, remarked Janczewski, the edges of leaves are at right angles to the sun's rays. Kudelka theorized that such a position would protect the leaves from excessive transpiration. This vertical leaf position is rare, however, and fully developed leaves in *Ribes* are almost always nearly horizontally orientated.

The ubiquitous presence of hydathodes in the leaves of *Ribes* has been all but ignored in previous studies, except those of Kudelka (1907). While it is true that hydathodes are of rather common occurrence throughout the angiosperms and that the process of guttation is a widespread phenomenon among vascular plants (Volkens, 1883; Haberlandt, 1894, 1928; Burgerstein, 1920; Johnson, 1937; Sperlich, 1939; Frey-Wyssling, 1941), their prominence in *Ribes* and possible significance in the comparative anatomy of the Saxifragaceae *sensu lato* should not be overlooked.

In a study of the petioles in seven species and several varieties of *Ribes*, Brienne (1948) concluded that the genus can be divided into two large groups based upon whether the petiolar bundles remain separate or become fused over the greater part of their course through the petiole. In the first group, with distinct petiolar bundles, he placed *R. aureum*, *R. nigrum*, *R. petraeum* and *R. rubrum*; in the second group, with fused petiolar bundles, he listed *R. alpinum*, *R. flavum* and *R. grossularia*. These groups bear no relationship to the taxa *Ribes sensu stricto* and *Grossularia*. Our examination of 21 species of *Ribes* (Table 2) confirmed that the configuration of vascular bundles in petioles is not correlated with these taxa, separate and united vascular tissue being found in species assigned to both *Ribes* and *Grossularia*. Brienne has remarked on the variability within the same petiole, indicating that entirely different vascular conditions can be seen over a length of 3–4 mm. We have found a similar situation in *R. speciosum*, where petioles of two different specimens have been examined. Table 2 shows that in one specimen the petiolar bundles at mid-section are more or less fused and in another specimen petiolar bundles in the same position are entirely fused. It is probable that developmental conditions influence the time at which coherence of bundles in petioles occurs.

Young stem

Both Janczewski (1907) and Kudelka (1907) thought that the presence of a collenchymatous or sclerenchymatous hypodermis in the young stem was a useful taxonomic indicator. Janczewski mentioned that the differences which he observed in stems of *Ribes* were purely quantitative and for this reason were not of any particular interest, with the exception of the tissue subtending the epidermis. He asserted that in all the unarmed currants and in the spinous members of the subgenus *Berisia* (*R. diacantha*, *R. giraldii* and *R. pulchellum*), the layers underlying the epidermis are collenchymatous, and that in the spinous currants of the subgenera *Grossularia* and *Grossularioides* the underlying layers are lignified. Janczewski ascribed the wrinkling of stems at the summit of vigorous shoots to the formation of this lignified layer between the cortex and epidermis. Accordingly, his key to the hermaphrodite species of *Ribes* is divided into two groups based upon whether the hypodermis is collenchymatous or lignified. Similarly, Kudelka established *R. nigrum* as the type for species bearing collenchyma under the epidermis of the shoot and *R. grossularia* as the type for species in which sclerenchyma is formed under the epidermis.

In our own studies we examined hand-cut sections from young stems of 12 species of *Ribes* assigned to both *Ribes sensu stricto* and *Grossularia*. Sections were treated with phloroglucinol and hydrochloric acid to test for the presence of lignin. In no case was a positive reaction noted in the walls of cells immediately subtending the epidermis.

Based on these tests, it can be assumed that these cell walls are not lignified or sclerified, as asserted by Janczewski and Kudelka.

What does seem to be present beneath the epidermis is a more or less well defined hypodermis of collenchyma cells. In *R. aureum*, for example, several cell layers immediately beneath the epidermis comprise thick-walled cells, and the layers are thus quite distinct. However, in other species, e.g. *R. bracteosum*, *R. lobbii*, *R. nevadense*, *R. speciosum* and *R. viscosissimum*, layers underlying the epidermis comprise cells with more or less thickened walls, and thus the layers are more or less well differentiated from the epidermis on one side and the cortex on the other. The lignified and sclerified tissues reported by Janczewski and Kudelka probably refer to the distinction or lack of distinction of the cell layers under the epidermis caused by greater or lesser thickening of the walls of constituent cells.

Secondary xylem

The presence of axial parenchyma in wood of *Ribes* has been variously reported: Record & Hess (1943) indicated that it was apparently absent or sparingly paratracheal; Tippo (1938) asserted that it was metatracheal forming bands one to six cells wide but quite rare; Greguss (1959) listed abundant diffuse parenchyma in *R. alpinum* and noted its presence also in *R. sativum* and *R. uva-crispa*; and Yatsenko-Khmelevsky (1954) wrote that poor diffuse and terminal axial parenchyma were present in *R. alpinum*, *R. biebersteinii* and *Grossularia reclinata*. In our own comprehensive study of *Ribes* woods we have recorded *bona fide* axial parenchyma only in single species: *R. americanum*. We suspect that reports of the presence of axial parenchyma in *Ribes* wood refer to what we have described as radially uniseriate rows of vascular ray parenchyma, and not to axial parenchyma derived through the activity of fusiform cambial initials. Indeed, viewed in transverse section this tissue does give the impression of diffuse axial parenchyma. Seen in radial section, however, its origin from ray initials of the cambium is evident.

Although Record & Hess, Greguss and Yatsenko-Khmelevsky report the presence of libriform wood fibres, that is, imperforate tracheary elements with simple pits, none has been observed in the woods which we have examined. Our position supports that of Tippo, who indicated that only tracheids and septate fibre-tracheids were present in the four *Ribes* species he studied. Likewise, only Tippo has reported the presence of simple perforation plates in conformity with our observations on the presence of these structures in *R. americanum* and *R. aureum*.

ACKNOWLEDGEMENT

This paper is part of a series based upon research supported by the National Science Foundation of the United States of America under grant number GB 7431 awarded to the University of Maryland under the direction of the first author.

REFERENCES

Arnott, H. J., 1959. Leaf clearings. *Turtox News*, **37**: 192–194.

Bates, J. C., 1933. Comparative anatomical research within the genus *Ribes*. *Kans. Univ. Sci. Bull.*, **21**: 369–398.

Berger, A., 1924. A taxonomic review of currants and gooseberries. *Tech. Bull. N. Y. St. agric. Exp. Stn.*, **109**: 1–118.

BRIENNE, J. R. DE, 1948. Étude anatomique des pétioles de diverses éspèces et variétés de *Ribes*. *Bull. Soc. linn. Normandie*, sér. 9, **5**: 141–144.

BUGNON, P., 1927. Différenciation de la trace foliaire trifasciculée du *Ribes sanguineum*. *Bull. Soc. bot. Fr.*, **73**: 1032–1038.

BURGERSTEIN, A., 1920. *Die Transpiration der Pflanzen* (*Ergänzungsband*). Jena: Gustav Fischer.

COMMITTEE ON NOMENCLATURE, INTERNATIONAL ASSOCIATION OF WOOD ANATOMISTS, 1933. Glossary of terms used in describing woods. *Trop. Woods*, **36**: 1–12.

COMMITTEE ON NOMENCLATURE, INTERNATIONAL ASSOCIATION OF WOOD ANATOMISTS, 1957. International glossary of terms used in wood anatomy. *Trop. Woods*, **107**: 1–36.

COVILLE, F. V. & BRITTON, N. L., 1908. Grossulariaceae. *North Am. Flora*, **22**: 193–225.

CRONQUIST, A., 1968. *The evolution and classification of flowering plants*. Boston: Houghton Mifflin.

DULAC, J., 1867. *Flore du département des Hautes-Pyrénées, plantes vasculaires spontanées, classification naturelle* . . . Paris: F. Savy.

ENGLER, A., 1928. Saxifragaceae. In Engler & Prantl, *Die natürlichen Pflanzenfamilien*, 2nd ed., **18a**: 74–226. Leipzig: Wilhelm Engelmann.

ERDTMAN, G. & METCALFE, C. R., 1963*a*. Affinities of certain genera *incertae sedis* suggested by pollen morphology and vegetative anatomy. I. The myrtaceous affinity of *Kania eugenioïdes* Schltr. *Kew Bull.*, **17**: 249–250.

ERDTMAN, G. & METCALFE, C. R., 1963*b*. Affinities of certain genera *incertae sedis* suggested by pollen morphology and vegetative anatomy. III. The campanulaceous affinity of *Berenice arguta* Tulasne. *Kew Bull.*, **17**: 253–256.

ETTINGSHAUSEN, C. VON, 1861. *Die Blatt-Skelete der Dikotyledonen* . . . Wien: Kais. Kön. Hof- und Staatsdruckerei.

FREY-WYSSLING, A., 1941. Die Guttation als allgemeine Erscheinung. *Ber. schweiz. bot. Ges.*, **51**: 321–325.

GREGUSS, P., 1959. *Holzanatomie der europäischen Laubhölzer und Sträucher*, 2nd ed. Budapest: Akadémiai Kiadó.

HABERLANDT, G., 1894. Ueber Bau und Funktion der Hydathoden. *Ber. dt. bot. Ges.*, **12**: 367–378.

HABERLANDT, G., 1928. *Physiological plant anatomy*. (Translated from the 4th German edition by Montagu Drummond.) London: Macmillan.

HOLLE, G., 1893. Beiträge zur Anatomie der Saxifragaceen und deren systematische Verwerthung. *Bot. Cbl.*, **53**: (1) 1–9; (2) 33–41; (3) 65–70; (4) 97–102; (5) 129–136; (6) 161–169; (7/8) 209–222.

HRYNIEWIECKI, B., 1913. Ein neuer Typus der Spaltöffnungen bei den Saxifragaceen. *Bull. int. Acad. Sci. Lett. Cracovie* (Sér. B, *Sci. nat.*), 1912: 52–73.

HUTCHINSON, J., 1959. *The families of flowering plants*, 2nd ed., **1**. *Dicotyledons*. Oxford: Clarendon Press.

HUTCHINSON, J., 1967. *The genera of flowering plants* (*Angiospermae*). *Dicotyledones*, **2**. Oxford: Clarendon Press.

HUTCHINSON, J., 1969. *Evolution and phylogeny of flowering plants*. London & New York: Academic Press.

JANCZEWSKI, E. DE, 1907. Monographie des groseilliers *Ribes* L. *Mém. Soc. Phys. Hist. nat. Genève*, **35**: 199–517.

JENSEN, W. A., 1962. *Botanical histochemistry*. San Francisco & London: W. H. Freeman.

JOHANSEN, D. A., 1940. *Plant microtechnique*. New York & London: McGraw-Hill.

JOHNSON, M. A., 1937. Hydathodes in the genus *Equisetum*. *Bot. Gaz.*, **98**: 598–608.

KUDELKA, M. W., 1907. Vergleichende Anatomie der vegetativen Organe der Johannisbeergewächse (*Ribes*). *Bull. int. Acad. Sci. Lett. Cracovie* (*Cl. Sci. math. nat.*), 1907: 24–40.

LAMARCK, J. B. A. P. M. DE, & DE CANDOLLE, A. P., 1805. *Flore française*, 3rd ed., **4** (2). Paris: H. Agasse.

LANJOUW, J. & STAFLEU, F. A., 1964. Index Herbariorum. *Regnum veg.*, **31**: 1–251.

METCALFE, C. R. & CHALK, L., 1950. *Anatomy of the dicotyledons*. Oxford: Clarendon Press.

MORVILLEZ, M. F., 1918. L'appareil conducteur des feuilles des Saxifragacées. *C.r. hebd. Séanc. Acad. Sci., Paris*, **167**: 555–558.

PAVLOVA, N. M., 1927. A survey of the literature on the genus *Ribes*. *Trudӯ prikl. Bot. Genet. Selek.*, **17**: 463–513. (Text in Russian; references in language of publication.)

PETIT, L., 1886. Le pétiole des dicotylédones au point de vue de l'anatomie comparée et de la taxinomie. *Mém. soc. Sci. phys. nat. Bordeaux* (*sér.* 3), **3**: 217–404.

PYYKKÖ, M., 1966. The leaf anatomy of East Patagonian xerophytic plants. *Ann. Bot. fenn.*, **3**: 453–622.

RECORD, S. J. & HESS, R. W., 1943. *Timbers of the New World*. New Haven: Yale University Press.

REHDER, A., 1940. *Manual of cultivated trees and shrubs hardy in North America*, 2nd ed. New York: Macmillan.

RICHARD, A., 1823. *Botanique médicale, ou histoire naturelle et médicale* . . . Pt. 2. Paris: Béchet.

SASS, J. E., 1958. *Botanical microtechnique*, 3rd ed. Ames: Iowa State University Press.

SOLEREDER, H., 1908. *Systematic anatomy of the dicotyledons*. (Translated by L. A. Boodle and F. E. Fritsch; revised by D. H. Scott.) Oxford: Clarendon Press.

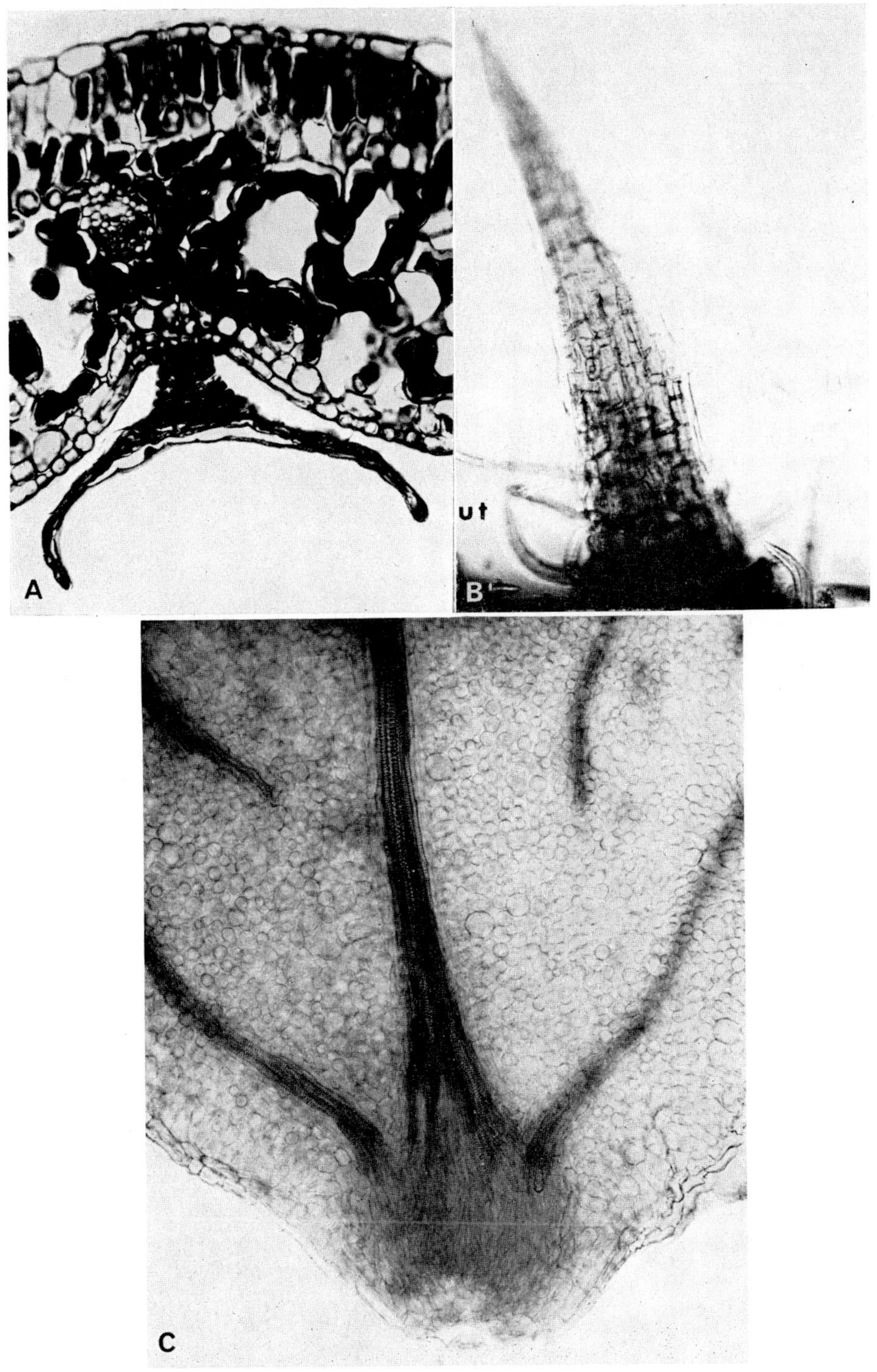
A
ut
B
C

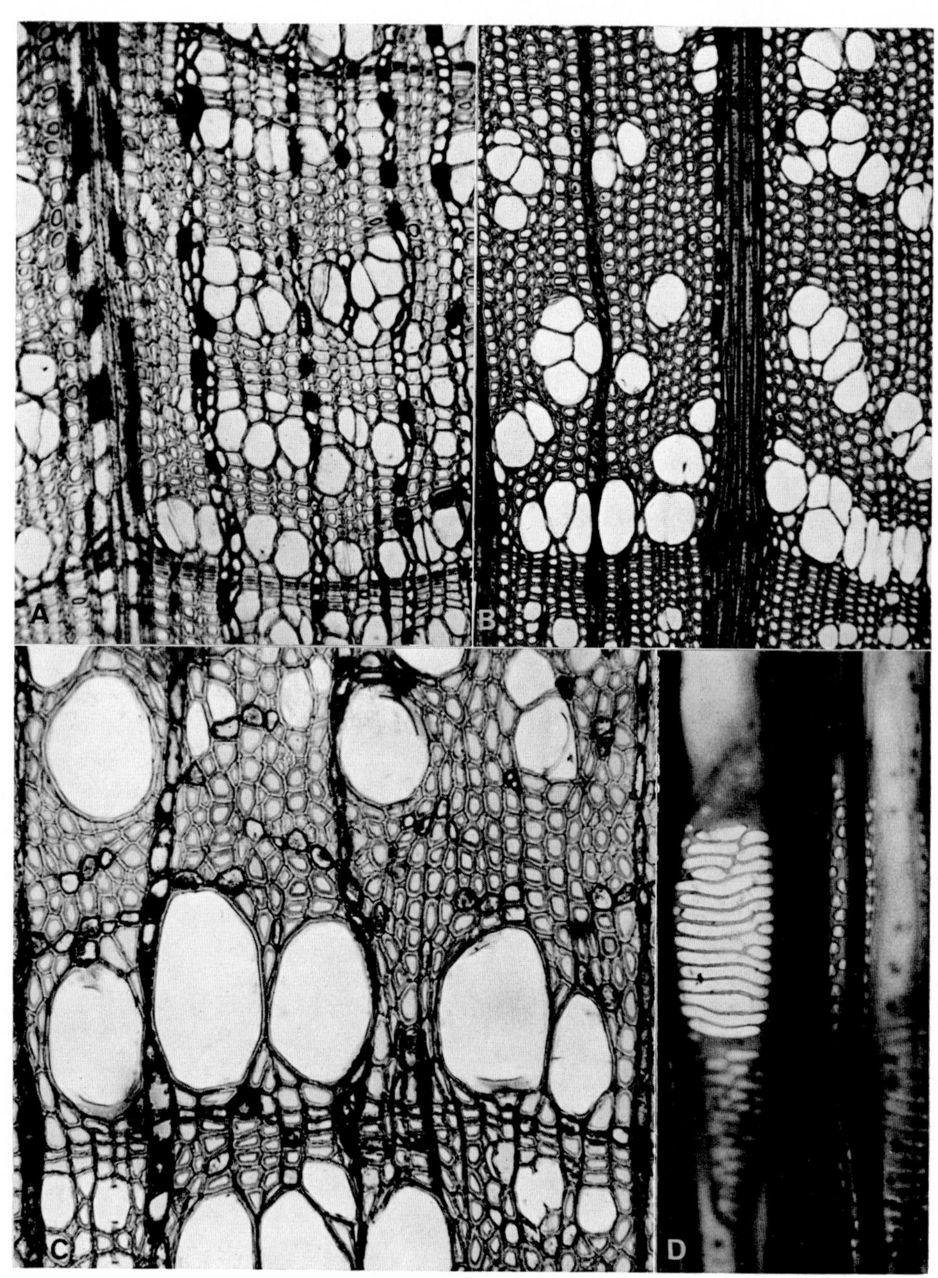

W. L. STERN, E. M. SWEITZER AND R. E. PHIPPS

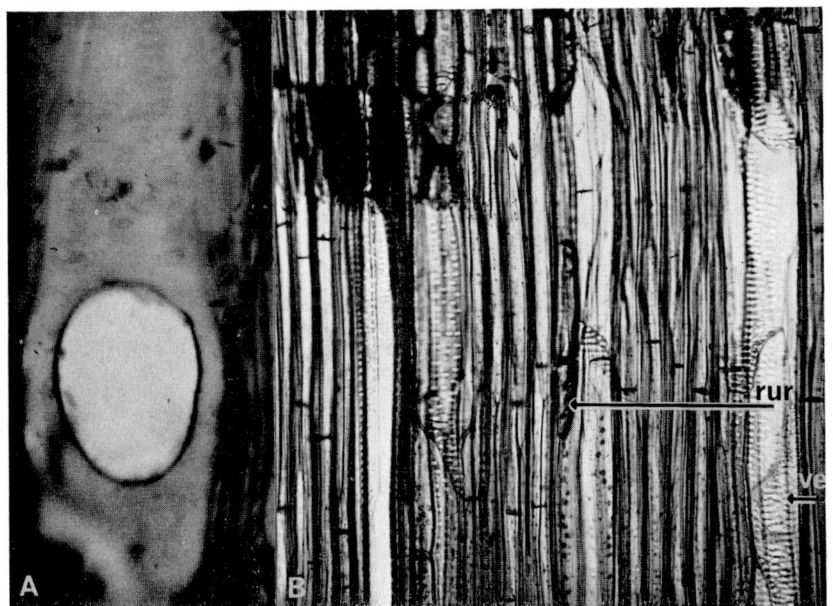

W. L. STERN, E. M. SWEITZER AND R. E. PHIPPS

SPERLICH, A., 1939. *Das trophische Parenchym. B: Exkretionsgewebe.* In Linsbauer, Handbuch der Pflanzenanatomie, **IV**. Berlin: Borntraeger.
STERN, W. L., 1967. Index Xylariorum. *Regnum veg.*, **49**: 1–36.
STERN, W. L., BRIZICKY, G. K. & EYDE, R. H., 1969. Comparative anatomy and relationships of Columelliaceae. *J. Arnold Arbor.*, **50**: 36–75.
STERN, W. L. & CHAMBERS, K. L., 1960. The citation of wood specimens and herbarium vouchers in anatomical research. *Taxon*, **9**: 7–13.
SWAIN, T., 1965. The tannins. In Bonner & Varner, *Plant biochemistry*. New York & London: Academic Press.
TAKHTAJAN, A., 1969. *Flowering plants, origin and dispersal.* (Translated by C. Jeffrey.) Edinburgh: Oliver & Boyd.
THORNE, R. F., 1968. Synopsis of a putatively phylogenetic classification of the flowering plants. *Aliso*, **6**: 57–66.
TIPPO, O., 1938. Comparative anatomy of the Moraceae and their presumed allies. *Bot. Gaz.*, **100**: 1–99.
VOLKENS, G., 1883. Ueber Wasserausscheidung in liquider Form an den Blättern höher Pflanzen. *Jb. K. bot. Gart. bot. Mus. Berlin*, **2**: 166–209.
WALKER, J. C., 1969. *Plant pathology*, 3rd ed. New York: McGraw-Hill.
WILLIS, J. C., 1966. *A dictionary of the flowering plants and ferns.* (Revised by H. K. Airy Shaw.) 7th ed. Cambridge: University Press.
YATSENKO-KHMELEVSKY, A. A., 1954. *Woody plants of the Caucasus*, **1**. Erevan: Botanical Institute, Academy of Science, Armenian SSR. (Text in Russian.)

EXPLANATION OF PLATES

PLATE 1

A, **B**. *Ribes viburnifolium.*
A. Peltate hair on lower surface of leaf. Note that the base of the hair is situated in an indentation of the leaf. × 260. **B**. Multicellular non-capitate hair bearing unicellular trichomes, ut. × 180.
C. Hydathode illustrated from cleared leaf material in *R. californicum*. The vascular supply is always trimerous; the large central vein is flared distally. × 200.

PLATE 2

A. *Ribes indecorum.* Transverse section of wood showing diffuse-porous condition, tangential arrangement of pores, pore clusters, solitary pores, and broad and narrow vascular rays. ×140.
B. *R. bracteosum.* Transverse section of wood to show ring-porous condition, tangential arrangement of pores, pore clusters, solitary pores, and broad and narrow vascular rays. ×130.
C. *R. americanum.* Transverse section of wood to show diffuse-in-aggregates axial parenchyma. ×300.
D. *R. valdivianum.* Radial section of wood showing scalariform perforation plate and circular bordered pits in vasicentric tracheids. ×280.

PLATE 3

A. *Ribes americanum.* Radial section of wood showing simple perforation. ×500.
B. *R. valdivianum.* Radial section of wood with radially uniseriate vascular ray, rur, and adjacent vessel element,ve. ×200.

The publications of C. R. Metcalfe

1930

Metcalfe, C. R. The 'shab' disease of lavender. *Jl. R. hort. Soc.*, **55**: 271–5.

1931

Metcalfe, C. R. The 'aerenchyma' of *Sesbania* and *Neptunia*. *Kew Bull.*, 151–4.

Metcalfe, C. R. The wood structure of *Fokienia hodginsii* and certain related Coniferae. *Kew Bull.*, 420–5.

Metcalfe, C. R. The breathing roots of *Sonneratia* and *Bruguiera:* a review of the recent work by Troll and Dragendorff. *Kew Bull.*, 465–7.

1933

Metcalfe, C. R. A note on the structure of the phyllodes of *Oxalis herrerae* R. Knuth and *O. bupleurifolia* St. Hil. *Ann. Bot.*, **47**: 355–9.

Metcalfe, C. R. The structure and botanical identity of some scented woods from the East. *Kew Bull.*, 3–15.

1935

Metcalfe, C. R. The structure of some sandalwoods and their substitutes and some other little known scented woods. *Kew Bull.*, 165–95.

1936

Metcalfe, C. R. An interpretation of the morphology of the single cotyledon of *Ranunculus ficaria* based on embryology and seedling anatomy. *Ann. Bot.*, **50**: 103–20.

Metcalfe, C. R. & Hill, A. W. The effect of atmospheric pollution on vegetation. *Proc. Conf. natn. Smoke Abatement Soc.*, Oct. 1936: 67–73.

1937

Metcalfe, C. R. The effect of atmospheric pollution on vegetation. *J. Park Administration*, **2** (July): 51–5.

Sprague, T. A. & Metcalfe, C. R. The taxonomic position of *Rhynchocalyx*. *Kew Bull.*, 392–4.

1938

Metcalfe, C. R. The morphology and mode of development of the axillary tubercles and root tubers of *Ranunculus ficaria*. *Ann. Bot.*, N.S., **2**: 145–57.

Metcalfe, C. R. Note on the anatomy of *Fraxinus oxycarpa* and *F. Pallisae*. Appendix to: Anderson, E. & Turrill, W. B. Statistical studies on two populations of *Fraxinus*. *New Phytol.*, **37**: 160–72.

Metcalfe, C. R. Extra-floral nectaries on *Osmanthus* leaves. *Kew Bull.*, 254–6.

Metcalfe, C. R. Anatomy of *Fraxinus oxycarpa* and *F. Pallisae*. *Kew Bull.*, 258–62.

1939

Metcalfe, C. R. The sexual reproduction of *Ranunculus ficaria*. *Ann. Bot.*, N.S., **3**: 91–103.

Metcalfe, C. R. & Templeman, W. G. Experiments with plant growth-substances for the rooting of cuttings. *Kew Bull.*, 441–56.

1941

MELVILLE, R. & METCALFE, C. R. The germination of belladonna seeds. *Pharm. J.* (ser. 4), **92**: 116.

METCALFE, C. R. Damage to greenhouse plants caused by town fogs with special reference to sulphur dioxide and light. *Ann. appl. Biol.*, **28**: 301–15.

1942

METCALFE, C. R. Economic value of the common stinging nettle. *Nature, Lond.*, **150**: 83.

METCALFE, C. R. A short history of the Jodrell Laboratory. *Chronica bot.*, **7**: 174–6.

1944

METCALFE, C. R. History and recent work of the Jodrell Laboratory, Kew. *Ann. appl. Biol.*, **31**: 166–7.

METCALFE, C. R. On the taxonomic value of the anatomical structure of the vegetative organs of the dicotyledons. 1. An introduction, with special reference to the anatomy of the leaf and stem. *Proc. Linn. Soc. Lond.*, 1942–3: 210–14.

1946

METCALFE, C. R. The systematic anatomy of the vegetative organs of the angiosperms. *Biol. Rev.*, **21**: 159–72.

1948

METCALFE, C. R. The elder tree (*Sambucus nigra* L.) as a source of pith, pegwood and charcoal, with some notes on the structure of the wood. *Kew Bull.*, 163–9.

METCALFE, C. R. Lesser rubber plants. *Research*, **1**: 438–46.

1949

METCALFE, C. R. & RICHARDSON, F. R. The use of polyvinyl alcohol and related compounds as a mounting medium for microscope slides. *Kew Bull.*, 569–71.

1950

METCALFE, C. R. & CHALK, L. *Anatomy of the dicotyledons*, 2 vols, 1500 pp. Oxford: Clarendon Press.

1951

METCALFE, C. R. The anatomical structure of the Dioncophyllaceae in relation to the taxonomic affinities of the family. *Kew Bull.*, 351–68.

1952

METCALFE, C. R. *Medusandra richardsiana* Brenan. Anatomy of the leaf, stem and wood. *Kew Bull.*, 237–44.

METCALFE, C. R. Notes on the anatomy of *Heptacodium*. *Kew Bull.*, 247–8.

METCALFE, C. R. Notes on the anatomy of the leaf and stem of *Anisophyllea guianensis* Sandwith. *Kew Bull.*, 291–3.

1953

METCALFE, C. R. Effects of atmospheric pollution on vegetation. *Nature, Lond.*, **172**: 659–61.

METCALFE, C. R. The anatomical approach to the classification of the flowering plants. *Sci. Prog., Lond.*, **41**: 42–53.

1954

METCALFE, C. R. An anatomist's views on angiosperm classification. *Kew Bull.*, 427–40.

1955

METCALFE, C. R. Recent work on the systematic anatomy of the monocotyledons (with special reference to investigations at the Jodrell Laboratory at Kew). *Kew Bull.* (1954), 523–32.

1956

METCALFE, C. R. Mr. James Pryde [Obituary]. *Proc. Linn. Soc. Lond.*, 1953–4: 38–9.

METCALFE, C. R. Papillae and fluted veins on the abaxial leaf surface of *Stephania zippeliana* Miq. (Menispermaceae). *Kew Bull.*, 71–2.

METCALFE, C. R. The taxonomic affinities of *Sphenostemon* in the light of the anatomy of its stem and leaf. *Kew Bull.*, 249–53.

METCALFE, C. R. *Scyphostegia borneensis* Stapf. Anatomy of stem and leaf in relation to its taxonomic position. *Reinwardtia*, **4**: 99–104.

METCALFE, C. R. Some thoughts on the structure of bamboo leaves. *Bot. Mag., Tokyo*, **69**: 391–400.

METCALFE, C. R. Gas and fog injury to plants. *Dictionary of Gardening* (suppl. vol.), pp. 219–21. Roy. hort. Soc. Oxford: Clarendon Press. [Also 2nd ed. (suppl. vol.), pp. 320–3, 1969.]

1957

BATE-SMITH, E. C. & METCALFE, C. R. Leuco-anthocyanins. 3. The nature and systematic distribution of tannins in dicotyledonous plants. *J. Linn. Soc.* (*Bot.*), **55**: 669–705.

1958

METCALFE, C. R. Cecil Prescott Hurst [Obituary]. *Proc. Linn. Soc. Lond.*, 1956–7: 242–3.

1960

METCALFE, C. R. *Anatomy of the monocotyledons.* **I.** *Gramineae*, 731 pp. Oxford: Clarendon Press.

1961

METCALFE, C. R. (ed.) *Anatomy of the monocotyledons.* **II.** *Palmae* by P. B. Tomlinson, 453 pp. Oxford: Clarendon Press.

METCALFE, C. R. The anatomical approach to systematics. General introduction with special reference to recent work on monocotyledons. *Recent Advances in Botany*, 146–50. Univ. Toronto Press.

1963

ERDTMAN, G. & METCALFE, C. R. Affinities of certain genera *incertae sedis* suggested by pollen morphology and vegetative anatomy. *Kew Bull.*, **17**: 249–56.

METCALFE, C. R. Comparative anatomy as a modern botanical discipline, with special reference to recent advances in the systematic anatomy of monocotyledons. *Adv. bot. Res.*, **1**: 101–47.

1964

METCALFE, C. R. Botany today. *Pharm. J.*, **139**: 103–4.

METCALFE, C. R. Systematic anatomy of monocotyledons. (Paper presented at X Int. Bot. Congr.) *Notes Jodrell Lab.*, **2**: 1–6.

METCALFE, C. R. & GREGORY, M. Comparative anatomy of monocotyledons. Some new descriptive terms for Cyperaceae with a discussion of variations in leaf form noted in the family. *Notes Jodrell Lab.*, **1**: 1–11.

1965

METCALFE, C. R. Royal Botanic Gardens, Kew. New Jodrell Laboratory. *Nature, Lond.*, **207**: 239–41.

1966

METCALFE, C. R. Opening of the new Jodrell Laboratory. *Kew Guild J.*, 593–7.

METCALFE, C. R. Prof. P. Maheshwari [Obituary]. *Nature, Lond.*, **211**: 804.

METCALFE, C. R. Distribution of latex in the plant kingdom. *Notes Jodrell Lab.*, **3**: 1–18.

METCALFE, C. R. Notes on the anatomy of leaf and stem of *Panda oleosa* and *Galearia celebica.* (Appendix I to paper by L. L. Forman.) *Kew Bull.*, **20**: 318–19.

PARAMESWARAN, N. & METCALFE, C. R. Notes on the wood structure of *Panda oleosa* and *Galearia celebica.* (Appendix II to paper by L. L. Forman.) *Kew Bull.*, **20**: 319–21.

1967

GREGORY, M. & METCALFE, C. R. Bibliography for the anatomy of the Cyperaceae. *Notes Jodrell Lab.*, **5**: 1–17.

METCALFE, C. R. Distribution of latex in the plant kingdom. *Econ. Bot.*, **21**: 115–27.

1968

METCALFE, C. R. Some current problems in systematic anatomy. *Phytomorphology*, **17** (1967): 128–32.

METCALFE, C. R. Current developments in systematic plant anatomy. Pp. 45–57 in *Modern methods in plant taxonomy* (ed. V. H. Heywood). London: Academic Press.

METCALFE, C. R. & CLIFFORD, H. T. Microhairs on grasses. *Kew Bull.*, **21**: 490.

METCALFE, C. R., LESCOT, M. & LOBREAU, D. A propos de quelques caractères anatomiques et palynologiques comparés d'*Allantospermum borneense* Forman et d'*Allantospermum multicaule* (Capuron) Nooteboom. *Adansonia* (ser. 2), **8**: 337–51.

1969

ERDTMAN, G., LEINS, P., MELVILLE, R. & METCALFE, C. R. On the relationships of *Emblingia. Bot. J. Linn. Soc.*, **62**: 169–86.

METCALFE, C. R. (ed.) *Anatomy of the monocotyledons.* **III**. *Commelinales—Zingiberales*, by P. B. Tomlinson, 446 pp. Oxford: Clarendon Press.

METCALFE, C. R. (ed.) *Anatomy of the monocotyledons.* **IV**. *Juncales*, by D. F. Cutler, 357 pp. Oxford: Clarendon Press.

METCALFE, C. R. Anatomy as an aid to classifying the Cyperaceae. *Am. J. Bot.*, **56**: 782–90.

In press

METCALFE, C. R. *Anatomy of the monocotyledons.* **V**. *Cyperaceae.* Oxford: Clarendon Press.

METCALFE, C. R. (ed.) *Anatomy of the monocotyledons.* **VI**. *Dioscoreales*, by E. S. Ayensu. Oxford: Clarendon Press.

Index